D0764120

UTSA LIBRARY RENEWALS 458-2440

WITHDRAWN
UTSA LIBRARIES

MUNICIPAL BENCHMARKS

Assessing Local Performance and Establishing Community Standards

SECOND EDITION

DAVID N. AMMONS

Sage Publications
International Educational and Professional Publisher
Thousand Oaks ▪ London ▪ New Delhi

Copyright © 2001 by Sage Publications, Inc.

All rights reserved. No part of this book may be reproduced or utilized in any form or by any means, electronic or mechanical, including photocopying, recording, or by any information storage and retrieval system, without permission in writing from the publisher.

For information:

Sage Publications, Inc.
2455 Teller Road
Thousand Oaks, California 91320
E-mail: order@sagepub.com

Sage Publications Ltd.
6 Bonhill Street
London EC2A 4PU
United Kingdom

Sage Publications India Pvt. Ltd.
M-32 Market
Greater Kailash I
New Delhi 110 048 India

Printed in the United States of America

Library of Congress Cataloging-in-Publication Data

Ammons, David N.
  Municipal benchmarks: Assessing local performance and establishing community standards / by David N. Ammons.— 2nd ed.
      p. cm.
Includes bibliographical references and index.
  ISBN 0-7619-2078-1 (cloth: alk. paper)
  1. Municipal services—United States—Evaluation. 2. Municipal government—United States—Evaluation. 3. Benchmarking (Management)—United States. I. Title.
  HD4605.A667   2001
  352.3′214′0973—dc21   00-011069

01 02 03 04 05 06 07 7 6 5 4 3 2 1

Acquiring Editor:	Marquita Flemming
Editorial Assistant:	MaryAnn Vail
Production Editor:	Diane S. Foster
Editorial Assistant:	Cindy Bear
Typesetter/Designer:	Janelle LeMaster
Indexer:	Molly Hall
Cover Designer:	Michelle Lee

Library
University of Texas
at San Antonio

Contents

In memory of Jan Ammons and Dot Hudson—

both benchmarks, themselves.

Preface

City governments need performance benchmarks, if they are serious about the efficient delivery of quality services. And their citizens need municipal benchmarks, if they are not!

Some years ago, as a local government practitioner, I fielded intermittent inquiries from council members and citizens who were concerned about our police department, our parks and recreation department, or our public works department, and who wondered whether local performance was up to par. Such questions were difficult to answer back then, except perhaps by anecdote —my tales of good performance balancing their bad ones—or by resorting to testimonials or evidence of "busy-ness." Not surprisingly, however, my assurances that the police department was making as many arrests as last year or that the streets department was laying as many tons of asphalt were hardly compelling to a caller who doubted that last year's performance was very good either! Such instances begged for a standard or benchmark against which local performance could be judged.

Now, as a university faculty member with a specialty in the management of local government, I still get those questions. My perspective might have changed a little because my relationship to the caller and the municipality in question is different and because the locales from which the calls come are more widespread, but the questions are basically the same. Mayors, council members, general citizens, and municipal administrators, too, want to know how to judge the service delivery performance of their local government. Often, the answers in the past were regrettably incomplete. "Hmm, I think typical performance is about . . ." or "I heard some time ago that professional standards were being considered. Maybe they have been established and would apply in this case." As one city official, contacted prior to the first edition of *Municipal Benchmarks,* said, "Unfortunately, there is no central repository of such standards."

Well, there is now.

Mayors, city council members, city managers, department heads, other municipal officials, and citizens who want a measuring rod for local government services can turn to the benchmarks presented in this book. They will find national standards, "engineered" standards, statistical norms, rules of thumb, selected statistics from comparative performance measurement projects, excerpts from actual performance records, and performance targets of a collection of respected cities.

The list of benchmarks is long—and with the second edition the list keeps growing and the benchmarks become more and more ambitious. That is the nature of benchmarks, and it is also the nature of the drive for continuous improvement.

Acknowledgments

The preparation of this book turned out to be a bigger project than I imagined at its beginning. I am indebted to many persons who have graciously assisted me along the way. Matt Bronson, Ben Canada, Larry DiRe, Randy Harrington, and Sonya Smith, graduate students at the University of North Carolina, helped me with this second edition—much like their fine group of predecessors, Jordan Davis, Jeremy Chen, Anne Lockwood, and Debra Hill at the University of Georgia, did do when I was working on the first edition. Each tackled various data collection tasks and doggedly pursued authorities who could answer the recurring questions "Are there standards on this?" and "What does this number mean?"

Alex Hess, Marsha Lobacz, and Edie Hughes of the Institute of Government Library at the University of North Carolina tracked down elusive bits of information for me time after time. My son, Drew, readied various figures for publication. Elaine Welch, Patt Dower, and Elizabeth Gardner were immensely helpful as I pursued the tedious tasks of cataloging information and developing tables for the manuscript. The project was probably bigger than they had imagined, too.

Richard W. Campbell, my friend and colleague at the University of Georgia, saw value in this project from the beginning and urged me on. I thank him again for that.

Above all, I am indebted to the many officials who each added bits and pieces to the growing collection of material that eventually became the heart of this volume. Officials in various professional associations affiliated with or relevant to municipal government responded to inquiries, provided documents, and offered helpful advice when my pursuits needed to be redirected. A host of municipal officials scattered across the United States and Canada were generous with their time and with pertinent documents. Most of all, they were generous with their words of encouragement, providing reassurance time and again that this product was eagerly awaited and would be worth the effort.

Introduction

"How'm I doin'?" was the trademark question of Ed Koch, mayor of New York City during the 1980s. In posing his question, Mayor Koch sought reassurances regarding the depth of his political base as well as feedback on his stewardship of municipal operations. As veteran mayors and city managers know, these two dimensions of municipal leadership are not independent of each other. Just as political waves can rock an administration, solid operational stewardship can undergird political strength and stability.

The performance of local government operations from animal control to zoning administration can affect the political health of mayors and other elected executives and the professional well-being of city managers and other appointed administrators in local government. Unless they—or trusted assistants—are "minding the store," executives' political or professional stock can quickly decline. Yet all too many officials have only a vague sense of how their own municipal operations are faring. "How are we doing?" is a question that should be asked—and deserves to be answered.

Gauging Municipal Performance

Surveys of local government officials indicate a broad awareness of the importance of performance measurement and monitoring. Tangible evidence, however, reveals the back-burner status of performance measurement in most communities relative to the hotter issues cooking at any given time.

By most counts, more than half of all U.S. cities collect performance measures of some type (Cope, 1987, 1992; Governmental Accounting Standards Board [GASB] & National Academy of Public Administration [NAPA], 1997; Poister & Streib, 1994, 1999). In most cases, however, the majority of the measures collected by a given municipality merely reflect workload—for example, the number of applications processed, arrests made, recreation class participants, or inspections completed.

1

In essence, workload measures are a form of "bean counting." Such a count is important. To anyone wanting to get ahead in the bean business, however, it is also important to know the *quality* of the beans and the *efficiency* with which they are grown and harvested.

A smaller number of local governments systematically assess performance quality and efficiency, meticulously gauging and reporting the effectiveness and unit costs for even a handful of major departments (Ammons, 1995; Grizzle, 1987; LeGrotte, 1987; MacManus, 1984; Usher & Cornia, 1981). Despite the existence of how-to manuals for local government performance measurement and encouragement from a host of professional associations, most local governments have not yet invested the resources and energy needed to advance beyond the bean-counting stage.

Advanced Performance Measures

Measures of efficiency, effectiveness, and productivity have been developed for local government services. Each reveals much more about municipal operations than does a simple workload figure. Efficiency measures—often expressed as unit costs or as units produced per employee hour—depict the relationship between services or products and the resources required to produce them. Effectiveness measures depict the quality of municipal performance or indicate the extent to which a department's objectives are achieved. Measures of timeliness and citizen satisfaction are among the common forms of effectiveness measures.

Productivity measures, although extremely rare in municipal performance monitoring systems, combine efficiency and effectiveness components in a single indicator. "Utility meters read *without error* per employee hour" would be an example.

Performance Compared With What?

For decades, municipal officials have been urged to measure performance and have been given advice on how to get started. Those who have followed the prescriptions have soon discovered that a performance measure is virtually valueless if it appears in the form of an isolated, abstract number. Its value comes in comparison with a relevant peg.

Several options exist. Authorities have long suggested that current marks could be compared with those from earlier periods, with other units in the same organization, with relevant outside organizations, with preestablished targets, or with existing standards (GASB, 1992; Hatry, 1980b, 1999; Hatry, Fountain, & Sullivan, 1990). In theory, such comparisons would provide municipal officials with both an internal gauge, marking year-to-year progress of a work unit or highlighting unit-to-unit performance differences, and an external gauge, showing how a municipality's operations stack up against other jurisdictions or professional ideals. Until recently, only the internal gauge comparing this year's performance to last year's had proven sufficiently practical to receive widespread application.

Many local governments have routinely reported year-to-year comparisons of performance indicators—although often they are merely workload measures. Few, however, reported external comparisons with standards or with performance indicators from other jurisdictions. Then, in the mid-1990s, changes began to occur, spurred principally by the urgings of GASB and the emergence of a handful of projects designed to yield reliable interjurisdictional performance comparisons.

GASB encouraged cities to measure their service efforts and accomplishments and, where possible, to compare their results with other cities. Portland, Oregon, and a few other cities began to do so and publicized their reports.

Meanwhile, a few groups of cities established cooperative projects to collect and share performance statistics. Projects administered by the International City/County Management Association (ICMA) and the Innovation Groups were national in scope (and with the participation of a few Canadian cities in the ICMA project, it became *international*), whereas a few others focused on selected cities within an individual state, for example North Carolina and South Carolina (Coe, 1999; Few & Vogt, 1997; Kopczynski & Lombardo, 1999). Each project grappled with the challenges of matching services and cost-accounting systems across its cities, but as these problems were resolved the participating cities began to report comparative statistics and, in some cases, to use comparative information to reduce costs or improve services (Ammons, 2000a; Ammons, Coe, & Lombardo, 2001; Jones, 1997; Rivenbark & Carter, 2000).

Why So Few External Comparisons?

Why have there been only a few comparative performance measurement projects and why have relatively few cities chosen to participate? Furthermore, why have so few individual municipalities attempted external comparisons on their own? Undoubtedly, some local government officials are happy with the status quo, preferring not to have the performance of their organization compared with others. After all, only in Garrison Keillor's fictitious village of Lake Wobegon can *everyone* be "above average."[1] In real life, only about half can achieve that status, and, for the rest, the desire to know one's ranking must be motivated not by a yen for publicity and praise but instead by a yearning to improve. Sadly, many of those jurisdictions mired at the bottom of the perfor-

mance scale—the ones, they might say, that make the upper 90% possible—
may prefer to remain oblivious to their status.

In contrast, some local government officials—including those that have
joined formal projects or developed their own comparative statistics—not
only are willing to engage in external comparisons but are eager to do so.
Driven by a desire to climb into the ranks of outstanding municipal performers or to be recognized for already being there, these officials long for external
benchmarks but often are disappointed to find no more than a handful having
any usefulness. Those hoping to spur greater accomplishments by urging their
organization to keep up with the "municipal Joneses" often are disappointed
to find that few relevant statistics on the Jones family even exist—at least in a
form usable for this purpose. When the most that the city manager can discover about the performance of other cities is a set of workload and expenditure levels, few useful comparisons are possible.

If good comparison statistics are so hard to find, why not rely on municipal
performance standards set by professional associations and others? It sounds
simple—and in some cases it is possible to make such comparisons against
standards (Ammons, 1994, 1996). Unfortunately, however, many so-called
standards are vague or ambiguous, have been developed from limited data or
by questionable methods, or may be self-serving. Furthermore, most standards are not widely known and often are difficult to track down. There had
been no single repository of standards relevant to municipal operations prior
to the first edition of *Municipal Benchmarks* (1996). Mayors, council members, city managers, department heads, other officials, and citizens wishing to
sort well-developed and usable standards for a particular operation from the
rest—or even to *find* those standards—previously faced the prospect of an
often difficult and time-consuming search. Understandably, few persevered.

Meeting the Information Needs of
Local Government Officials and Citizens

The practical reality of competing demands, limited time, and scarce resources severely restricts the viability of using external gauges for performance assessment unless those gauges are readily available. Few local government officials or interested citizens can afford to spend the hundreds of hours
necessary to track down applicable standards or suitable interjurisdictional
performance indicators. Desire to make external comparisons—or lack of
that desire—is not necessarily the issue; time, resources, and other practical
constraints may simply constitute too much of a hurdle if a reservoir of relevant comparison information is not at hand.

That is where this volume comes into play. Between the covers of this book,
local government managers, elected officials, and citizens will find standards

and comparison statistics intended for ready use. Countless contacts with lo-cal government officials, professional associations, and trade organizations, coupled with careful scouring of budgets, financial statements, and other performance-reporting documents from local governments across the United States, along with a few from Canada, have produced a collection of munici-pal benchmarks that will enhance a community's ability to answer the ques-tion "How are we doing?"

Included in the pages that follow are "standards," "norms," and "rules of thumb" offered by professional associations, trade organizations, and other groups with a stake in local government. Readers are cautioned here—and will be reminded elsewhere—that the motives of such groups in prescribing standards may range from civic-minded service to self-serving protection of the status and working conditions of association members. It is also important to realize that standards may be intended in some instances to represent mini-mum acceptable levels of performance, in other cases to designate norms, and in still other cases to identify targets toward which local governments should aspire.

Collections of actual performance indicators have been gleaned from the documents of more than 150 local governments of various sizes scattered across the nation. The set of municipalities chosen for this project does not constitute a random sample; each municipality was included because it mea-sures performance and because its reporting documents were made available. Most of the municipalities included were selected from lists of recipients of the Government Finance Officers Association's (GFOA, 1993, p. 4) Distin-guished Budget Presentation Award—an award that includes as a nonmanda-tory criterion the reporting of performance results. The input of GFOA award recipients was supplemented by that of other cities that were also found to document performance in a manner conducive to cross-jurisdictional com-parison.

The performance indicators reported in this volume are neither workload measures nor unit costs. Workload measures were excluded because of their limited usefulness in cross-jurisdictional comparisons and their inability to an-swer the "how well" and "how efficiently" questions. Unit costs have been omitted because of their extreme vulnerability to inflation (e.g., during peri-ods of high inflation, unit costs may become quickly outdated unless "con-stant dollar" calculations are made), economic differentials (e.g., regional variations in the cost of labor could produce erroneous judgments on the rela-tive efficiency of labor-intensive operations), and accounting vagaries (e.g., inconsistent accounting practices across jurisdictions for overhead, employee benefits, capital acquisition, and depreciation, to name a few, could distort comparisons). Retained is a collection of indicators that easily traverse juris-dictional contexts or that otherwise have been standardized in a manner that removes the most obvious effects of differences in population. Measures such as "average response times to public safety emergencies," "average return on municipal investments relative to U.S. treasury bill rates," "inspections per electrical inspector," "park acreage per 1,000 persons," and "percentage of recreation program expenses recovered through fees" are examples.

Where performance indicators from actual cities conform in format to the standards promulgated by professional associations, they offer readers a "reasonableness review" for those standards. Actual performance of respected cities may influence the interpretation attached to a given standard—that is, whether it should be regarded as a minimum acceptable level, a norm, or a target of excellence. The performance records of recognizable and respected cities will help answer the practitioner's pragmatic questions:

"Is everybody else meeting this standard?"
"Is *anybody* meeting it?"

Contents, Format, and Uses

Following this introduction is a second chapter that encapsulates the major messages of some of the most prominent books, articles, and how-to manuals on performance measurement for local government. Although these publications instruct local officials on the development of ideal performance measures, they typically offer few, if any, relevant comparisons for use once those measures are crafted and operating statistics are compiled. This book takes a different approach.

Following brief instructions on the development of good performance indicators, the focus of this volume turns toward the interpretation of performance information once it has been collected. Officials who have devoted a year or more to the development of a set of performance measures and the collection of relevant data often are disappointed to find no other jurisdictions reporting the same measure in the same fashion. They understandably may be frustrated to learn that they must collect the measure a second year before any relevant comparisons can be made—and then only with their own city's performance in the earlier year. With the information in this volume, officials will be able to make more immediate comparisons that will place local performance in an external context.

Following the chapter on performance measurement is a series of 30 chapters devoted to performance standards and cross-jurisdictional performance indicators pertinent to major functions common to municipal government. The concluding chapter, "Performance Milestones," will offer a concise set of suggested performance targets drawn from the standards developed by professional associations and performance indicators reported by cities included in this book.

A Few Comments Before Proceeding

The collection of benchmarks found in this volume can be a useful tool for gauging and improving municipal performance. Like other tools, the benchmarks are most effective in the hands of a craftsperson who knows how to use them and who understands their limitations.

On Benchmarks and Benchmarking

Comparing local performance statistics with selected benchmarks is a valuable step in evaluating municipal operations, but simple comparisons are not the same as full-fledged, corporate-style benchmarking. Corporate-style benchmarking entails the analysis of performance gaps between one's own organization and best-in-class performers, the identification of process differences that account for the gap, and the adaptation of key processes for implementation in one's own organization in an effort to close the gap.

A simple comparison of local performance with selected municipal benchmarks is more limited than corporate-style benchmarking, but it is also a less expensive and more practical form of benchmarking for a general assessment of a broad range of functions. Such a comparison places local performance in context and, where major performance gaps are detected, may suggest the need for additional analysis, perhaps leading to a full-scale, corporate-style benchmarking project focusing on narrower areas of concern.

Reactive Versus Proactive Management

Many of the performance statistics included in this book reflect a reactive approach to service delivery—reporting, for example, how quickly potholes are filled or how well firefighters limit the spread of fire. In such cases, the emphasis is on speed or skill in responding to problems. As more local governments set their sights on proactive management that anticipates problems before they occur, we can expect in the future more performance statistics that measure effectiveness in preventing undesirable occurrences. Even then, however, high-performance municipalities will still be adept at responding to problems as they happen.

Aggregate Statistics as Camouflage

A municipality that performs well overall may nevertheless be deficient in delivering services to some neighborhoods or to a particular category of customers (e.g., senior citizens, youth, minorities, or businesses). Simply comparing citywide performance statistics with municipal benchmarks would hide those deficiencies. Disaggregating statistics to reflect performance as it relates to various groups of customers often is more revealing.

Unaudited Data

Much of the performance information in this book comes directly from municipal reporting documents. As such, it is largely unaudited; but it is information that each municipality has confidently reported of its own volition rather than in response to some questionnaire that could evoke as many estimates as real numbers. Of potentially greater concern to consumers of these municipal benchmarks is the possibility of inconsistent definition of terms from one city to another. Statistics are reproduced as reported. In some cases, analysts may wish to dig deeper—for instance, if a benchmark seems too good to be true.

Time Frame Differences

Performance targets and statistics have been collected from different fiscal years and reporting periods.[2] Because inflation-sensitive expenditure statistics (e.g., total costs and unit costs) have been excluded, the measures that remain are, for the most part, little affected by reporting period variations or the passing of time. A library boasting a circulation of six items per capita in 1993, for example, was performing admirably then and could still be proud of that mark today—much like a university that still has emblazoned on its stadium, "National Champions, 12–0," a decade or more after the accomplishment. On the other hand, performance worthy of the label "commendable" in a few services may experience more rapid evolution. Especially in services such as information systems, in which changing technology is a major factor, appropriate caution should be exercised in the application of benchmarks more than 5 or 6 years old.

Audiences for This Book

This volume is intended to serve three audiences. It is designed (a) for elected officials who desire a context within which to judge municipal operations, (b) for local government managers and department heads who wish to improve their performance measurement systems and who want to supplement internal with external performance gauges, and (c) for citizens who wish to know whether their community's public services really measure up.

Notes

1. Keillor, author of *Lake Wobegon Days* (1985), *Leaving Home* (1987), and other works, fondly describes Lake Wobegon as a place "where all the women are strong, the men are good-looking, and all the children are above average" (1987, p. xvii).

2. Hyphenated fiscal years (e.g., FY 1997–1998) are reported in this book simply as the year in which the fiscal period ends (e.g., 1998).

2

Performance Measurement and Benchmarking

Aggressive businesses and industries measure performance meticulously. In their popular book on excellence in corporate America, Tom Peters and Bob Waterman (1982) characterized top companies as "measurement-happy and performance-oriented" (p. 240). They found that the *very best of the best* complement their measurement systems with an action orientation that separates them from the pack. Outstanding companies carefully monitor and record the relevant dimensions of performance—they know the important facts and figures—and they act on that knowledge.

Municipalities and other public sector agencies confront a far different circumstance. Without the pressure of competition or the unforgiving bottom line of profit or loss, governmental units are apt to neglect performance measurement as they focus on more pressing matters. Carefully conceived systems of performance measurement are more than a little complex; often they are more than a little threatening to the status quo; and they do impose expenses on the organization from conception through administration. Consequently, performance measurement often is allowed to slip in priority, except among municipalities that have an extraordinary commitment to management information.

A decade ago, federal researchers examined successful private corporations in hopes of gleaning lessons for possible application in government. They found that most industry leaders were distinguished not only by management continuity and consistency, long-range vision, and customer orientation, but also by their "systematic strateg[ies] for measuring performance" (Thompson, 1991, p. 4). Osborne and Gaebler (1992) declared in their book *Reinventing Government* that entrepreneurial governments measure their performance "focusing not on inputs but on *outcomes*" (p. 19). Similarly, performance measurement and related matters are prominent in management expert Peter Drucker's (1975) list of characteristics common to successful service institutions:

- They clearly define the nature and scope of their function, mission, and activities.
- They set clear objectives.
- They set priorities, concentrate on the most important objectives, and set standards of performance for themselves.
- They measure performance, analyze the results, and work to correct deviations from established standards.
- They regularly conduct audits of performance to ensure that the management system continues to function properly (pp. 158–159).

Why Measure Performance?

Properly developed and administered, a performance measurement and monitoring system can offer important support to a host of management functions (Broom, Jackson, Vogelsang Coombs, & Harris, 1998; Glover, 1993; Hatry, 1980a, 1999; Osborne & Gaebler, 1992, p. 156; Poister & Streib, 1999; Thompson, 1991, p. 14; U.S. General Accounting Office, 1993, p. 1). Each of those functions has important ramifications for cities that aspire to be excellent.

Accountability. Managers in top-performing organizations insist on accountability from their subordinates, and, in turn, expect to be held accountable by their organizational superiors. Performance measures document what was done by various departments or units and, ideally, how well it was done and what difference it made. Through such documentation, outstanding departments and entire municipalities earn the trust of their clients and citizens as they demonstrate a good return in service provided for tax dollars received.

Planning/Budgeting. Cities with an objective inventory of the condition of public services and facilities, a clear sense of service preferences among their citizens, and knowledge of the cost of providing a unit of service at a given level are better equipped to plan their community's future and to budget for that future. Again, performance measurement—incorporating unit costs and indicators of citizen demand or preference—is key.

Operational Improvement. Municipalities that measure performance are more likely to detect operational deficiencies at an early stage. Furthermore, performance records enhance their ability to confirm the effectiveness of corrective actions.

Program Evaluation/Management-by-Objectives/Performance Appraisal.
Carefully developed performance measures often provide valuable information for the systematic evaluation of program effectiveness. Management-
by-objectives (MBO) programs and pay-for-performance systems for managerial employees, where they exist, typically are tied to performance measures.
In some cases, employee performance appraisals or other forms of systematic
performance feedback have been based, at least in part, on individual performance relative to established measures.

Reallocation of Resources. A clear indication of program effectiveness and
unit costs—in essence, a scorecard on tax dollar investments and returns—can
aid decision makers in reallocation deliberations, especially in times of financial duress.

Directing Operations/Contract Monitoring. Managers equipped with a
good set of performance measures are better able to detect operational
strengths and weaknesses, to provide relevant feedback to employees and
work units, and to deploy close supervision where it is needed most. Performance measures also provide evidence useful in determining whether the service quality specified in contractual arrangements is, in fact, being achieved.

Stated simply, performance measurement provides local governments with
a means of keeping score on how their various operations are doing. As noted
by Harry Hatry (1978), that score-keeping function is vital:

> Unless you are keeping score, it is difficult to know whether you are win-
> ning or losing. This applies to ball games, card games, and no less to gov-
> ernment productivity. . . . Productivity measurements permit govern-
> ments to identify problem areas and, as corrective actions are taken, to de-
> tect the extent to which improvements have occurred. (p. 28; reprinted by
> permission)

Types of Performance Measures

Efforts to identify different types of performance measures sometimes have
yielded lengthy lists. Often, however, such lists are inflated by including measures of resource input and other pseudomeasures that reflect performance
only indirectly, if at all (Hatry, 1980b).[1] True performance measures in local
government generally may be categorized as one of four types:

1. Workload (output) measures
2. Efficiency measures
3. Effectiveness (outcome) measures
4. Productivity measures

Workload measures, also called *output measures,* indicate the amount of work performed or the amount of services received. By comparing workload measures reporting, for example, the number of applications processed by the human resource department, the number of sets of city council minutes prepared by the city clerk, the number of arrests by the patrol division of the police department, and the number of trees planted by parks crews with the corresponding records from a previous year, a city official or citizen can see whether workload volume is up or down. Although that information can be of some value, it reveals only how much work was done—not how well or how efficiently it was done.

Efficiency measures reflect the relationship between work performed and the resources required to perform it. Unit costs are an obvious example, but efficiency measures can take other forms, too.

Unit costs are calculated by dividing total costs of a service or function by the number of units provided. For example, if 2,000 feet of 8-inch sewer line are installed by municipal crews at a total cost of $100,000, then the unit cost of sewer line installation is $50 per foot. A reversal of the ratio—dividing the number of units by the resources consumed—reveals the number of units produced per dollar or per $1,000 and is also an efficiency measure (e.g., 20 feet of sewer line per $1,000). Other forms of efficiency measures reflect alternative types of resource input (for example, units produced per labor hour)[2] or production relative to an efficiency standard. If meter readers complete only one half of their assigned rounds, if repairs by municipal auto mechanics take twice as long as private garage manuals say they should, or if expensive maintenance equipment is operated only 10% of the time it is available, questions of efficiency relative to a prescribed or assumed standard may be raised.

Effectiveness measures, also called *outcome measures,* depict the degree to which performance objectives are achieved or otherwise reflect the quality of local government performance. Meter reading error rates of less than 0.5%, a consistent record of fire suppression with only minimal spread, and low return rates on auto repairs reflect effective operations. Response times and other measures of service quality sometimes are only indirectly related to effectiveness but typically are included among effectiveness measures for their presumed linkage.

Productivity measures combine the dimensions of efficiency and effectiveness in a single indicator. For example, whereas "meters repaired per labor hour" reflects efficiency and "percentage of meters repaired properly" (e.g., not returned for further repair within 6 months) reflects effectiveness, "unit costs (or labor hours) per *effective* meter repair" reflects productivity. The costs (or labor hours) of faulty meter repairs as well as the costs of effective repairs are included in the numerator of such a calculation, but only good repairs are counted in the denominator—thereby encouraging efficiency and effectiveness by meter repair personnel.

Table 2.1   Examples of the Four Principal Types of Performance Measures

Municipal Function	Workload Measure	Efficiency Measure	Effectiveness Measure	Productivity Measure
City clerk	number of sets of city council meeting minutes prepared	employee hours per set of city council minutes prepared	percentage of city council minutes approved without amendment	percentage of city council minutes prepared within 7 days of the meeting and approved without amendment
Library	total circulation	circulation per library employee	circulation per capita	circulation per $1,000
Meter repair	number of meters repaired	cost per meter repair	percentage of repaired meters still functioning properly 6 months later	cost per properly repaired meter (i.e., total cost of all meter repairs divided by number of meters needing no further repairs within 6 months)
Personnel	job applications received	cost per job application processed; cost per vacancy filled	percentage of new hires/promotions successfully completing probation and performing satisfactorily 6 months later	cost per vacancy filled successfully (i.e., employee performing satisfactorily 6 months later)

Examples of the four types of performance measures for several common municipal functions are provided in Table 2.1. The value of workload measures is limited. Much greater insight into the performance of municipal operations may be gained from efficiency, effectiveness, and productivity measures.

Criteria for a Good Set of Performance Measures

Properly developed sets of performance measures possess several distinctive characteristics (Bens, 1986; Broom et al., 1998; Hatry, 1980b, 1999; Hatry et al., 1992, pp. 2–3). Good sets include measures that are:

- *Valid.* They measure what they purport to measure—that is, a high score on a given measure does, in fact, reflect possession of the underlying dimension or quality.
- *Reliable.* The measure is accurate and exhibits little variation due to subjectivity or use by different raters (e.g., a measuring tape is a reliable instrument in that it is highly objective, and two different persons using the same instrument are likely to get similar measurements).
- *Understandable.* Each measure has an unmistakably clear meaning.
- *Timely.* The measures can be compiled and distributed promptly enough to be of value to operating managers or policymakers.
- *Resistant to Undesired Behavior.* The development of a performance measure raises the profile of the performance dimension being examined. A higher profile sometimes brings unintended consequences or even strategies designed to "beat the system"—for instance, overzealous ticket writing if the police department is measured by that activity alone or the watering down of garbage to increase its weight if collection crews are rated solely by tons collected. The best sets of performance measures have little vulnerability to such actions because they have been devised carefully and because they typically include multiple measures that address performance from several dimensions and thereby hold potentially perverse behavior in check.
- *Comprehensive.* The most important performance dimensions are captured by a good set of measures. Some minor facets of performance might be overlooked, but the major elements are addressed.
- *Nonredundant.* By favoring unique measures over duplicative measures, the best sets of performance measures limit information overload for managers, other decision makers, and consumers of municipal reports. Each measure contributes something distinctive.
- *Sensitive to Data Collection Cost.* Most dimensions of municipal performance can be measured either directly or through proxies. In some cases, however, measurement costs may exceed their value. Good sets of performance measures include the best choices among *practical* measurement options.
- *Focused on Controllable Facets of Performance.* Without necessarily excluding important, overarching, and perhaps relatively uncontrollable characteristics relevant to a particular function, good sets of performance measures emphasize outcomes or facets of performance that are controllable by policy initiatives or management action. For example, although a police department's set of performance measures might include the rate of domestic homicides in the jurisdiction, a good set of measures would also include indicators of public safety more widely considered controllable by police efforts.

Sources of Performance Data

Typically, performance data are secured from various combinations of the following sources:

- Existing records
- Time logs
- Citizen/client surveys
- Trained observer ratings
- Specially designed data collection processes

The simplest and most desirable source of performance measurement data is the set of records already maintained by a municipality. Workload counts, complaint records, and response times for various services are common, even among cities with modest performance-reporting practices. Such information can serve as the foundation of a performance measurement system, and sometimes can be converted to higher-level measures. Depending on its level of precision, for example, workload data might be combined with expenditure information to yield efficiency measures.

Time logs completed on either a comprehensive or random basis provide resource input information for labor-related efficiency measures (e.g., work units per employee hour) or for measures comparing actual work completed in a given amount of time with the amount of work expected on the basis of engineered work standards. Apart from vehicle maintenance and perhaps a handful of other services, however, engineered work standards are rarely available for common municipal functions.

The surveying of users of particular services or citizens in general is a feedback mechanism used by many city governments. Surveys typically tap respondent perceptions regarding the adequacy of selected services, the nature of any perceived deficiencies, and the extent to which respondents avail themselves of various services and facilities. For a survey to contribute meaningfully to a city's performance measurement system, it must be conducted with sufficient scientific rigor to produce reliable results. Such rigor typically introduces costs that exceed those of tempting, low-cost alternatives, but surveys conducted "on the cheap" frequently are misleading and can be easily discredited. Carefully designed and properly randomized telephone surveys have been found to be widely accepted, moderately priced alternatives to more expensive face-to-face interviews and to low-cost, low-response-rate mail surveys.

Trained observer ratings have been used successfully in several cities for evaluating the condition of facilities and infrastructure. Typically, persons employed in positions outside the department responsible for facility maintenance are trained to serve on an intermittent basis as raters of street cleanliness or parks maintenance, for example. Instructed on the finer points of distinguishing between various grades of facility condition and customarily armed with photographic depictions of those grades, trained observers can provide a more systematic evaluation of condition than can usually be drawn from records of citizen compliments or complaints, or even from citizen surveys. The effective administration of trained observer programs requires careful attention to training and issues of reliability. Supervisors often spot-check facilities to corroborate ratings and, where two or more raters assess the same facility within a short period, supervisors usually reinspect if the grades of raters are more than one unit apart.[3]

Unfortunately, information already on hand or readily available through customary means does not always meet current performance measurement needs. Analysts might discover, for example, that existing workload counts and random time logs have been based on estimates, rather than on precise figures. The citizen survey that everyone in town likes to quote might turn out to be suspect because it was conducted using questionable methods—perhaps relying on a tear-out questionnaire in the local newspaper. Even federal figures, presumed to be rock solid, might be less comprehensive and conclusive than originally thought; FBI crime statistics, for example, include only reported crimes—not all crimes actually committed. Typically, at least a few new data collection procedures must be introduced to support the development of a good set of performance measures.

Status of Performance Measurement in City Government

More than a half century ago, a pair of esteemed observers noted the availability of a number of "rough and ready measurement devices," but conceded the limited development of performance measurement among city governments.

> It is probably true that in the present state of our knowledge the citizen can often better judge the *efficiency* of his local government by the political odor, be it sweet or foul, which emanates from the city hall than by any attempt at measurement of services. (Ridley & Simon, 1943, p. ix; reprinted by permission)

Through the years, various movements to improve management practices, rationalize decision making, and enhance accountability have called for improved performance measurement. Respected professional associations, including the American Society for Public Administration, GASB, GFOA, and ICMA, have promoted the practice. Some cities have responded to that encouragement. Many of the writings on performance measurement in local government repeat the names of familiar municipalities as exemplars: Sunnyvale and Palo Alto, California; Phoenix, Arizona; Charlotte, North Carolina; Dayton, Ohio; Savannah, Georgia; Randolph Township, New Jersey; Dallas, Texas; Aurora, Colorado; New York City; Alexandria and Charlottesville, Virginia; and Portland, Oregon (see, for example, Bens, 1986, p. 1; Hatry et al., 1992, p. 1; Osborne & Gaebler, 1992, p. 349; Smith, 1992, p. 8). These cities, however, are far from typical in the advanced status of their performance measurement systems. Despite survey responses suggesting widespread and fairly sophisticated performance measurement in local government, more exacting research involving the examination of actual performance reporting documents reveals far more limited development (Ammons, 1995; Grizzle, 1987; LeGrotte, 1987; MacManus, 1984; Usher & Cornia, 1981).[4] Not all cities even engage in performance measurement. Among those that do, most rely heavily on workload measures. Some report a few efficiency or effectiveness measures for various functions. Rarely are productivity measures reported.

Although performance measurement is well developed and still improving in some city governments, most cities have a long way to go. Speaking of his own city in the not-too-distant past, one Philadelphia official remarked, "If you were to ask how much it cost to pick up a ton of trash, I hope you weren't waiting for an answer the same year" (Barrett & Greene, 1994, p. 44). The situation, he noted, had improved greatly in recent years. Perhaps the remarks of an Indianapolis official put the development of municipal performance measurement and program evaluation systems in proper perspective: "We're coming from an era of cleaning our clothes on rocks, and now we have come up with a hand-cranked washing machine. It's a much better life, but of course we'd rather have an automated washing machine, eventually" (Barrett & Greene, 1994, p. 44).

Overcoming Resistance to Performance Measurement

The likelihood is great that some form of resistance will be encountered in efforts to develop or enhance performance measurement systems. Although the precise source often is not predictable, some resistance almost inevitably will come from *somewhere*. The proper starting point for developing a coping strategy is to try to understand the reasons for that resistance.

Typically, performance measurement is seen by various groups or individuals in an organization as a threat to their status. Some employees may fear that it is the first step in a process that will lead to tougher work standards and a forced speedup of work processes, perhaps to be followed by the layoff of workers who are no longer needed or who cannot keep up.

Supervisors and managers may feel threatened by what they perceive to be the insinuations of a performance measurement drive. Believing that executive and legislative satisfaction with their performance would generate no such movement, they may sense an accusation of poor performance. They may fear loss of status or even loss of employment if performance measures confirm those accusations, and they feel certain of a loss of discretion as the upper echelons gain yet another means of looking closely over their shoulders.

Even some top city officials, who seem to have so much to gain from improved performance measures, may resist their development. Especially prone to do so are those who believe that, by virtue of their own political adeptness or membership in the dominant political coalition, their preferences are more likely to prevail in political negotiations without the influence of performance facts and figures.

Usually, such fears are overblown. Rarely are performance measurement efforts inspired by sinister or mean-spirited motives. Although detractors often fear worker exploitation or loss of supervisory discretion, the most commonly prescribed blueprints for measurement system development call for the involvement of those parties, rather than for their exclusion. Well-designed measurement systems are more often heralded for their usefulness in supporting the exercise of discretion by operating officials than criticized for any tendency to curb it. Furthermore, measurement initiatives typically are most successful when they secure the input and support of frontline employees and supervisors, thereby increasing the likelihood that the right things are being measured and that they are being measured fairly. When that happens, a new measurement and reporting system—and the clarification of work unit objectives and priorities that accompanies the system's development—is likely to earn accolades from frontline employees and supervisors, rather than charges of exploitation.

Persons who fear that politics will be displaced in local decision making through the rationalizing influence of objective performance measures need not be overly concerned. Advances in management practices may, at most, supplement political considerations in decision making. In a democratic setting, politics can never be displaced—nor should it be.

Resistance to advances in performance measurement may emanate from a variety of concerns and may manifest themselves in different forms. A few things in this regard, however, are fairly predictable. Proponents of improved performance measures should expect three common declarations from opponents.

"You can't measure what I do!" If the performance of a department or office has not been measured in the past, it should not be surprising that incum-

bents might reason that their activities must be unmeasurable. If their performance could have been measured, it already would have been. Offices or departments making this declaration often are characterized by nonroutine work and the absence of an existing data collection system. Both factors can make measurement difficult. Rarely is it impossible.

Sometimes a bit of creativity is necessary to devise either a suitable direct measure or a proxy that gauges performance indirectly. Often, interviewing skills are helpful. "If your office closed shop for a few weeks, I know you would be missed," the interviewer might suggest. "But who would suffer the greatest impact, and what aspect of your work would they miss the most?" Gradually, with cooperation from employees and departmental officials, a widely accepted set of performance measures can emerge.

"You're measuring the wrong thing!" Again, the involvement of service providers is a key to resolving their complaints or calming their fears. It could be that relatively insignificant performance dimensions are being measured and that more important dimensions are being ignored. If so, the remedy is to replace the former with the latter. The involvement of service providers can ensure corrective action.

On the other hand, the debate over appropriate performance measures may have uncovered a fundamental problem. A difference of opinion may exist between upper management and the service delivery unit over desired elements of service, or the unit may simply have labored under a long-standing misconception about what management wants. Such disagreements or misunderstandings should be resolved.

Yet another group of stakeholders in measuring the right thing is the citizen-customers. Their involvement, typically through strategic planning processes, focus groups, and citizen commissions, can enhance the likelihood that the right things are being measured.

"It costs too much, and we don't have the resources!" Understandably, departmental officials who already feel that their resources are stretched too thin may be reluctant to tackle new, time-consuming measurement and reporting tasks that may siphon program resources. Those officials are a little like the overburdened logger who, facing stacks of uncut logs, felt he could not spare the time to sharpen his dull saw. Well-meaning but reluctant officials must be persuaded of the value of performance measurement as a tool for improving services and making better use of scarce resources.

There is no reason to embark on performance measurement improvements unless better measures are expected to lead to improved services, to make services more efficient, or to make them more equitable. Service providers must be assured of management's commitment to those ends and be convinced of management's resolve to use performance measurement to improve, rather than to drain resources from, services to the public. It may be helpful to point out to skeptics that most local government officials in jurisdictions that have

Table 2.2   Steps in the Development and Administration of a Performance
Measurement and Monitoring System

1. Secure managerial commitment.
2. Assign responsibility (individual or team) for spearheading/coordinating departmental efforts to develop sets of performance measures.
3. Select departments/activities/functions for the development of performance measures.
4. Identify goals and objectives.
5. Design measures that reflect performance relevant to objectives:
    *   Emphasize service quality and outcomes rather than input or workload.
    *   Include neither too few nor too many measures.
    *   Solicit rank-and-file as well as management input/endorsement.
    *   Identify the work unit's customers and emphasize delivery of services to them.
    *   Consider periodic surveys of citizens, service recipients, or users of selected facilities.
    *   Include effectiveness and efficiency measures.
6. Determine desired frequency of performance reporting.
7. Assign departmental responsibility for data collection and reporting.
8. Assign centralized responsibility for data receipt, monitoring, and feedback.
9. Audit performance data periodically.
10. Ensure that analysis of performance measures incorporates a suitable basis of comparison.
11. Ensure a meaningful connection between the performance measurement system and important decision processes (e.g., goal setting, policy development, resource allocation, employee development and compensation, and program evaluation).
12. Continually refine performance measures, balancing the need for refinement with the need for constancy in examining trends.
13. Incorporate selected measures into public information reporting.

performance measurement systems report that their systems have proven to be worthwhile (McGowan & Poister, 1985; Poister & Streib, 1999).

Developing a Performance
Measurement and Monitoring System

Performance measures will not lead inevitably to improved performance. Although the act of measuring an operation for its results draws attention to that function and may thereby inspire greater efforts and improved performance, such results cannot be guaranteed. Performance measurement is merely a tool. If wielded properly, it can identify areas of performance adequacy and areas of performance deficiency; however, it can neither explain the former nor prescribe remedies for the latter. Reliable explanations and appropriate prescriptions require subsequent analysis of those targeted operations.

Performance measurement and monitoring systems may be developed in a variety of ways. The steps outlined in Table 2.2 have been gleaned from the experience of several municipalities and distilled from writings on the topic, but many variations on this pattern are possible and, indeed, perhaps desirable in a given setting.

Securing management's commitment—at the highest level possible—is an important first step. Not only will that commitment help to overcome resistance at lower levels, but also management interest will increase the likelihood that resulting measures will be used in subsequent decisions, thereby reinforcing and reinvigorating the process. Although the commitment of the city's chief executive is highly desirable, the bare essential for performance measurement at the service unit level is the commitment of that service unit's administrator or supervisor.

Decisions regarding functional responsibility for coordinating the development of performance measures and identifying functions to be measured should be made at an early stage. Depending on the measurement foundation already in existence, some cities may be capable of establishing a solid performance measurement system covering all departments simultaneously; however, some experts recommend a more gradual approach, beginning with one or two departments or activities as demonstration or pilot projects (Bens, 1986). Whichever strategy is adopted, it is reasonable to anticipate that some departments will be more receptive to performance measurement than others. It is also important to recognize that some departments perform functions more conducive to easy measurement than others do.

Performance measures should be consistent with and pertinent to a department's goals and objectives. If such goals and objectives have not been articulated, their development or clarification should become an early step in the process. Mission statements and goals are broad expressions of general purpose. Objectives are more precise, declaring specifically what is going to be accomplished and when. Performance measures track progress in achieving those objectives, ideally documenting the efficiency with which they are pursued and the extent of their achievement.

Performance measures should emphasize the quality of services and the outcomes that those services produce rather than merely reporting the resources consumed or providing workload or activity figures. Indicators that simply record compliance with prescribed processes rather than service quality or results should be deemed acceptable substitutes only when more direct measures of effectiveness are impractical.

Measurement systems should focus on neither too few nor too many performance indicators. Having too few can cause important service dimensions to be overlooked or can inadvertently encourage perverse behavior in the overzealous pursuit of high marks on the featured dimension. Having too many measures can contribute to information overload and disuse of the system.

The proper approach may be the development of a system with a moderate number of performance indicators to be tracked by departmental officials, with only a subset of key measures published in performance reporting documents designed for higher-level officials or the public (Hatry et al., 1992, p. 3). To reduce the tendency to swamp readers by reporting relatively unimportant as well as vital information, for instance, the city of Milwaukee limits departments to reporting only their five most important outcome measures in the budget (Barrett & Greene, 1994, p. 42).

Performance measures should be customer sensitive, emphasizing effectiveness in meeting customer expectations as well as efficiency in service delivery. The early involvement of rank-and-file employees and service recipients along with management in the design of appropriate performance indicators will minimize the likelihood of overlooking important performance dimensions.

Establishing the logistics of performance measurement and monitoring is important. Assigning responsibility within the service delivery units for data collection and reporting, determining the frequency of reporting (e.g., monthly, quarterly, or semiannually), and assigning centralized responsibility for data receipt, monitoring, and feedback are essential for ensuring the viability of the system and sustaining interest in its continuation. Maintenance of the system should also include provisions for periodic audit of performance data to ensure their accuracy.

For performance data to be useful, periodic analysis should examine the performance of service delivery units in comparison with performance in previous periods, standards, targets, comparable jurisdictions, or other suitable bases of comparison. Although continuity of performance measures is desirable for identifying trends, performance measures should be refined or replaced whenever necessary.

A meaningful connection should be established between the performance measurement system and the organization's decision processes. For example, the design of performance measures should be influenced by a municipality's previous goal-setting efforts. Performance measurement results may, in turn, influence future goal-setting processes, policy development, resource allocation, and employee development and compensation decisions. Furthermore, a performance measurement system may be instrumental in a municipality's subsequent program evaluation efforts, as well as its accountability and public information initiatives.

Benchmarking

To determine whether a municipality's performance is favorable or unfavorable, it is necessary to compare that jurisdiction's performance marks against some relevant peg. Among city governments that monitor their own performance, many compare current performance with figures for the same measures in previous reporting periods. Some compare the performance measures of different units in the same jurisdiction providing similar services or compare performance records with predetermined targets. Until recently, relatively few have used national or state standards, private sector performance, or the performance records of other jurisdictions as benchmarks for gauging their own jurisdiction's performance (Hatry, 1980b).

In the private sector, benchmarking has been widely acclaimed as a technique that contributes to performance improvement. By identifying best-in-class performers and the practices that make them so, industries may refine their own processes in a quest to meet or exceed the benchmarks set by outstanding performers.

Public sector application of the term *benchmarking* has been a bit broader. In some cases, objectives or targets set arbitrarily by a state or local government itself have been labeled *benchmarks*. In other cases, the term has been used in a manner more consistent with private sector application. In this volume, the term *benchmark* will be reserved for anticipated or desired performance results anchored either in professional standards or in the experience of respected municipalities.

Municipalities desiring to identify suitable benchmarks for their operations confront two major issues. One is data availability. This volume will overcome that problem by providing information on standards and the experience of several other municipalities for a host of local government functions. The second issue is comparability. In assembling suitable benchmarks, each jurisdiction must be vigilant in identifying factors that make some jurisdictions suitable for comparison and others unsuitable. Cost comparisons are especially vulnerable; differences in reporting periods, accounting practices, and cost of living may confound simple comparison. Social and economic factors that may influence the difficulty of a given jurisdiction's incoming workload may similarly set it apart from its counterparts and limit the value of comparisons. Care is therefore in order as communities select their benchmarks.

Cautionary notes are appropriate for jurisdictions embarking on benchmarking, but so are words of encouragement. The principle of accountability demands that department heads and managers be able to tell city councils how their operations are doing, and that city councils, in turn, be able similarly to inform the public. How a city government is doing in comparison with last year is interesting, but not as interesting as how it is doing in comparison with national standards or with others in the same field. Such comparisons are sometimes difficult and often require explanations. The willingness of some public officials to grapple with that difficulty and to provide explanations when they are needed may well distinguish those truly committed to the principle of accountability from others who are not.

Notes

1. Hatry (1980b) identifies 10 types of performance measures: cost measures, workload-accomplished measures, effectiveness/quality measures, efficiency/productivity measures, actual unit cost–to–workload standard ratios, efficiency measures and effectiveness quality, resource utilization measures, productivity indices, pseudomeasures, and cost-benefit ratios. Although this is an excellent list of the types of indicators presented as performance measures in various reports, some may be dismissed as impostors (e.g., cost measures and pseudomeasures), and others may be incorporated into the four major types of performance measures identified in this chapter.

2. *Labor hour* is one of several terms used in this volume to depict the equivalent of an hour of work performed by one employee or a combination of employees (e.g., two employees working $\frac{1}{2}$ hour apiece on a given project is also 1 labor hour). *Man hour,* the term historically used in local government, will be used interchangeably in this volume with *labor hour, staff hour,* and other similar terms.

3. For more details on the use of trained observer ratings, see Hatry (1999, pp. 86–93), Hatry et al. (1992), and O'Connell (1989).

4. When simply asked in surveys if they measure performance, local government officials tend to respond affirmatively, with high percentages claiming to use not only workload measures but also efficiency and effectiveness measures (see, for example, O'Toole & Marshall, 1987; O'Toole & Stipak, 1988). More exacting research, requiring actual evidence of performance measurement, often reveals less extensive measurement activity and only limited use of efficiency, effectiveness, and productivity measures.

3

Animal Control

A nimal control problems may not be the number one aggravation of every mayor and city manager in America, but they easily make most "top ten" lists. In many cases, problems erupt over matters that seem trivial in relation to much bigger issues confronting the community. A neighbor's dog barks too much; someone's roaming cat assaults songbirds at the backyard feeder; wandering dogs—or even dogs being walked by their owners—foul the lawn; a dog "protecting" its owner's front yard frightens bicyclists and pedestrians; someone's flower bed was destroyed last night; the garbage was tipped over, *again*. Monumental crises? In most instances, no; but on some occasions, animal control problems rise beyond the nuisance and aggravation level to pose serious threats to health and safety. Dog attacks are the greatest reportable childhood public health problem in the nation—greater than measles, mumps, and whooping cough combined (Handy, 1993, p. 2).

Whether threatening or merely annoying, disputes over pets can be serious and often quite emotional for the persons involved. Prompt and effective action is important.

Evaluating Animal Control Operations

Communities secure animal control services in a variety of ways. Some contract for these services from other local government units or from a local humane society. Others provide animal control using their own employees, sometimes organized as a separate animal control department but often as a component of the public health or police department.

Whether secured by contract or through the efforts of local government workers hired for that function, a city that *pays* for animal control services needs some means of judging the adequacy of these services. Typically, a city will simply compare workload statistics for 2 or 3 years and assume that the operation is doing all right if the numbers are up for licenses, apprehensions, and adoptions and if complaints to the mayor and city council members are down. As informative as these numbers may be, they compare the jurisdiction only with itself in earlier periods. If it was performing well in the past, it is doing even better today. But what if previous performance was woefully inadequate? Even if performance is better today, is it where it should be? The answer to that question requires comparison with relevant external pegs.

Officials wishing to assess the adequacy of an animal control operation should consider criteria focusing on two important performance dimensions:

1. Compliance with standards for the proper operation of an animal control service, especially as they pertain to humane animal treatment
2. Operational efficiency and effectiveness

Standards of Proper
Operation and Humane Treatment

The Humane Society of the United States (HSUS) has been among the national leaders in the development of standards for animal control operations. HSUS standards address topics ranging from animal shelter operating hours, fees, and policies (Table 3.1) to operating procedures (Table 3.2), estimated kenneling needs (Table 3.3), and shelter size and design. Local policies, procedures, and facilities may be assessed by comparing them with HSUS guidelines, if deemed applicable and appropriate by community officials.[1] Local compliance with such standards may be a more revealing indicator of program adequacy than year-to-year comparisons of workload statistics.

HSUS contends that most successful animal control programs emphasize legislation, education, and sterilization. Common elements include

- An ordinance establishing differential licensing with lower fees for spayed or neutered pets
- A low-cost spay/neuter clinic or equivalent program
- Mandatory sterilization of animals adopted from the public shelter
- A public education program (Handy, 1993, p. 2)

Sterilization is an important component of animal control programs not only because it limits the animal population but also because it tends to curb animal aggression. According to HSUS, unsterilized animals are twice as likely as spayed or neutered animals to bite humans and, furthermore, account for 95% of all fatal maulings (Handy, 1993, p. 11). Some communities require that animals impounded more than once be spayed or neutered; some require that cats allowed to roam outdoors be sterilized; and many follow the HSUS recommendation of offering lower licensing fees for such pets (Handy, 1993).

Table 3.1   Policy Standards for Animal Control Operations

Animal Shelter Operating Hours
- Shelter should receive sick or injured animals 24 hours per day.
- Shelter should be open for adoption/redemption Monday through Friday, with extended hours to 7 p.m. at least one weekday and open at least 4 hours on Saturday and/or Sunday.

Adoption Standards
- Adoption programs should strive to find responsible lifelong homes for animals.
- Animals should be placed only with adults who intend to keep them as household pets; never as a gift to another person; never to a transient individual; never to an adopter without a properly fenced yard or suitable means of exercise.
- Puppies/kittens should be at least 4 months old before being placed in a home with children less than 6 years of age.
- Only domestic animals should be placed as pets.

Sterilization of Animals Placed by Shelter
- All animals placed by the shelter should be sterilized prior to release or within 30 days of adoption. Any purebred animal should be sterilized before leaving the shelter. Shelters should actively follow up on defaulted sterilization contracts.

Nonadoption Policy for Fighting Dogs
- Any dog bred and conditioned for fighting purposes should be euthanatized as soon as legally possible rather than placed for adoption.

Policies on Fees
- There should be no fee for picking up or receiving animals.
- Adoption fee for purebred or especially desirable animals should be no higher than for others; adoption fees for dogs should be no higher than for cats.

SOURCES: Adapted by the Humane Society of the United States (HSUS) in May 2000 from the following publications to reflect current HSUS policy: *HSUS Guidelines for Animal Shelter Policies* (Washington, DC: HSUS, April 1982); *HSUS Shelter Guidelines Pertaining to Potentially Dangerous Dogs* (Washington, DC: HSUS, November 1988); *HSUS Guidelines for Responsible Pet Adoptions* (Washington, DC: HSUS, n.d.). Used by permission.

Operational Efficiency and Effectiveness

In addition to standards for facilities and recommended policies, national performance statistics and rules of thumb developed by professional associations and municipal leagues provide pegs for comparison with local performance marks. One such rule of thumb, useful for estimating shelter requirements or for documenting a greater-than-normal animal control problem, suggests that the number of animals entering the shelter during a year will equal approximately 7% to 10% of the human population of the community (HSUS, 1990, p. 1).

Useful benchmarks for assessing operational effectiveness may be drawn from national studies and from the performance targets and experience of respected municipalities. On a national basis, a little more than half of the dogs

Table 3.2   Summary of Guidelines for Animal Shelter Operations

Facility
- Facility size and design should permit necessary animal separation (e.g., dogs from cats, sick or injured from healthy, puppies from adults, aggressive from others, etc.) and minimize stress.
- Sealed concrete (or other nonporous) floors should slope toward drains.
- Nonporous walls at least 4 feet high and topped with at least 2 feet of chain link or wire mesh should separate kennels.
- Runs should be covered with fence fabric or wire mesh.
- Heavy-duty drainage and plumbing system should include drainage for each run to prevent cross contamination.
- Facility design should provide adequate heating, cooling, humidity control, and air circulation (guidelines: floor temperature of at least 65°F for adult animals, at least 75°F for infant animals; circulation sufficient to permit exchange with outside air 8 to 12 times per hour).
- Cages for indoor holding of dogs should have floor space of at least 12 square feet for small breeds, 20 square feet for medium breeds (36–50 pounds), and 24 square feet for large breeds.
- Dog kennels with runs should include at least 56 square feet combined (24 square feet for kennel and 32 square feet for run). For shared kennels, allow a similar amount of space per dog.
- Cages for cats should provide floor space of at least 9 square feet—with assignment of 1 cat per cage, except for mothers with kittens.
- Euthanasia room and dead-animal storage should be accessible from the kennel but separated from public view.

Procedures
- Feeding 3 times per day for puppies/kittens less than 12 weeks of age; twice per day if 12 weeks to 1 year old. Make allowances for individual animals.
- Potable water should be available at all times.
- Self-feeders, if used, should be mounted to prevent contamination.
- Each cat cage should contain a litter pan and a perch (or resting shelf).
- Animals should be inspected and treated for disease on arrival and daily thereafter.
- All kennels, cages, and runs should be scrubbed daily with hot water and disinfectant—with animals placed in separate holding area and remaining there until enclosure is completely dry.
- Dogs confined in cages should be exercised at least twice a day.
- Euthanasia should be performed only by trained personnel using humane methods, preferably intravenous (IV) injection of sodium pentobarbital or pentobarbital plus local anesthetic. However, bottled carbon monoxide (CO) is conditionally acceptable under specified conditions and application. Hot, unfiltered CO from an engine is not acceptable, and CO is not acceptable on old or injured animals or those 16 weeks or younger. Dogs and cats should never be placed in a CO chamber together, nor should like species be overcrowded in the chamber.
- Disposal of dead animals should be by incineration, burial in a landfill, or another method approved by the community.

Records
- Thorough and easily accessible records should be maintained for all animals entering the shelter and for all receipts (e.g., donations and fees).

Vehicles
- Vehicles should be well marked and fully equipped with animal control equipment and should include a protected, well-ventilated, and separate enclosure for each animal carried.

SOURCE: Adapted by the Humane Society of the United States (HSUS) in May 2000 from *HSUS Guidelines for the Operation of an Animal Shelter* (Washington, DC: HSUS, September 1990), to reflect current HSUS policy. Used by permission.

Table 3.3   Projected Kenneling Needs for Dogs and Cats

Human Population	Minimum Kenneling Needed for Number of Animals to Be Handled	
	Dogs and Puppies	Cats and Kittens[a]
Under 50,000	0–50	0–20
50,000–100,000	50–75	20–30
100,000–200,000	75–100	30–35

SOURCE: Phyllis Wright, Barbara A. Cassidy, and Martha Finney, "Local Animal Control Management," *Management Information Service Report, 18*(7) (Washington, DC: International City/County Management Association, July 1986), p. 12. Reprinted by permission.

a. The calculations given for cat kenneling do not take into account the increasing number of laws that include cat licensing and rabies vaccinations. If there is a local ordinance that requires cats be held for 5 days or more, more kenneling for the species would be needed.

and cats brought to animal shelters are captured as strays (Moulton, Wright, & Rindy, 1991). More than half of the dogs and more than 70% of the cats eventually are euthanized, with the remainder either redeemed by their owners or placed in new homes (National Council on Pet Population Study and Policy, 2000, p. 3). A shelter that avoids high euthanization rates by returning to their owners or placing in new homes approximately two out of every five dogs and cats it receives is achieving the national average; a shelter that exceeds this rate of placements is performing quite well by national standards. The shelter in Germantown, Tennessee, for example, boasts a redemption-by-owner rate of more than 25% and an adoption rate of more than 50%, forcing the euthanization of less than one fourth of the animals entering that shelter (Table 3.4).

Raw counts of animal apprehensions and other shelter intakes should be interpreted with care. Rising numbers indicate greater animal control activity and may suggest increased effectiveness. At some point, however, increased effectiveness in animal control—including reduction in the number of the animals at large and reduction of unwanted pets as a result of effective education and sterilization programs—should *reduce* the number of animals entering the animal shelter.

Responsiveness is an important dimension of public service. How prompt should responses to animal control calls be? The response times and response time targets of several cities compiled for this chapter suggest that an average emergency response time of 20 minutes or less is achievable by many agencies (Table 3.5). A set of standards developed by the League of California Cities (LCC) distinguishes between different degrees of response urgency (Table 3.6). Calls about threats to human life or safety should receive highest priority

---

Table 3.4   Animal Recovery/Adoption Statistics for Local Shelters: Selected Cities

Germantown, TN
 Redeemed: 25.3%
 Adopted: 52.6%
 Euthanized: 22.1% (1996)

Palo Alto, CA
 Targets: Reunite at least 65% of stray dogs with owners; at least 10% of stray cats (1999)

Plano, TX
 Redeemed: 30%
 Adopted: 30% (1997)

Boca Raton, FL
 Percentage redeemed or adopted: 58% (1997)

Lubbock, TX
 Percentage recovered: 28% (1996), 41% (1997)

Grand Prairie, TX
 Percentage adopted: 27% (1996)

Denton, TX
 Redeemed: 20%
 Adopted: 15%
 Euthanized: 65% (1997)

Wichita, KS
 Reclaimed: 15%
 Adopted: 11%
 Euthanized: 74% (1998)

Chesapeake, VA
 Percentage of impounded animals redeemed or adopted: 24.5% (1997)

---

and immediate response; slower response times for routine calls are generally acceptable.

Top-notch programs recognize that effective animal control involves much more than efficient apprehension of strays. Many set ambitious targets for key aspects of an effective program, including high rates of animal licensing (Table 3.7). Other points of emphasis range from animal health and sterilization to cost recovery and animal control officer efficiency (Table 3.8).

The California standards noted previously (LCC, 1994, p. 20) offer rules of thumb for judging animal control officer workload. Field officers in high-service departments may be expected to handle fewer than 1,000 calls per year, those in medium-service departments are likely to deal with 1,000 to 2,500 calls each, and officers in low-service departments might be required to respond to more than 2,500 calls for service annually. According to the LCC's judgment, field officers in high-service agencies are likely to spend more than half their time on patrol, compared with 30% to 50% in medium-service departments and less than 30% in low-service departments.

Table 3.5   Prompt Response to Animal Control Calls:
Performance Targets and Experience of Selected Cities

Duncanville, TX
Average response time to emergency calls: 6 minutes (1998)
Average response time to nonemergency calls: 15 minutes (1998)

Irving, TX
Target: Average of less than 20 minutes to animal bite incidents
Actual: 13-minute 45-second average (1998)

Prairie Village, KS
Average response time to animal bite calls: 15.5 minutes (1995)

Lubbock, TX
Actual: 16-minute average for emergency responses; 37-minute average for routine responses (1997)

College Station, TX
Percentage of responses within 15 minutes during work hours: 77% (1998)

Overland Park, KS
Target: Average response time of 20 minutes to calls for service concerning animal control problems

Denton, TX
20.59-minute average (1997)

Athens–Clarke County, GA
Targets: Respond to emergency calls within 15 minutes; all calls within 30 minutes
Actual: 29-minute average for emergency calls; 144-minute average for all calls (1997)

College Park, MD
Targets: Respond within 30 minutes 90% of the time during business hours; within 2 hours 90% of
    the time during nonbusiness hours

Charlotte, NC
Targets: Respond to (a) 90% of all emergency calls within 45 minutes and (b) 100% of all emergency
    calls within 1 hour; and respond to, investigate, and resolve (c) 98% of nonemergency complaints
    within 1 workday and (d) 100% of all nonemergency complaints within 2 workdays
Actual: (a) 94%; (b) 100%; (c) 97%; (d) 100% (1987)

Palo Alto, CA
Target: Respond to at least 85% of calls for service within 45 minutes (1999)

Reno, NV
Targets: Respond to all emergency calls within 1 hour; 95% of nonemergency calls within 8 hours
    (1999)

Corpus Christi, TX
Targets: Respond to bite and attack reports within 1 hour; injured animal calls within 1 hour; animals
    in traps picked up within 8 hours; complaints, including stray animals, answered within 24 hours

Boca Raton, FL
Targets: Respond to calls for pickup within 1 hour during the day; within 2 hours at night (1999)

Kansas City, MO
Average response time to high-priority calls: 76 minutes (1995)

Boston, MA
Actual: 75% of responses within 24 hours of emergency complaint (1992)

Table 3.6   Response Time Standards: League of California Cities

Incident	Agency Service Level		
	High	Medium	Low
Nuisance animal	Less than 4 hours	4 to 72 hours	More than 3 days
Aggressive animal	Less than 20 minutes	20 to 120 minutes	More than 2 hours
Neglect or cruelty	Less than 1 hour	1 to 24 hours	More than 24 hours
Confined or trapped animal	Less than 90 minutes	1.5 to 24 hours	More than 24 hours
Dead animal pickup	Less than 4 hours	4 to 36 hours	More than 36 hours
Sick or injured animal	Less than 15 minutes	15 minutes to 4 hours	More than 4 hours
Animal endangering human life or safety	Immediate action is the only acceptable performance		

SOURCE: League of California Cities (LCC), A "How To" Guide for Assessing Effective Service Levels in California Cities (Sacramento, CA: LCC, 1994), pp. 18–19. Reprinted by permission.

Table 3.7   Licensing of Animals: Selected Cities

Oak Ridge, TN
  Registered animals per 1,000 residents: 161.4 (1997)

San Clemente, CA
  Dog licenses issued per 1,000 residents: 104.4 (1998)

Jacksonville, FL
  License tags dispensed per 1,000 residents: 81.4 (1998)

Long Beach, CA
  Dog licenses issued per 1,000 residents: 54 (1994), 52 (1995), 58 (1996)

Fort Worth, TX
  Registered animals per 1,000 residents: 52.9 (1997)

User fees for public services—primarily license, shelter, and adoption fees in the case of animal control operations—are important policy issues in many communities. How much of the expense of animal control operations should, for instance, be recovered through license fees? In some cities, such as Eugene, Oregon (Table 3.8), a substantial portion of expenses are recouped, but those cities stand as exceptions. A 1990 survey conducted by the Colorado Municipal League revealed that license fees covered at least three quarters of the ani-

Table 3.8   Odds and Ends in Animal Control: Selected Cities

SPAY/NEUTER

Reno, NV
Target: At least 78% of adopted animals spayed/neutered (1999)

SCREENING OF ANIMALS FOR ADOPTION

Oklahoma City, OK
Target: Not more than 2% of adoptions returned for medical reasons (i.e., unhealthy animals)
Actual: 1.91% (1999)

ANIMAL BITE INCIDENCE RATE

Duncanville, TX
Persons per 1,000 population bitten by dogs at large: 0.46 (1997)

PROMPT QUARANTINE

Palo Alto, CA
Target: Quarantine at least 95% of domestic animals involved in a bite within 3 days of receiving bite
   report
Actual: 92% (1997)

COST RECOVERY THROUGH FEES

Eugene, OR
Percentage of service budget offset by revenue: 86% (1997)

PRODUCTION RATIOS

Duncanville, TX
Number of calls for service per animal control full-time equivalent (FTE): 1,828 (1997)

Chesapeake, VA
Number of calls per animal control officer: 1,546 (1997)

Oklahoma City, OK
Average daily intake of animals to shelter per on-duty field officer: 5.76 (1999)

mal control costs in only 11% of the responding cities in that state and less than one fourth of the cost in most others (Handy, 1993, p. 6).

Note

1. A model ordinance for the control and regulation of animals has been developed by the Humane Society of the United States (HSUS, *Responsible Animal Regulation,* 1986). For more information, contact the HSUS, 2100 L Street, NW, Washington, DC 20037.

4

City Attorney

Standards addressing the responsible and ethical conduct of attorneys have been articulated by the American Bar Association. Standards of *performance,* however—benchmarks that would aid a mayor, city manager, or citizen wishing to assess the performance of the city attorney's office—are more elusive.

Benchmark Possibilities

Performance standards are more easily developed for routine functions than for nonroutine work. Because so much of a city attorney's role falls into the latter category, the prospects for precise standards of output that would be both credible and broadly applicable are limited. Nevertheless, careful scrutiny of performance-reporting documents of selected municipalities reveals performance targets and actual experience that offer an external perspective on several key dimensions of attorney performance. That perspective may be helpful to cities wishing to evaluate their own operations.

Preparation of Legislation and Legal Opinions

As the municipality's legal expert, the city attorney is expected to prepare— or at least to approve as to form—proposed ordinances placed before the city council for adoption. The attorney also is expected to render legal opinions on all manner of issues coming before the city government. Cities tend not to report on the quality of the ordinances crafted by the city attorney or on the quality of opinions rendered, but several do report on the speed at which those services are provided.

Among the examined municipal documents that report on city attorney responsiveness, typical turnaround times for requested ordinances and resolu-

Table 4.1   Prompt Preparation of Ordinances and Resolutions: Performance
Targets and Experience of Selected Cities

San Antonio, TX
  Average turnaround for ordinance preparation: 3 days (1995)

San Diego, CA
  Average review time for ordinances: 7 days (1998)
  Average review time for resolutions: 1 day (1998)

Glendale, AZ
  Percentage of ordinances/resolutions prepared within 1 week: 90% (1997)

Orlando, FL
  100% of ordinances/resolutions prepared within 10 workdays of request when accompanied by
    backup material (1991)

Macon, GA
  Percentage of ordinances and resolutions drafted within 10 days: 95% (1995)

Grand Prairie, TX
  Percentage of ordinances and resolutions drafted or reviewed within 10 days: 88% (1996)

Dayton, OH
  Target: Complete 95% of legislation within 2 weeks or within the time designated

Cary, NC
  Percentage of ordinances drafted by other departments that are reviewed within 30 days: 98%
    (1999)

tions range from a few days to as many as 30 days (Table 4.1). Legal opinions
typically are expected within 2 or 3 weeks (Table 4.2). A few expect more
rapid response—notably, San Antonio's 3-day turnaround for ordinance
preparation and informal legal opinions; San Diego's 1-day review of resolutions; and the 4-day turnaround for legal opinions in Chandler, Arizona, and
Fort Collins, Colorado.

Responsiveness to Requests for Legal Action

How quickly does the city attorney file legal documents, review pending
contracts, and perform other important tasks? Municipal attorneys in Ann Arbor, Michigan, attempt to process all requests for criminal charge authoriza-
tions within 3 days and to file lawsuits within 15 days (Table 4.3). The city attorney in Decatur, Illinois, files demolition cases for unsafe structures within
30 days and most environmental and nuisance cases within 21 days. Reported
review times for municipal contracts ranged from 24 hours to an average of 9

Table 4.2 Prompt Preparation of Legal Opinions and Miscellaneous Legal Services: Performance Targets and Experience of Selected Cities

San Antonio, TX
Percentage of informal opinions issued within 3 days: 95% (1995), 96.1% (1997)
Percentage of formal opinions issued within 10 days: 90% (1995), 90.7% (1997)

Chandler, AZ
Target: 4 workdays for routine legal opinions
Actual: 4-day average (1998)

Fort Collins, CO
4-day average turnaround on legal service requests (1991)

Fayetteville, AR
Response to 100% of requests for legal services within 5 days (1991)

Glendale, AZ
Percentage of legal opinions and contract reviews provided within 1 week: 75% (1997)

Ann Arbor, MI
Target: Respond to all requests for written information or opinions within 7 days

Boston, MA
Percentage of requests for legal opinions answered within 10 days: 100% (1996)

Bakersfield, CA
Target: Respond orally or in writing to all requests for legal advice immediately or within 10 days if research is required

Tucson, AZ
Percentage of mayoral, council, city manager, and departmental requests for legal opinions completed within 2 weeks of request or approved extension: 95% (1997)

Oakland, CA
Target: Respond to at least 85% of requests for legal assistance within 21 days
Actual: 85% within 21 days (1992)

Decatur, IL
Target: Respond to at least 85% of written legal inquiries within 30 days
Actual: 75% within 30 days (1995)

Overland Park, KS
Target: Oral opinions on routine matters within 24 hours, on nonroutine matters within 30 days; written opinions on routine matters within 15 days, on nonroutine matters within 30 days

Wilmington, DE
Target: Respond within 30 days to at least 90% of requests for legal opinions

Milwaukee, WI
Target: Complete at least 25% of opinions and assignments within 30 days; 50% within 90 days
Actual: 15% completed within 30 days; 40% within 90 days (1992)

days. The time required to prepare a contract is longer—for example, 15 workdays in Orlando.

Table 4.3   Responsiveness to Requests for Legal Action: Selected Cities

PROMPT FILING OF ARREST DOCUMENTS

Ann Arbor, MI
Targets: Process all requests for authorization of criminal charges within 1 day if they do not require preparation of a complaint; within 3 days if they require preparation of a complaint

Orlando, FL
100% of charging documents for city code violations filed within 15 days (1991)

PROMPT FILING OF CASES/LAWSUITS

Ann Arbor, MI
Target: File lawsuits within 15 days after all necessary information has been received

Decatur, IL
100% of demolition cases (unsafe structures) filed within 30 days of council approval; 89% of environmental and nuisance cases filed within 21 days of request for prosecution (1995)

PROMPT REVIEW/PREPARATION OF CONTRACTS

Fayetteville, AR
100% reviewed within 24 hours (1991)

Winston-Salem, NC
Target: Review contracts within 2 days

Ann Arbor, MI
Targets: Review all standard contracts within 3 days; review or prepare all nonstandard contracts within 15 days after all necessary information is received

Cary, NC
Percentage of contracts reviewed within 5 workdays of receipt: 100% (1999)

Glendale, AZ
Percentage reviewed within 1 week: 75% (1997)

Boston, MA
9 days (average) for review of noncompetitively bid contracts; 5 days (average) for competitive contracts (1992)

Charlottesville, VA
Target: Review at least 75% of all contracts within 5 workdays of receipt
Actual: 59% (1998)

Grand Prairie, TX
Percentage of contracts reviewed or drafted within 10 days: 92% (1996, estimated)

Denver, CO
Average number of days for drafting and processing contracts: 11.0 (1998)

Orlando, FL
100% prepared within 15 workdays of request when accompanied by backup material (1991)

PROMPT PREPARATION OF CITY CODE MATERIAL

Orlando, FL
100% prepared within 60 workdays of ordinance adoption (1991)

PROSECUTION

Oklahoma City, OK

Target: Review and file, or decline to file, court charges within 48 hours of receipt of reports (2000)

TORT CLAIMS

Oklahoma City, OK

Target: Review and offer recommendations on all tort claims within 90 days of receipt (2000)

OTHER

Ann Arbor, MI

Target: review within 1 day all Freedom of Information requests, domestic restraining orders, and subpoenas received from the police department

Grand Prairie, TX

Percentage of open records requests handled within 10 days: 96% (1996, estimated)
Percentage of ordinances and resolutions reviewed or drafted within 10 days: 88% (1996, estimated)

Palo Alto, CA

Target: Respond to at least 70% of city advisory board or commission conflict-of-interest inquiries within 10 workdays
Actual: 80% (1997)

Attorney Workload

The array and volume of tasks an attorney is expected to perform is likely to vary from city to city—and even within a single city large enough to allow specialization among members of the legal staff. Nevertheless, the attorney workloads reported by Shreveport, Louisiana, and San Antonio, Texas, offer a general frame of reference for others to consider when reviewing their own performance (Table 4.4).

Success Ratios

Several aspects of the city attorney's performance lend themselves to the calculation of success rates: convictions as a percentage of all cases, collection rates on claims and other funds owed to the city, recovery rates for delinquent taxes, percentage of lawsuit settlements, percentage of litigation avoided, favorable disposition rates, successful subrogation rates, amount of judgment or settlement as a percentage of the amount demanded of the municipality by claimants, payout ratios, and cases affirmed on appeal, to name several (Table 4.5). If the number of cases used in the calculation of a given ratio is large, that ratio may be adopted as a benchmark more confidently than if the number of cases is small. If, for example, a conviction rate is based on a large number of cases—as the 96% mark in San Diego and the 95% in San Antonio undoubt-

(text continues on page 42)

Table 4.4   Production Ratios for Attorneys: Shreveport and San Antonio

Shreveport, LA
 Lawsuits assigned per attorney: 60 (1994)
 Contracts reviewed per attorney per year: 32 (1994)

San Antonio, TX
 Average number of lawsuits per attorney: 25.6 (1997)
 Number of formal opinions rendered per attorney: 17.5 (1997)
 Number of informal opinions rendered per attorney: 181 (1997)
 Number of ordinances prepared per attorney: 97 (1997)

Table 4.5   Success Ratios for City Attorneys: Performance Targets and Experience
              of Selected Cities

CONVICTION RATE

San Diego, CA
 Percentage of criminal cases resulting in conviction or favorable finding: 96% (1998)

Fayetteville, AR
 Conviction rate for DWI:[a] 98%; warrant charge: 96%; carrying a weapon: 85%; battery: 73% (1998)

San Antonio, TX
 95% conviction rate on traffic court cases litigated (1995)
 86% conviction rate on all cases tried in municipal courts (1997)

Shreveport, LA
 Actual: 95% traffic conviction rate; 95% conviction rate for DWI;[a] 95% conviction rate for municipal
   ordinance violations (1994)

Overland Park, KS
 Actual: 94.7% DUI[a] conviction rate (1991)

Alexandria, VA
 Actual: 94% conviction rate for drug cases (1995)

Dayton, OH
 Target: 90% in housing cases; 85% in prostitution-related cases

Calgary, Alberta
 Conviction rate of 85% (1996)

New York, NY
 Conviction rate in criminal juvenile cases: 78% (1998)

RECOVERY OF FUNDS OWED

Denton, TX
 Actual: 92% of claims collected (1997)

Denver, CO
 Percentage of referred accounts collected: 75% (1998)

Macon, GA
Percentage of outstanding debt in referred accounts collected: 75% (1995)

Lubbock, TX
Actual: 60.2% of claims collected (1991)

Boston, MA
Recovered funds as percentage of program costs: 693% (1996)

RECOVERY OF DELINQUENT TAXES

Nashville–Davidson County, TN
Collection of delinquents as a percentage of all delinquent taxes: 87.5% (1997)

Blacksburg, VA
Collection rate for delinquent real estate tax referrals: 85% (1997)
Collection rate for delinquent business license tax referrals: 100% (1997)

PROMPT SETTLEMENT OF LAWSUITS

Oakland, CA
Actual: 33% of lawsuits resolved within 1 year of filing (1996)

Fayetteville, AR
Actual: 19% of current year's litigation cases closed; 72% of prior year's litigation cases closed (1991)

PROMPT ACTION ON SMALL CLAIMS CASES

Decatur, IL
Actual: Action taken on 57% of cases within 30 days of receipt (1995)

LITIGATION AVOIDANCE

Bellevue, WA
Target 1: Among cases set for jury trial, resolve at least 99% without jury trial
Actual: 98.8% (1997)
Target 2: Among cases set for bench trial, resolve at least 62% without trial
Actual: 61.1% (1997)

Charlotte, NC
Actual: 67% of cases settled by negotiation (1987)

Milwaukee, WI
Actual: 54% of cases resolved by pretrial evaluations (1994)

Oakland, CA
Target: Settle at least 30% of prelitigation claims
Actual: 30% (1992), 5% (1996)

FAVORABLE DISPOSITIONS

Fort Worth, TX
Percentage of cases successfully litigated or dismissed in the city's favor: 97.4% (1997)

San Diego, CA
Percentage of cases seeking monetary damages settled favorably to the city: 90% (1998)
Percentage of land use cases resulting favorably to the city: 95% (1998)

(continued)

Table 4.5   Success Ratios for City Attorneys: Performance Targets and Experience of Selected Cities (continued)

Ann Arbor, MI
Success rate on ordinance enforcement cases: 90% (1998)

Denver, CO
Target: Maintain at least an 85% win record in suits filed against the city
Actual: City prevailed in 85% of claims cases (1998)

San Antonio, TX[b]
Ratio of lawsuit wins to losses: 40 to 3 (1991)
Favorable dispositions: 90% (1995), 85% (1997)

Oakland, CA
Percentage of lawsuits resolved with no monetary payout: 45% (1996)

Bellevue, WA
Percentage of cases resolved at not more than 10% above estimated case value: 88.9% (1998)

SUCCESSFUL SUBROGATION
Alexandria, VA
Percentage of city property damage losses collected from third parties: 68% (1990), 72% (1991), 79% (1992), 81% (1993)

JUDGMENT/SETTLEMENT AS PERCENTAGE OF DEMAND
Shreveport, LA
15% (1994)

LOW PAYOUT RATIO
Cincinnati, OH
Payouts as a percentage of cases closed: 9.4% (1995)

CASES AFFIRMED ON APPEAL
Bellevue, WA
96.4% (1996), 100.0% (1997)

---

a. Driving while intoxicated (DWI); alternatively, driving under the influence (DUI).
b. According to San Antonio, "favorable dispositions" include "favorable rulings administered in court" as well as instances in which the city "filed appropriate motions and case was dismissed or the case was settled favorably" (City of San Antonio [TX], *Adopted Annual Budget, Fiscal Year 1998–1999*, p. 181).

edly are—that rate is less likely to be uncharacteristically high or low simply as the result of a streak of good or bad courtroom luck. Such distortions can more easily occur when the percentage is based on a relatively small number of cases.

Should every city attorney be expected to meet or exceed the ratios reported in Table 4.5? Probably not. Several factors beyond the control of the city attorney could intervene to influence the success ratio in one direction or the other—for example, the difficulty of the cases, the practice of contracting out the most difficult ones, the city council's willingness to accept the city attorney's recommendations regarding litigation, and the municipality's insurance coverage and its ramifications for litigation or settlement. Furthermore, without a larger set of reporting municipalities, it is impossible to know whether these numbers are at the high or low end of a range of reasonable performance expectations. Even so, they offer a useful external context for performance assessments. When the local city attorney is boasting about having won 20 out of 30 cases for a 67% success rate, it might be useful to know that the city attorney's office in San Antonio won 40 out of 43 cases in 1991 and Fort Worth's attorneys successfully litigated or secured favorable dispositions in 97% of that city's cases in 1997—whether this potentially deflating information is shared with the local city attorney or not.

5

City Clerk

Duties of city clerks—often titled "city secretaries"—are varied. In some communities, the city clerk is the chief appointed official and serves as coordinator or manager of most municipal functions. In other communities, the chief executive or administrative director role is performed by another official—typically the mayor, city manager, city administrator, or executive assistant to the mayor—and the primary duties of the city clerk are more narrowly defined. In most cases, basic responsibilities of the office include providing secretarial services to the mayor and city council, preparing minutes of city council meetings, serving as custodian of official records and ensuring access to those records, and serving as a principal contact for citizen and business inquiries. In many municipalities, the city clerk has a major role in coordinating the assembly and delivery of city council agenda packets that provide background information on items being considered at upcoming city council meetings.

Performance Benchmarks From Other Cities

Perhaps because the role of city clerk varies from one community to another, incumbents in that office sometimes assume that meaningful performance comparisons across cities are impossible. Variations in population and differences in the scope and scale of duties assigned to the city clerk reduce the likelihood that the sum of work done in one community closely matches the work done in another.

Although it is true that the performance of most municipal clerks typically is reported only in the form of workload statistics (e.g., city council meetings attended, minutes prepared, and number of requests for official records) and that raw figures from one community are of little use or interest elsewhere, several cities do report information that *is* of use as potential benchmarks.

That information focuses on the fundamental aspects of the job of city clerk—in other words, the responsibilities that tend to fall to the city clerk whether the remainder of assigned duties are narrow or broad—and the information is presented in a standardized way that neutralizes much of the significance of community size and other variations. In such cases, the performance targets and experience of other municipalities can help a city frame its own expectations as it ponders answers to these and other questions regarding the performance of the city clerk:

- How far in advance of a meeting should the city council reasonably expect to receive the meeting agenda and background information?
- How soon after a meeting should the city council expect to receive the minutes?
- How can the quality of the minutes be judged?
- If one assumes that custodial duties for official records imply ensuring general accessibility, does that mean that official action and documents should be indexed? If so, how up-to-date should the index be?
- When the city council takes action that must then be processed by the city clerk, what level of promptness is reasonable to expect?
- When inquiries are made that involve official city records, how quickly should the city clerk be expected to respond?

In some cases, the answers to these questions are established by state law or local ordinance. In other communities, affected parties debate these and other questions of responsiveness and proceed to establish or reaffirm local expectations—often without the benefit of relevant information from other jurisdictions. "It's always taken about a month to prepare the minutes; it's unreasonable to expect them sooner," for example, is an argument based entirely on local experience, deriving no benefit from external information.

Advance Material for Upcoming Meeting

City clerks responsible for assembling and distributing city council agenda packets often take the heat when a last-minute agenda announcement reveals that a controversial or otherwise unwelcome topic will be taken up at the next meeting or when insufficient lead time is provided for reading and considering critical background information and staff analysis. How much lead time is reasonable? Preparing background information for delivery too far in advance of a meeting may deny council members important late-breaking information or may encourage the practice of tacking on one agenda addendum after another as the meeting date approaches. On the other hand, failure to provide adequate lead time restricts the ability of the city council to perform its role. Information reported by some cities suggests a preference for provid-

Table 5.1   Timely Notice of Council Agenda and/or Issuance of Agenda Packet:
Selected Cities

Cary, NC
Percentage of agendas and supporting documents available the week prior to council meetings: 95%
(1999)

Charlottesville, VA
Target: Deliver agenda packets on the Wednesday prior to Monday meetings
Actual: 100% (1998)

Norman, OK
Percentage of agendas provided to city council 5 days in advance of meeting: 100% (1998)

San Clemente, CA
Percentage of agenda packets provided 5 days in advance of meeting: 100% (1998)

Anaheim, CA
Target: Provide notice of council agenda by noon of the Thursday preceding the meeting

Bellevue, WA
Percentage of agenda packets delivered to council at least 4 days before meeting: 98% (1997)

Palo Alto, CA
Target: Prepare and distribute council agendas and accompanying backup material 4 days prior to
the regular meeting (1999)

Sunnyvale, CA
Target: Deliver agenda packets to council members at least 4 days prior to meeting

Grants Pass, OR
Target: Distribute materials on the Thursday prior to the meeting date at least 90% of the time
Actual: 50% (1995)

Cambridge, MA
Target: Agenda ready for distribution at least 72 hours prior to regular city council meetings
Actual: 95% (1996, estimated)

Edmonton, Alberta
Target: Agenda packages provided to councilors at least 3 days prior to meetings

Boston, MA
Target: Distribute council meeting agendas 24 to 48 hours before all meetings

ing advance notice of at least 3 or 4 days, ideally extending over a weekend
(Table 5.1).

Promptness of Minutes Preparation

Considerable variation exists in the turnaround time for city council meeting minutes. Cities such as Boston, Massachusetts; Duncanville, Texas;
Sunnyvale, California; Edmonton, Alberta; and Oak Park, Michigan, expect
minutes to be prepared within 2 days following a meeting, whereas some

cities apparently are comfortable with much longer turnaround periods—even a full month or longer (Table 5.2). If prompt preparation of city council minutes is a high priority only to the city clerk and only as a matter of professional pride, the wisdom of that priority might be debatable in light of other demands. If, on the other hand, city council members and other consumers of those minutes desire more rapid turnaround, preparation time of, say, 2 weeks is well within the band of reasonableness. Turnaround that is even quicker may be difficult but is not unprecedented.

Quality of Minutes

Many city clerks aspire to produce minutes that are error free and are approved by the city council without amendment. The minutes of the town council in Cary, North Carolina, achieved that standard in one recent year, and the cities of Duncanville, Texas, and Overland Park, Kansas, came within a whisker (Table 5.3).

Indexing of Council Documents and Actions

Many progressive city clerks consider the indexing—sometimes the *electronic indexing*—of city council documents and actions to be part of their responsibilities as custodian of official records. Indexing increases the likelihood that persons seeking information on a given topic will find that information and will find it quickly. Oklahoma City set the speed record among cities examined, reportedly indexing council actions within 2 or 3 hours of their occurrence (Table 5.4). Norman, Oklahoma, and Hurst, Texas, attempt to index council items within 2 and 3 days, respectively. Other cities that reported prompt indexing as a priority established targets in the 1-week to 6-week range.

Prompt Processing of Official Documents

Once action has been taken at a city council meeting, the responsibility for seeing that official documents are executed, perhaps published, and copies distributed often rests with the city clerk. How quickly should those duties be performed? Processing times among reporting cities ranged from a prompt 1 or 2 days in Alexandria, Virginia; Lubbock, Texas; Boston, Massachusetts; and Cary, North Carolina, to as many as 30 days as the targeted maximum (Table 5.5).

Retrieval of Records and Information

Speed of information retrieval depends in part on the diligence of clerical employees but also on the adequacy of filing, storage, indexing, and information retrieval systems. Rapid retrieval often is a testimony to those systems. Among cities reporting targets for actual retrieval times, a few reported average retrieval in a matter of minutes—typically 5, 10, or 30 minutes—but several

Table 5.2   Prompt Preparation of City Council Minutes: Performance Targets and
                  Experience of Selected Cities

Boston, MA
  Percentage of minutes distributed within 48 hours: 100% (1996)

Duncanville, TX
  Target: Within 2 workdays of meeting
  Actual: 98% within 2 workdays (1996); 99% (1998)

Sunnyvale, CA
  Target: Within 2 days of meeting
  Actual: 89% within 2 days (projected, 1994)

Edmonton, Alberta
  Target: Draft minutes provided electronically to councilors within 2 days after meeting

Oak Park, MI
  Target: Within 2 business days (1999)

Ann Arbor, MI
  Target: Within 72 hours of council meeting
  Actual: 95% (1998)

Norman, OK
  Percentage of council minutes prepared within 3 days following meeting: 95% (1998)

Reno, NV
  Target: Within 5 working days (1999)

Scottsdale, AZ
  Percentage of minutes transcribed within 7 days: 100% (1999)

Grand Prairie, TX
  Percentage of minutes completed within 1 week following council meeting: 90% (1996)

Plano, TX
  Percentage of minutes prepared within 7 days: 90% (1998, estimated)

Hurst, TX
  Target: Within 7 workdays (1999)

Irving, TX
  Actual: 100% recorded within 8 days of meeting (1998)

Chapel Hill, NC
  Target: Within 10 business days
  Actual: 100% (1995), 100% (1996)

College Park, MD
  Percentage of minutes distributed to council for review within 10 days of meeting: 95% (1999)

St. Petersburg, FL
  Target: By Friday of the following week
  Actual: 100% by Friday of the following week (1990)

Winston-Salem, NC
Target: Prepare summary of minutes within 2 weeks of meeting
Actual: 90% (1998)

Bellevue, WA
Percentage of minutes presented for council approval within 2 weeks: 86% (1997)

Ocala, FL
Target: Within 2 weeks of meeting

Overland Park, KS
Target: Within 4 weeks of meeting (1999)

Anaheim, CA
Target: Available by the second subsequent meeting

Tucson, AZ
Target: Complete minutes and special verbatim transcripts within 30 days of the meeting
Actual: 90% (1997)

Palo Alto, CA
Target: Finalize council minutes within 8 weeks of the meeting (1999)

Table 5.3   Quality of City Council Minutes: Selected Cities

Cary, NC
100% error free (1999)

Duncanville, TX
97% of minutes approved without amendment (1996); 99% (1998)

Overland Park, KS
99% error free (1997)

Corpus Christi, TX
88% of minutes required no corrections (1991)

San Clemente, CA
Target: 100% submitted to council by next meeting with 100% accuracy
Actual: 87% (1998)

Palo Alto, CA
Target: Ensure council minutes require no more than 5 corrections per set (1999)

---

targeted their efforts toward guaranteeing the availability of requested records within 1 or 2 working days (Table 5.6). In some instances, allowances
were made for longer response time in the event that requested information

---

Table 5.4   Prompt Indexing of Council Documents and Actions:
Selected Cities

Oklahoma City, OK
Target: Within 2 hours of adjournment
Actual: 2.5-hour average (1992)

Norman, OK
Percentage of council items indexed and properly distributed within 2 days following council action: 100% (1998)

Hurst, TX
Targets: Index minutes within 3 workdays following approval; index and file official
documents within 2 weeks of final action (1999)

Duncanville, TX
Target: Within 5 workdays
Actual: 95% (1996), 100% (1998)

Anaheim, CA
Target: Within 5 workdays

Oakland, CA
Target: 80% within 7 working days

Ocala, FL
Target: Within 2 weeks

Lubbock, TX
Actual: 80% within 1 month (1991)

Tallahassee, FL
Actual: 35% within 6 weeks (1991)

---

was of a sensitive nature (e.g., falling under Freedom of Information Act provisions) or stored in a remote or otherwise less accessible location.

Other Possible Benchmarks

The availability of comparable statistics from several cities on a given performance dimension lends a degree of confidence to the reasonableness of a
reported performance range. Although less compelling, the reported performance of three, two, or even a single jurisdiction may sometimes be of use.
The city clerk's office in Blacksburg, Virginia, for example, responds to all citizen requests within 24 hours; Palo Alto, California, publishes the council
agenda digest in the local newspaper at least one day before regular meetings;
Sunnyvale attempts to prepare its council action digest within 24 hours of
council meetings; and Oakland, California, reports that 80% to 90% of all
necessary referrals are made within 2 business days following council action

Table 5.5   Prompt Processing of Official Documents: Selected Cities

Alexandria, VA
Target: Complete resolutions within 2 days following council meeting; submit ordinances to city attorney for completion within 1 day following meeting; complete action docket within 1 day following meeting
Actual: 100% (1998)

Lubbock, TX
Actual: 100% of documents processed within 1.5 days of council meeting (1991)

Boston, MA
Percentage of documents processed within 48 hours: 99% (1996)
Percentage of filings processed within 48 hours: 90% (1996)
Percentage of campaign reports processed within 48 hours: 100% (1996)

Cary, NC
Percentage of documents processed within 2 workdays: 100% (1999)

Oakland, CA
Actual: 85% of legislation numbered and distributed within 3 days (1992)

Ocala, FL
Target: Execution of legal documents within 3 days of official action

Reno, NV
Actual: 95% of documents executed within 5 days (1990)

Ann Arbor, MI
Percentage of documents processed within 5 days of receipt: 50% (1998)

Decatur, IL
Actual: 100% of documents processed within 7 days (1995)

Grand Prairie, TX
Percentage of ordinances/resolutions filed within 1 week of adoption: 90% (1996)

Anaheim, CA
Target: Transmit documents within 8 days

Oak Ridge, TN
Target: Execute, publish, or file 90% of official documents within 10 days of adoption, receipt, or
authorization
Actual: 85% within 10 days (1991); 70% (1992)

Irving, TX
Actual: 85% of documents processed and distributed within 14 workdays (1995); 100% (1998)

Tucson, AZ
Percentage of official city documents processed and filed within 30 days of receipt, as required by
law: 100% (1997)

(Table 5.6). Other possible benchmarks address an array of city clerk duties
ranging from the handling of correspondence to the microfilming of records.

Table 5.6   Prompt Retrieval of Municipal Records and Information: Selected Cities

Irving, TX
  Actual: 100% within 5 minutes (1995)

Norman, OK
  Percentage of requests for information from central files handled within 5 minutes: 98% (1998)

Overland Park, KS
  Actual: 10-minute average (1997)

Duncanville, TX
  Actual: 12-minute average

Nashville–Davidson County, TN
  Actual: 98% within 30 minutes (1998)

St. Petersburg, FL
  Target: Available the same day requested (records requested after 3 p.m. available by 8 a.m. the next workday)

Bellevue, WA
  Percentage of record retrieval requests responded to within 24 hours: 100% (1997)

Norfolk, VA
  99% of requests for information (on-site records) processed within 1 day of request (1997)

Denton, TX
  Percentage of information requests from citizens fulfilled within 1 day: 97% (1997)

Oakland, CA
  Target: 90% within 1 working day

Peoria, AZ
  Target: 80% within 24 hours

Tallahassee, FL
  Actual: 100% within 2 days (1991)

Chandler, AZ
  Target: Process at least 95% of all requests for records within 2 working days
  Actual: 99% (1998)

Tucson, AZ
  Percentage of citizen and staff requests for information processed within 2 workdays: 98% (1997)

Santa Ana, CA
  Percentage of records requests processed within 2 days: 90% (1995)

Edmonton, Alberta
  Target: 100% retrieval of records from records center within 48 hours of request

Orlando, FL
  Target: Within 48 hours

Sunnyvale, CA
  Target: 95% within 2 days; microfilmed or inactive records within 10 working days

Palo Alto, CA
  Target: Ensure that requests for current, microfilmed, or inactive records are filled within 5 working days (1999)

Alexandria, VA
  Target: Process requests under Freedom of Information Act within 5 days
  Actual: Average of 10 days (1990); 5 days (1991–1993)

Table 5.7 Odds and Ends for City Clerks: Miscellaneous Benchmarks From Selected Cities

PROMPT RESPONSE TO CITIZEN REQUESTS

Blacksburg, VA
Response to citizen requests and/or referral within 24 hours: 100% (1997)

Duncanville, TX
Percentage of citizen requests resolved within 3 workdays: 98% (1996), 100% (1997)

ACTION/COMMENT LINE

Tucson, AZ
Target: Transcribe and distribute comments received through the mayor and council's Citizen Comment Line within 10 work hours
Actual: 100% (1997)

CORRESPONDENCE

Oak Park, MI
Target: Correspondence follow-up to be completed within 6 days (1999)

PROMPT RESPONSE TO MAYOR/COUNCIL

Oak Park, MI
Target: Complete response to requests for information from the mayor and council within 7days (1999)

NEWSPAPER PUBLICATION

Palo Alto, CA
Target: Publish council agenda digest in the newspaper at least 1 day before regular meetings (1999)

PROMPT PREPARATION OF COUNCIL ACTION DIGEST FOR NEWSPAPER PUBLICATION

Sunnyvale, CA
Target: Prepare at least 95% within 24 hours following meeting (1994)

PROMPT ISSUANCE OF COUNCIL REFERRALS

Cary, NC
Percentage of action agendas (synopsis of council action) prepared the day following council meeting: 100% (1999)

Cambridge, MA
Target: Notification of council actions completed 38 hours after meeting
Estimate: 90% (1996)

Oakland, CA
90% within 2 working days of meeting (1991); 80% (1992)

Plano, TX
Percentage of agenda items processed within 2.5 days: 100% (1998, estimated)

Winston-Salem, NC
Percentage of adopted documents distributed to departments within 4 days of meeting: 95% (1998)

(continued)

Table 5.7   Odds and Ends for City Clerks: Miscellaneous Benchmarks From
Selected Cities (continued)

Oak Ridge, TN
80% of council actions followed up within 5 days (1997)

LEGISLATIVE ACTIONS COMPUTERIZED

Palo Alto, CA
Target: Provide that adopted legislative actions are computerized within 60 days of the city council
meeting (1999)

POSTING OF COUNCIL ACTIONS

Oakland, CA
Percentage of legislative actions posted to the Legislative Information System (LIS) within 72 hours:
95% (1996)

CODIFICATION

Palo Alto, CA
Target: Transmit newly adopted ordinances electronically to publisher for codification within 2 business days after receipt of ordinance from the city attorney (1999)

Flagstaff, AZ
Percentage of ordinances codified within 1 week of their effective date: 100% (1997)

BOUND OFFICIAL RECORD

Cambridge, MA
Target: Permanent bound record produced within 6 months after completion of legislative year
Estimate: 95% (1996)

RECORDS STORAGE

Tucson, AZ
Percentage of records processed and stored within 1 week of receipt from city departments: 35%
(1997)

BOARDS/COMMISSIONS

Palo Alto, CA
Target: Ensure vacancies on a board/commission have at least 2 applicants (1999)

OUTPUT STATISTICS: PROCESSING TIME FOR VITAL STATS

Boston, MA
17 minutes per record (birth, marriage, or death certificate) (1992)

OUTPUT STATISTICS: MICROFILMING

Reno, NV
333 pages microfilmed per hour (1990)

6

Courts

Be careful talking about performance standards and performance measurement around municipal court judges. Courts, they are likely to declare, are about due process and justice; they are *not* about revenues and efficiency. And who wants to argue with a judge?

Courts *are* about justice and due process, but measures of effectiveness can be devised that are consistent with that mission. Furthermore, measures of efficiency need not be considered an affront to the court's higher mission. Prompt—that is, efficient—judicial action backed by efficient support services is a worthy objective.

Possible Benchmarks

The nature of the municipal court function renders quite slim the prospect of extremely precise and widely accepted standards of performance, at least insofar as the nonroutine elements of the court's work are concerned. The more routine support services attached to court operations are another matter. Many of those services can be the focus of engineered standards or performance benchmarks. Unfortunately, few cities report municipal court performance statistics in a fashion that makes them viable candidates for performance benchmarks.

The relative scarcity of measures of efficiency and effectiveness from individual municipal courts increases the difficulty of identifying a reasonable performance range, but the small number does not render the reported measures valueless. They simply must be applied more cautiously.

Prompt Scheduling and Disposition

The right to a speedy trial is a fundamental principle of the U.S. judicial system. Some trials, however, are much speedier than others. For example, most of the cases in the municipal courts of Fort Worth and College Station, Texas, and all of the cases in Reno, Nevada, were set for trial within 60 days of arrest or request for trial (Table 6.1). Even speedier court action could be found in Savannah, Georgia, where the average time between the filing and disposition of criminal accusations was 24 days in 1997. A more common interval for case disposition among cities reporting this statistic was 90 days or less.

Processing Paperwork

Administrative and clerical tasks associated with municipal court operations—from inputting case information into the court's record system to the processing of appeals—are important components of court workload. The performance targets and experience of other jurisdictions in completing these tasks offer an external context in which to assess local performance (Table 6.2).

Court Employee Production Rates

Selected production benchmarks for court employees are provided in Table 6.3. Because of the limited number of cities reporting such information, these benchmarks should be applied cautiously.

Warrants Served and Cases Closed

When courts function effectively, warrants are served promptly and cases are closed in a timely fashion, with few lingering unresolved. Lubbock, Texas, led all other reporting cities with 97% of its warrants cleared in 1997 (Table 6.4). Chesapeake, Virginia, reported that 99.7% of its General District Court cases in 1997 were concluded within 12 months to lead other cities reporting performance in that category (Table 6.5).

Collection Rate for Fines

What is a reasonable collection rate for fines imposed by the municipal court? Figures provided by selected cities suggest that a realistic rate lies somewhere between 40% and 90% (Table 6.6).

Other Indicators of Effectiveness

Courts that hear high percentages of the cases filed and manage to handle other aspects of their work promptly may look to still other indicators for additional signs of effectiveness (Table 6.7). The municipal court in Duncanville, Texas, for example, reported that only 5% of its decisions were appealed in 1998. The court in Prairie Village, Kansas, had none of its decisions reversed on appeal in 1995. Compliance with terms of probation was not a problem in Norman, Oklahoma, where in 1997 only 1% failed to comply.

Table 6.1   Prompt Scheduling and Disposition of Court Cases: Selected Cities

SCHEDULING

Reno, NV
  Target: Schedule all trials within 60 days of arrest
  Actual: 100% (1990)

College Station, TX
  Percentage of contested court cases set for hearing within 60 days of request: 90% (1998)
  Percentage of noncontested hearings set within 30 days of request: 85% (1995)

Fort Worth, TX
  Target: Schedule 85% of cases on a court docket within 60 days of receipt of the request for a court
    setting
  Actual: 55% docketed in less than 60 days (1997)

Irving, TX
  Percentage of pretrial hearings scheduled within 4 weeks: 100% (1998)
  Percentage of trials scheduled within 90 days: 100% (1998)

Fayetteville, AR
  Target: Set criminal cases for trial within 90 days

DISPOSITION

Savannah, GA
  Target: Average time of 21 days between filing and disposition on all criminal accusations
  Actual: 24 days for criminal defendants (1997)
  Target: Average of 40 days for processing traffic citations from filing to disposition
  Actual: 35 days for traffic citations (1997)

Prairie Village, KS
  Average case processing time (from random sample): 6 weeks (1995)

Overland Park, KS
  Target: Dispose of minor traffic cases within 45 days; major traffic cases within 60 days; misde-
    meanor and driving under the influence (DUI) cases within 75 days; code violation cases within 45
    days; and sentencing (when pre-sentence report or an evaluation is required) within 30 days (1996)

Norman, OK
  Cases disposed within 90 days: 95% (1997), 94% (1998)

Oklahoma City, OK
  Percentage of criminal cases disposed within 90 days: 86% (1991), 88% (1992)
  Percentage of jury cases disposed within 90 days: 84% (1991), 85% (1992)
  Percentage of traffic cases disposed within 90 days: 91% (1991), 100% (1992)

Milwaukee, WI
  83% of nonpriority cases tried within 90 days after intake date (1997)

Grand Prairie, TX
  40% of cases disposed within 90 days (1996, estimated)

Eugene, OR
  Average number of working days from citation to trial date: 114 (1997)

Table 6.2   Prompt Preparation and Processing of Court Paperwork: Selected
Cities

INPUTTING OF CASE INFORMATION

Oklahoma City, OK
  Percentage of criminal case information input within 24 hours of receipt: 100% (1992)
  Percentage of jury case information input within 24 hours of receipt: 100% (1992)
  Percentage of traffic case information input within 24 hours of receipt: 95% (1992)

Norman, OK
  Percentage of case information input within 24 hours of receipt: 99% (1997), 98% (1998)

College Station, TX
  Percentage of parking citations entered within 24 hours of being filed in court: 95% (1998)
  Percentage of nonparking citations entered within 24 hours of being filed in court: 98% (1998)

Hurst, TX
  Percentage of citations input within 1 day: 80% (1997)

Houston, TX
  Target: Enter 95% of tickets within 2 days of receipt

Fort Worth, TX
  Target: Within 7 days (criminal cases)

UPDATING OF PARKING CASE RECORDS

Oklahoma City, OK
  Percentage updated within 24 hours of receipt: 100% (1992)

Orlando, FL
  Percentage of newly issued tickets input within 2 workdays: 100% (1991)

Reno, NV
  Target: Enter traffic citations into court's computer database within 48 hours of receipt
  Actual: 100% (1990)

Boston, MA (Department of Transportation)
  Percentage of parking tickets processed within 72 hours of receipt from ticket-issuing agencies: 100%
    (1992)
  Percentage of payments entered into system within 48 hours of receipt: 75% (1992)

Fort Worth, TX
  Target: Within 7 days

CORRESPONDENCE/NOTIFICATION OF COURT DATE

Oklahoma City, OK
  Percentage of notifications mailed within 12 hours: 100% (1992)

College Station, TX
  Percentage of case paperwork prepared with notification to all parties ready at least 3 weeks prior
    to hearing: 90% (1998)
  Percentage of attorney correspondence processed within a week after request receipt: 80% (1995)

PROVIDING TRIAL DATA TO CITY ATTORNEY

Reno, NV
  Target: Provide scheduled trial data to city attorney within 5 days of arraignment
  Actual: 95% (1990)

UPDATING OF CASE RECORDS

Oklahoma City, OK
Percentage updated within 12 hours: 97% (1992)

BOND REFUNDS/DISTRIBUTIONS

College Station, TX
Percentage of cases requiring a refund processed within 72 hours: 92% (1995)
Percentage of cases requiring a bond distribution processed within 72 hours: 92% (1995)

Scottsdale, AZ
Percentage of bond refunds processed within 5 business days: 59% (1999)

PROCESSING OF BOND FORFEITURE NOTICES

Oklahoma City, OK
Percentage processed within 8 hours: 98% (1992)

Scottsdale, AZ
Percentage of bond forfeitures processed within 5 business days: 70% (1999)

ISSUANCE OF WARRANTS

Tempe, AZ
Percentage issued within 48 hours when defendant fails to appear for criminal arraignment: 91%
   (1997)

Oklahoma City, OK
Average number of days between request for arrest warrant and issuance: 3 days (1992)
Average number of days between failure to appear and issuance of bench warrant: 6 days
   (1999)

Fayetteville, AR
Target: Within 10 workdays

Overland Park, KS
Target: Within 10 days

Fort Worth, TX
Target: Issue at least 80% of warrants within 90 days from the date cases become eligible for
   warrant issuance
Actual: 50% issued within 90 days (1997)

ISSUANCE OF OVERDUE NOTICE ON
UNPAID PARKING CITATIONS

Oklahoma City, OK
Within average of 15 days following due date (1992)

FAILURE TO APPEAR

College Station, TX
Percentage of final notices issued by the 18th day to defendants who fail to appear on citation: 90%
   (1995)

DEFAULT JUDGMENTS

Scottsdale, AZ
Target: Process 75% of all civil traffic defaults within 15 days of court date
Actual: 39% (1998)

(continued)

Table 6.2   Prompt Preparation and Processing of Court Paperwork: Selected
                      Cities (continued)

Tucson, AZ
   Percentage of civil traffic default judgments entered within 30 days for defendants who fail to pay:
   96% (1997)

RESTITUTION

Scottsdale, AZ
   Percentage of restitution disbursements processed within 2 business days: 76% (1998)

PROCESSING OF APPEALS

Oklahoma City, OK
   Percentage of appeals processed within 24 hours of receipt: 98% (1992)

Overland Park, KS
   Target: Forward notice of appeal within 10 days of disposition

Reno, NV
   Target: File appeals in district court within 10 days of municipal court conviction date
   Actual: 95% (1990)

RECORDING OF CASE DISPOSITION
AND COMPLETION OF PAPERWORK

Overland Park, KS
   Target: Within 3 workdays

Reno, NV
   Target: Enter sentencing data within 5 days of trial

DEPOSIT/POSTING OF COURT REVENUES

Reno, NV
   100% deposited daily (1990)

Milwaukee, WI
   100% posted within 24 hours of receipt (1997)

Norman, OK
   Percentage of revenues posted within 24 hours of receipt: 100% (1998)

College Station, TX
   Percentage of payments processed within 24 hours: 99% (1998)
   Percentage of cash bonds from the jail posted by noon of the same day received: 99% (1998)

Houston, TX
   Target: Process 95% of mailed payments within 1 day of receipt

Orlando, FL
   Target: Process mailed parking fines within 2 workdays of receipt
   Actual: 81% (1991)

Table 6.3   Court Employee Production Ratios for Selected Cities

CASE WORKLOAD

Eugene, OR
Number of cases processed per FTE, by type: ordinances/traffic, 1,404; parking, 3,527 (1997)

Alexandria, VA
Number of cases heard and processed per employee year: 1,039 (1995), 1,099 (1998)

CLERICAL WORKLOAD

College Station, TX
Number of nonparking citations entered per worker hour: 30
Number of parking citations entered per worker hour: 41 (1998)

Alexandria, VA
Number of documents recorded and processed per year per staff member: 8,754 (1994), 6,108 (1995), 8,837 (1998)

WARRANT WORKLOAD

Oklahoma City, OK
Average number of warrant cases served and cleared per day per employee: 14 (1998), 11 (1999)

Table 6.4   Warrants Served/Cleared: Selected Cities

Lubbock, TX
Percentage of warrants cleared: 93% (1995), 96% (1996), 97% (1997)

Norman, OK
Ratio of warrants served to warrants issued: 85% (1998)

San Antonio, TX
Percentage of warrants cleared: 84% (1995)

Grand Prairie, TX
75% of warrants served and cleared (1996, estimated)

Duncanville, TX
Percentage of warrants cleared: 54% (1997), 69% (1998)

Fort Worth, TX
Target: 95% warrant clearance rate
Actual: 62% (1997)

Table 6.5   Cases Closed/Cleared: Selected Cities

Chesapeake, VA
  Percentage of cases concluded within 12 months (General District Court): 99.7% (1997)

Lubbock, TX
  Percentage of nonparking violations cleared: 88% (1995), 78% (1996), 92% (1997)
  Percentage of parking violations cleared: 72% (1995), 84% (1996), 80% (1997)
  Percentage of violations cleared overall: 80% (1995), 80% (1996), 87% (1997)

Oklahoma City, OK
  Percentage of parking citations disposed within 90 days: 82% (1998), 75% (1999)

College Station, TX
  Percentage of cases resolved: 80% (1995)

Duncanville, TX
  Percentage of cases disposed: 71% (1997), 61% (1998)

San Antonio, TX
  Percentage of cases closed: 33.6% (1997)

Table 6.6   Collection Rates for Fines Imposed by Court: Selected Cities

Norman, OK
  Collection rate for adjudicated traffic and non–traffic cases: 99% (1998)
  Collection rate for parking cases: 86% (1998)

Boston, MA
  Collection rate on parking violations up to 6 months old: 81.5%; from 6 to 12 months old: 85.7%;
    from 12 to 24 months old: 88.2% (1996)

Cambridge, MA
  Collection rate for parking tickets to in-state registrants: 85% (1995); to out-of-state registrants: 60%
    (1995)

Savannah, GA
  Actual: 82% (1993)

Fayetteville, AR
  Actual: 81% (1997)

San Luis Obispo, CA
  Collection rate for parking citations: 81% (1994)

Greeley, CO
  Collection rate for parking tickets: 75% (1997); for other fines: 81% (1997)

Chapel Hill, NC
  Collection rate for parking tickets: 80% (1999)

Eugene, OR
  Actual: 69.8% (1997)

Shreveport, LA
  Actual: 43% of all parking tickets (1994)

Table 6.7   Court System Odds and Ends: Selected Cities

PROMPT ARRAIGNMENT

New York, NY
Average time from arrest to arraignment: 21.4 hours (1998)

CASES HEARD

Shreveport, LA
Criminal cases heard as a percentage of cases filed: 70% (1994)
Civil cases heard as a percentage of total filed: 60% (1994)
Traffic cases heard as a percentage of total filed: 51% (1994)

APPEALS

Duncanville, TX
Percentage appealed: 7% (1997), 5% (1998)

REVERSALS

Prairie Village, KS
Percentage of cases reversed on appeal: 0% (1995)

Chandler, AZ
Among cases appealed, percentage overturned: 8% (1995), 5% (1998)

DEFAULT OF DRIVER'S LICENSE

Tempe, AZ
Percentage of driver's license defaults issued within 48 hours when defendant fails to appear for civil
arraignment: 92% (1997)

GARNISHMENTS

Shreveport, LA
Garnishments satisfied as a percentage of total filed: 34% (1994)

COMPLIANCE WITH TERMS OF PROBATION/COMMUNITY SERVICE

Norman, OK
Percentage of defendants complying with probation: 99% (1997), 97% (1998)

Greeley, CO
Compliance with community service requirements: 87% (1997)

Oklahoma City, OK
Percentage of successful terminations of probation: 73% (1999)

7

Development Administration

Although most municipal departments have a role to play in their community's development, a handful are crucial members of the cast for local development activities—especially the community's physical development. Department labels differ from place to place, but the functions clustered in this chapter under the heading *Development Administration* commonly are called *planning, inspection, code enforcement,* and *zoning.*

Development Climate

Some communities enjoy reputations for favorable development climates. Others sometimes find themselves under attack from real estate persons, builders, land developers, bankers, economic development specialists, or the local chamber of commerce for their *unfavorable* climate for development. In part, a community's development climate is determined by official policies that encourage or restrict industrial, commercial, or residential growth. Policies adopted to preserve the heritage of the community, to protect the integrity of neighborhoods, to rebuff smokestack industries, to encourage upscale development, or to prevent spot zoning or strip shopping centers often are criticized as antidevelopment—even if proponents of those policies prefer that they be cast more favorably as controlled growth or quality development policies.

In addition, administrative and operational factors shape the climate for local development. The responsiveness of departmental employees to the needs of developers and builders for prompt action on plans, inspections, and a host of other regulated activities influence development costs and profitability. Even in a community with restrictive development policies, a highly responsive bureaucracy can mitigate many developer complaints. Conversely, the development climate in a community with strong pro-growth policies may be dampened by inspection delays, slow approvals, and a general lack of responsiveness among operating personnel.

Benchmarks

Some cities set performance targets for various aspects of development administration. These targets, plus each city's record of success or failure in attempting to meet them, offer useful guidance to communities attempting to establish their own performance expectations.

Community Planning

Many work elements of community planning departments are nonroutine in nature and therefore difficult—although not necessarily impossible—to measure. Workload counts for one community—the number of inquiries received, the number of planning commission meetings, the number of zoning map updates, and so forth—are of little relevance as benchmarks for another community. More useful as benchmarks are indicators that measure the quality of various planning actions. For example, Portland, Oregon, reported that 95% of its planning staff decisions in one recent year were upheld on appeal by various boards and hearing officers. As an indication of the thoroughness of its planning staff work, Raleigh, North Carolina, reported that during a recent year its city council took action on 42% of all zoning cases and 73% of all site plan or subdivision cases as consent agenda items, meaning that little or no elaboration beyond that provided in advance by the staff was needed to reach a decision.

Processing speed is a much more commonly reported aspect of performance quality. Promptness is documented by several cities for a host of functions:

- During one recent year, the planning department of Duncanville, Texas, processed every zoning change application in 21 days or less (Table 7.1). Nearly comparable speed was found in places like Norman, Oklahoma; Nashville–Davidson County, Tennessee; and Corpus Christi, Texas. Many other communities, perhaps emphasizing qualities other than speed, required longer processing times—sometimes much longer.

- College Station, Texas, reported an average turnaround time of just 4 workdays for the review of development plans (Table 7.2). Duncanville reviewed most subdivision plats within 5 workdays. Many of their counterparts, however, consume 20 to 45 days in site development and subdivision plan review. The experience reported by the cities in Tables 7.1 and 7.2 generally validates the standards for the processing of proposed zone changes, variances, and development permits recommended by a League of California Cities (LCC) panels (1994, p. 32). High-service-level departments, the panel contends, process these applications and

Table 7.1   Time Frame for Planning Department Review and Recommendation on
                  Zoning Applications: Selected Cities

Duncanville, TX
  Target: Process zoning change applications within 21 days
  Actual: 100% (1998)

Norman, OK
  Items processed by planning commission within 24 days of receipt: 86% (1998)

Nashville–Davidson County, TN
  Target: Review all zoning applications within 28 days
  Actual: 100% (1998)

Corpus Christi, TX
  Targets: Process all regular zoning applications within 30 days; process at least 66% of special permit
    zoning applications within 30 days, and 100% within 44 days
  Actual: 100% of regular applications processed within 30 days; 100% of special permit zoning
    applications processed within 30 days (1996)

Oakland, CA
  Percentage of major zoning permit applications processed within 45 days: 40% (1996)
  Average processing time per major zoning application permit: 33 days (1996)
  Percentage of minor zoning permit applications processed within 45 days: 73% (1996)
  Average processing time per minor zoning application permit: 35 days (1996)

High Point, NC
  100% of rezoning applications evaluated within 35 days; 100% of special-use permit applications
    evaluated within 35 days; 100% of appeals to board of adjustment evaluated within 25 days (1998)

Charlotte, NC
  Target: Review rezoning applications within 30 days
  Actual: 100% within 30 days (1987); 100% of rezonings within 45 days (1994, 5-month record)

Hurst, TX
  Target: Completion time of 6 weeks or less for subdivision plats and zoning cases
  Actual: 6-week average (1997)

Chesapeake, VA
  Average number of days to review public hearing items: 45 days (1997)
  Average number of days to review non–public hearing items: 15 days (1997)

Iowa City, IA
  Target: Review rezoning applications, subdivision plats, and site development plans within 45 days

San Antonio, TX
  Average number of days from zoning application to council action: 73.5 (1995), 60 (1997)

Raleigh, NC
  Zoning cases—median days from application to decision: 74 days (1997), 82 days (1998)

San Jose, CA
  45% of conventional zoning applications processed within 90 days; 72% of planned development zon-
    ing applications processed within 180 days (1995)

Cincinnati, OH
  Target: Process at least 90% of zoning application requests within the following time frames:
    rezoning, 16 weeks; special use, 30 days; subdivision, 30 days; deeds, 7 days

Table 7.2 Time Frame for Planning Department Review and Recommendation on Development Applications: Selected Cities

Duncanville, TX
Target: Review subdivision plats within 5 workdays
Actual: 97% (1998)

College Station, TX
Percentage of development plans reviewed within 5 working days: 95% (1998)
Average turnaround time: 4.0 days (1998)

Lubbock, TX
Percentage of plats processed within 10 days: 90% (1998, estimated)

Overland Park, KS
Target: Review at least 75% of all subdivision, commercial site, detention, non–capital improvement public street and storm sewer, traffic control, and streetlighting plans within 10 workdays of receipt (1999)

Blacksburg, VA
Percentage of site plans and subdivision plans reviewed within 14 days on first submission: 100% (1997)
Percentage of site plans and subdivision plans reviewed within 10 days on second submission: 100% (1997)
Percentage of site plans and subdivision plans requiring more than 2 submissions: 14% (1997)

Chandler, AZ
Target: Perform site development plan reviews within 20 days
Actual: 95% (1999, 6-month year-to-date)

Tucson, AZ
Percentage of subdivision/development plans reviewed within 3 weeks: 100% (1997)

Charlotte, NC
Percentage of subdivision plans processed within 23 days: 100% (1999)

Norman, OK
Items processed by planning commission within 24 days of receipt: 86% (1998)

Nashville–Davidson County, TN
Target: Review all development applications within 28 days
Actual: 100% (1998)

Oak Ridge, TN
Average processing time for subdivision plats: 28 days (1997)
Percentage of engineering reviews of subdivision plans completed within 1 week: 70% (1997)

Charlotte, NC
Target: Review subdivision plans, final plats, multifamily projects, rezoning applications, and special use permits within 30 days
Actual: 100% within 30 days (1987); 100% of subdivision plans and final plats within 30 days (1994, 5-month record)

Phoenix, AZ
Target: Review at least 85% of all preliminary subdivision plats within 30 days

(continued)

TABLE 7.2 Time Frame for Planning Department Review and Recommendation on Development Applications: Selected Cities (continued)

Corpus Christi, TX
Target: Process within 35 days all plats requiring public notice; all others within 21 days (1999)

Raleigh, NC
Site plans—median days from submittal to decision: 43 days (1997), 40 days (1998)
Subdivision plans—median days from submittal to decision: 36 days (1997), 32 days (1998)

Hurst, TX
Target: Completion time of 6 weeks or less for subdivision plats and zoning cases
Actual: 6-week average (1997)

Chesapeake, VA
Average number of days to review public hearing items: 45 days (1997)
Average number of days to review non–public hearing items: 15 days (1997)

Iowa City, IA
Target: Review rezoning applications, subdivision plats, and site development plans within 45 days

Oakland, CA
Target: Secure 90% of subdivision application decisions within 45 days; 100% within 50 days

San Jose, CA
72% of planned development permit applications processed within 90 days; 92% of site development permits processed within 180 days; and 89% of conditional use permits processed within 180 days (1995)

make recommendations within 30 days; medium-service-level departments, within 45 days; and low-service-level departments require more than 60 days.

- The planning department of Scottsdale, Arizona, routinely reviewed sign permit applications within 3 days following submittal (Table 7.3). Orlando, Florida, reviewed most within 4 days.

- The planning departments of Charlotte, North Carolina, and Palo Alto, California, strive to update zoning maps within 30 days of each zoning change.

Differences in the nature and complexity of planning cases cause substantial variation in the amount of staff time each case requires. Nevertheless, the professional staff hour averages tabulated by a pair of municipalities in the early 1990s for a variety of planning cases offers a general point of reference for others. Wichita, Kansas, found that approximately 17 professional staff hours were required for zoning cases, 20 hours for plats, 8 hours for lot splits, and approximately 12 hours for planning cases in general. Reno, Nevada, reported that approximately 36 staff hours were required for each planning commission application, 25 hours for each board of adjustment application, 6

Table 7.3   Prompt Review of Sign Plans: Selected Cities

Scottsdale, AZ
  Percentage of sign permit applications reviewed within 3 days: 99% (1999)

Orlando, FL
  Percentage of sign permit applications reviewed within 4 days: 88% (1994), 85% (1996)

Greensboro, NC
  Percentage of sign permit applications reviewed within 5 days: 100% (1997)

Corpus Christi, TX
  Target: Process for planning commission action 100% of all conditional sign permit applications
  within 20 days (1999)

---

hours per final map, 3 hours per parcel map, 25 hours per site plan review, 6 hours per multiresidential, commercial, or industrial building permit, 45 minutes per single-family residential building permit, and 30 minutes per sign permit.

Plan Review and Permits

Careful review of construction plans prior to the issuance of a building permit is an important protection of the public interest. From the perspective of builders and their clients, however, a slow review can also mean expensive delays in the construction process. Some communities are more adept than others at minimizing such delays and their associated expenses to builders.

In some departments, plan review time statistics are reported only as the average for all plan reviews, making no distinction between plan reviews for commercial structures and the review of generally less complicated residential buildings. Many cities, however, distinguish between commercial and residential reviews in the performance targets they set and the statistics they report (Table 7.4). Most of the cities examined for this volume perform building permit reviews within 3 weeks for commercial projects and within 1 week for residential buildings. Review time in some cities is notably shorter, with reviews of commercial plans requiring only about a week and residential plans only a day. Although many cities declare their processing speed, few report error rates. Two cities that do track and report errors are Plano, Texas, and Winston-Salem, North Carolina. Plano reported that 98% of all building permit applications and site plans were reviewed without error in 1997. Winston-Salem reported that only 0.1% of its building permits were issued in violation of the zoning ordinance in 1996 and none at all in 1997.

Review of construction plans for fire safety purposes may in some cases be performed by building officials but often is performed by fire department

(text continues on page 73)

Table 7.4   Review Time for Building Permits: Selected Cities

COMMERCIAL BUILDINGS

Orlando, FL
  Percentage of major plans reviewed within 4 days: 92% (1994), 68% (1996)
  Percentage of minor plans reviewed within 3 days: 95% (1994), 92% (1996)

Denton, TX
  Target: 4 workdays
  Actual: 61% within 4 workdays (1997)

High Point, NC
  Target: Within 5 business days
  Actual: 90% (1998)

Ocala, FL
  Target: Within 1 week

Plano, TX
  Actual: 6-day average (1997)

Norman, OK
  Percentage reviewed within 20 days: 100% (1998)
  Average: 6.73 days (1997)

Savannah, GA
  Target: 7 days (if 3 stories or less); 10 days (4–7 stories); 14 days (if high-rise)
  Actual averages: 7, 10, and 14 days (1990); 7, 10, and 14 days (1991)

Tallahassee, FL
  Target: Within 8 days
  Actual: 71% (1991)

Winston-Salem, NC
  Target: Within 9 days
  Actual: 69% (1997), 67% (1998)

Cary, NC
  Target: Within 9 workdays
  Actual: 15% (1999)

Lubbock, TX
  Actual: 100% within 10 working days (1991)

Irving, TX
  Target: Within 10 working days
  Actual: 95% (1991)

Oklahoma City, OK
  Percentage reviewed within 10 working days: 83% (1999)
  Percentage of code reviews for remodels within 5 working days: 89% (1999)

Alexandria, VA
  10-day average (1998)

Grand Prairie, TX[a]
  Actual: 100% reviewed within 14 days; 85% issued within 14 days (1998)

Ames, IA
  Percentage of all building permits (commercial and residential) issued within 15 days: 85% (1997)

Raleigh, NC
  Actual: small commercial—5.8 days (1997), 13.0 days (1998); medium commercial—11.2 days (1997),
    19.0 days (1998); large commercial—17.6 days (1997), 26.0 days (1998)

Tempe, AZ
  Target: Within 15 days
  Actual: 21-day average (1997)

Fort Collins, CO
  Target: Within average of 3 weeks
  Actual: 3-week average (1991)

Glendale, AZ
  Percentage reviewed within 22 workdays: 97% (1997)

Reno, NV
  Target: Review permit applications for at least 60% of new commercial buildings and 90% of
    remodelings within 4 weeks (1999)

San Diego, CA
  Actual: 90% of commercial plan checks within 30 days; 92% of multifamily plan checks within 30 days
    (1998)

Eugene, OR
  Target: Within 63 calendar days (8 weeks)

RESIDENTIAL BUILDINGS

Anaheim, CA
  Target: Same-day turnaround on minor residential and tenant improvement building permits

Fayetteville, AR
  Actual: 100% within 8 working hours (1991)

Savannah, GA
  Target: Within 1 day for single-family and duplex; 7 days for triplex and quad; 10 days for apartments
  Actual: 99%, 99%, and 100% (1991); 99%, 85%, and 85% (1992)

College Station, TX
  Percentage of single family permits issued within 24 hours: 95% (1998)
  Percentage within 5 days: 100% (1998)

Denton, TX
  Target: Within 2 workdays
  Actual: 98% (1997)

Ocala, FL
  Target: Within 2 days

Orlando, FL
  100% within 2.1 days (1994); 100% within 2.3 days (1996)

High Point, NC
  Targets: single- and two-family within 2 business days; multifamily within 5 business days
  Actual: 90% within targeted time frame (1998)

Charlotte, NC
  Cycle time for normal permit review: 2.5 days (1998)

Winston-Salem, NC
  Target: Issue at least 90% of all building permits for single-family construction within 3 working days
  Actual: 98% (1997), 100% (1998)

(continued)

Table 7.4   Review Time for Building Permits: Selected Cities (continued)

Irving, TX
   Target: Within 3 workdays
   Actual: 98% (1991)

Grand Prairie, TX[a]
   Actual: 100% reviewed within 5 days; 92% issued within 3 days (1998)

Hurst, TX
   Target: Review plans and issue permit within an average of 3 days
   Actual: 3-day average (1997)

Plano, TX
   Actual: 3-day average (1997)

Cary, NC
   Target: Within 72 hours
   Actual: 20% (1999)

Durham, NC
   Percentage reviewed within 4 workdays: 90% (1999)

Grants Pass, OR
   Target: At least 90% processed within 1 week (1997)

Norman, OK
   Percentage reviewed within 5 days: 66% (1998)
   Average: 4.53 days (1997)

San Diego, CA
   Actual: 59% of residential plan checks within 8 days (1998)

Cincinnati, OH
   Target: Within 10 workdays, if 20 dwelling units or less
   Actual: 100%, with an average review time of 7 workdays (1995)

Alexandria, VA
   10-day average (1998)

Tempe, AZ
   Target: Within 8 days
   Actual: 11-day average (1997)

Raleigh, NC
   Actual: 14-day average (1997), 9-day average (1998)

Glendale, AZ
   Percentage reviewed within 16 workdays: 95% (1997)

Reno, NV
   Target: Review permit applications for at least 90% of new residential units and 95% of remodelings
      within 4 weeks (1999)

Eugene, OR
   Target: Within 39 calendar days (6 weeks)

a. Statistics for Grand Prairie, TX, are taken from Shelton and Bolen (1999).

Table 7.5   Prompt Fire Safety Review of Building Plans: Selected Cities

Cary, NC
Percentage of building and site plans reviewed within 3 days: 100% (1999)

Victoria, TX
Percentage of building plan reviews conducted within 5 days: 84% (1996)

Oak Ridge, TN
Target: Review by fire department within 7 days
Actual: 95% (1992), 97% (1997)

Corpus Christi, TX
Target: Review at least 90% of all building and fire protection system plans within 7 working days of receipt (1999)

Charlotte, NC
Percentage of plans reviewed within 10 days: 99.7% (1999)

Denver, CO
Percentage of plan reviews conducted within 10 working days: 100% (1998)

Flagstaff, AZ
Target: Perform all plan checks within 10 working days
Actual: 100% (1998)

San Clemente, CA
Actual: 90% of plan checks completed within 10 workdays (1998)

San Jose, CA
Targets: Review routine building plans within 14 days, routine fire protection system plans within 14 days, and at least 90% of hazardous materials plans within 45 days
Actual: 62%, 72%, and 99% (1995)

Santa Ana, CA
Actual: 90% of plan reviews completed within 15 days (1995)

Greeley, CO
Percentage of plans reviewed within 21 days: 83% (1997)

Palo Alto, CA
Target: Fire department review of plans, consultation with developers, and issuance of permits within 28 days at least 90% of the time
Actual: 100% (1997)

---

personnel. Most reporting municipalities indicated completion of most such reviews within 2 weeks (Table 7.5).

Some planning and development departments make prompt service for walk-in customers a priority. San Clemente, California, reported serving 94% of walk-ins within 10 minutes of arrival (Table 7.6). In Cincinnati, Ohio, the waiting times for most was a mere 2 minutes.

Table 7.6   Prompt Customer Service: Selected Cities

Cincinnati, OH
  Percentage of service desk clients receiving service within 2 minutes of arrival: 81% (1995)

San Clemente, CA
  Percentage of counter customers served within 10 minutes: 94% (1998)
  Percentage of responses to phone inquiries within 24 hours: 98% (1998)

Palo Alto, CA
  Percentage of working days when persons at the planning walk-in counter wait less than 15 minutes
    to be helped: 100% (1997)

Kansas City, MO
  Percentage of permit/plans review customers served within 30 minutes: 99% (1995)

Inspections

Ideal benchmarks for the building inspection function would document the effectiveness of inspections. Were building flaws or noncomplying construction practices spotted by inspectors and thus prevented? A few cities do report the number of inspections rejected by their inspectors, but how many others should have been rejected that slipped by? Inspectors in Raleigh, North Carolina, for example, in one recent year rejected an average per inspector of 448 building inspections, 366 electrical inspections, 297 fire inspections, 249 mechanical inspections, and 255 plumbing inspections. Knowing Raleigh's rejection rates may help officials in other cities judge whether charges of over-zealous enforcement against one of their own inspectors are credible or not, but the applicability of one city's rates to other communities is questionable, and the extent to which those rates represent the detection of all construction flaws in Raleigh remains unknown. In the absence of more desirable measures of effectiveness, other dimensions of inspection service quality and quantity may be gauged. The performance targets and experience of several cities, for example, indicate that it is reasonable to expect building inspections to be performed within 1 or 2 workdays from the time requested (Table 7.7). More aggressive inspection operations often can perform building inspections on the day requested.

Several cities report performing electrical, plumbing, and other specialty inspections within 1 workday of the request (Table 7.8). Lubbock, Texas, reported that during one year most of its electrical, mechanical, and plumbing inspections were performed within 4 work hours of the request.

Average inspector workload varies by inspection specialty among the municipalities reporting that statistic (Table 7.9). On the basis of reported figures,

Table 7.7   Prompt Response to Requests for Building/Construction Inspections: Selected Cities[a]

Fort Collins, CO
Actual: 3.5-hour average response time; 100% completed on the day requested (1991)

Oak Ridge, TN
Target: Within 4 hours of request (1997)

Flagstaff, AZ
Same-day service: 99% (1997)

Ocala, FL
Target: Same-day service

Corpus Christi, TX
Actual: 99% performed on the day requested (1991)

San Jose, CA
Actual: 99% on the day requested (1995)

Alexandria, VA
Percentage performed on the day requested: 98% (1998)

Raleigh, NC
Actual: 88% on the date requested (1997); 89% (1998)

Duncanville, TX
Actual: 97% within 8 hours of request (1998)

Hurst, TX
Target: Within 8 work hours of notice
Actual: 95% within 8 work hours (1997)

Norman, OK
Percentage performed on the same day (within 8 hours): 60% (1998); percentage performed within 48 hours: 90% (1998)

Chandler, AZ
Actual: 99% within 24 hours (1998)

San Clemente, CA
Percentage within 24 hours: 99% (1998)

College Station, TX
Percentage within 24 hours of request: 98% (1998)

Cary, NC
Actual: 97.5% within 24 hours of request (1999)

Glendale, AZ
Percentage performed within 24 hours of request: 97% (1997)

Tucson, AZ
95% of commercial inspections within 24 hours of request; 99% of residential inspections within 24 hours of request (1998, estimated)

(continued)

Table 7.7   Prompt Response to Requests for Building/Construction Inspections: Selected Cities[a] (continued)

Portland, OR
Actual: 95% within 24 hours of request (1998)

Durham, NC
Percentage performed within 24 hours: 90% (1999)

Grants Pass, OR
Target: At least 90% within 24 hours of request (1997)

Palo Alto, CA
Target: At least 90% within 1 working day
Actual: 99% (1997)

San Antonio, TX
Percentage of inspections completed by the end of the next business day: 99% (1997)

Kansas City, MO
Target: At least 95% within 1 day of request

Eugene, OR
Percentage of inspections performed within 1 working day of request: 95% (1995), 92% (1996)

Reno, NV
Target: At least 95% within 1 day of request (1999)

Winston-Salem, NC
Target: At least 90% within 1 day of request
Actual: 93% (1997), 88% (1998)

High Point, NC
Actual: 90% within 1 day of request (1998)

Scottsdale, AZ
Percentage performed by the end of the next workday: 99% (1999)

San Diego, CA
Percentage of combination inspections performed by the next working day: 89% (1998)

Oklahoma City, OK
Percentage performed within 16 working hours: 75% (1999)

a. Includes response statistics when "building/construction" is mentioned and also general inspection statistics when no other specialty (e.g., plumbing or electrical) is designated.

a workload of 9 to 16 general building inspections per day seems reasonable, as do slightly lower numbers of electrical or mechanical inspections.

The amount of time an inspector spends on each inspection, of course, influences daily workload. By holding down the inspector time per inspection, inspector workload may be increased. Quantity gains, however, may be negated by quality losses if inspector speed results in faulty inspections. It might

(text continues on page 80)

Table 7.8 Prompt Response to Requests for Electrical, Plumbing, and Other Specialty Inspections: Selected Cities

ELECTRICAL INSPECTIONS

Lubbock, TX
95% within 4 work hours of request (1991)

San Jose, CA
99% on the date requested (1995)

Raleigh, NC
90% on the date requested (1991)

Winston-Salem, NC
Within 1 day of request: 94% (1992), 89% (1997), 82% (1998)

San Diego, CA
84% performed by the next working day (1998)

FIRE INSPECTIONS

Raleigh, NC
93% on the date requested (1991)

Chandler, AZ
90% within 48 hours of request (1991)

San Clemente, CA
87% within 2 workdays (1998)

MECHANICAL/HVAC[a] INSPECTIONS

Lubbock, TX
80% within 4 work hours of request (1991)

Raleigh, NC
100% on the date requested (1991)

San Jose, CA
99% within 1 day of the request (1995)

Winston-Salem, NC
Within 1 day of the request: 95% (1992), 93% (1997), 92% (1998)

San Diego, CA
84% performed by the next working day (1998)

PLUMBING INSPECTIONS

Lubbock, TX
80% within 4 work hours of the request (1991)

Cincinnati, OH
Target: 100% on the date requested

Raleigh, NC
92% on the date requested (1991)

San Jose, CA
99% within 1 day of the request (1995)

Winston-Salem, NC
Within 1 day of request: 98% (1997), 98% (1998)

a. HVAC is heating, ventilation, and air conditioning.

Table 7.9   Inspector Workload: Inspections per Inspector in Selected Cities

Average for All Types	Building/Construction Inspector	Electrical Inspector	Fire Inspector	Mechanical/HVAC Inspector	Plumbing Inspector	General Code Enforcement Inspector
			Inspections per Day			
Sterling Heights, MI 17 (1995)	**Shreveport, LA** 16 (1991), 19 (1994)	**Irving, TX** 12.3 (1991)	**Savannah, GA** 8 (1997)	**New Hanover County (Wilmington), NC** 14.9 (1998)	**New Hanover County (Wilmington), NC** 24.1 (1998)	**Savannah, GA** Target: 15 code inspections/reinspections per day Actual: 18 (1990), 19 (1991)
Chesapeake, VA 17 (1997)	**Irving, TX** 16.3 (1991)	**Lubbock, TX** 11 (1995), 12 (1997)	**New York, NY** 5.7 (1996), 5.6 (1997)	**Shreveport, LA** 11 (1991), 12 (1994)	**Irving, TX** 17.5 (1991)	**San Antonio, TX** 11 (1995), 17.2 (1997)
San Antonio, TX 15.8 (1995), 16.5 (1997)	**Lubbock, TX** 15 (1995), 16 (1997)	**New Hanover County (Wilmington), NC** 12.0 (1998)	**Portland, OR** 2.12 (1991), 3.16 (1992, estimated)	**Greenville, SC** 6.7 (1989), 6.3 (1990)	**Lubbock, TX** 17 (1997)	**Shreveport, LA** 12 (1991), 15 (1994)
Savannah, GA 7 (1997)	**New Hanover County (Wilmington), NC** 15.1 (1998)	**Shreveport, LA** 10 (1991), 11 (1994)	**Bellevue, WA** 3 (1997)		**Shreveport, LA** 8 (1991), 12 (1994)	**Charlotte, NC** Target: 10 per inspector-day Actual: 10.4 (1987)
	San Clemente, CA Target: 12–14 (2000)	**New York, NY** 10.2 (1996), 9.3 (1997)			**New York, NY** 8.7 (1996), 11.1 (1997)	**Chesapeake, VA** 9 (1997)
	Greenville, SC 8.2 (1989), 12.4 (1990)	**Greenville, SC** 9.3 (1989), 7.0 (1990)			**Greenville, SC** 4.3 (1989), 5.9 (1990)	**Greenville, SC** General Inspector: 8.3 (1989), 7.7 (1990) Sign Inspector: 7.5 (1989), 6.7 (1990) Zoning Inspector: 17 (1990)
	College Station, TX 10.3 (1998)					**San Clemente, CA** Target: 4.5 (1999)
	New York, NY 8.9 (1996), 8.8 (1997)					

Inspections per Year

Eugene, OR 5,162 (1997)	**Long Beach, CA** 3,750 (1995)	**Raleigh, NC** 1,858 (1991)	**Shreveport, LA** 4,037 (1991)	**Raleigh, NC** 2,072 (1991)	**Raleigh, NC** 2,363 (1991)
Corpus Christi, TX 3,354 (1991)[a]	**Alexandria, VA** 4,566 (1991), 3,487 (1992), 3,512 (1993)		**Phoenix, AZ** 1,376 (1990, estimated)		
	Corpus Christi, TX 3,354 (1991)		**Raleigh, NC** 987 (1991)		
	Raleigh, NC 3,049 (1997), 3,085 (1998)		**Greensboro, NC** 964 (1997)		
	College Station, TX 2,460 (1998)		**Tempe, AZ** 385 (1997)		

a. Average per inspector includes building trade inspectors and general code enforcement.

be reassuring, therefore, to fall within the band of average inspection times among other cities (Table 7.10). Lying outside those bounds at either extreme could be a cause for concern—too slow and perhaps wasteful at one end or too fast and perhaps error-prone at the other.

The LCC (1994) contends that an inspector of single-family residential units working for a high-service-level department should be able to complete 12 framing or foundation inspections per day (pp. 21–22). Fewer than 12—or more than 20—might signal problems and a low service level.

Many cities attempt to recover their costs for permit and inspection activities through the fees they charge for these services. Several of the cities examined for this volume have been successful in that regard (Table 7.11).

Code Enforcement Activities

General code enforcement activities typically focus on existing neighborhoods, commercial areas, and other already developed properties rather than on new construction projects. Attention often is directed toward such offenses as illegal materials on private property or at curbside, overgrown vacant lots, and abandoned autos, but the range of code enforcement activities is not confined to these problems. The condition of housing and other structures can also be a major point of attention. Consider, for instance, the programs in Savannah, Georgia, and Cincinnati, Ohio.

In Savannah, the condition of housing stock is assessed through a structural condition survey. Each structure is classified using the following five categories:

- Standard—No problems.
- Minor Problem—Chipped or peeling paint; broken windows; minor replacement of wood siding or shingles needed; shutters that are broken or in need of paint; curling roof shingles; fascia repair needed; eave work needed.
- Moderate Problem—Complete repainting needed; damaged and unsafe steps; major replacements of shingles, siding, or wood needed; defective porch; or three or more minor problems.
- Major Problem—Defect in a major component of the building, such as a sagging or cracked load-bearing wall, a serious defect in the foundation, missing steps, caved-in roof, deteriorated windows; major replacement of wood siding or shingles needed.
- Dilapidated—Two or more major problems (City of Savannah [GA], *Neighborhood Quality Benchmarks: 1996 Report to the Community*, pp. 67–68).

Table 7.10 Inspector Speed: Average Inspector Time per Inspection in Selected Cities

Building/Construction Inspector	Electrical Inspector	Fire Inspector	Mechanical/HVAC Inspector	Plumbing Inspector	General Code Enforcement Inspector	Average (Trade Undesignated)
Fort Collins, CO 24 minutes (1991)	**Wichita, KS** 14 minutes (1990), 13 minutes (1991)	**Winston-Salem, NC** 0.63 inspector-hours per nonresidential inspection	**Houston, TX** 37 minutes (1992)	**Houston, TX** 21 minutes (1992)	**Wichita, KS** Zoning Inspection: 20 minutes (1990), 29 minutes (1991) Housing inspection: 51 minutes (1990), 60 minutes (1991)	**Plano, TX** 20 minutes (1997)
Wichita, KS 40 minutes (1990), 24 minutes (1991)	**Houston, TX** 22 minutes (1992)	**Charlotte, NC** 1 inspection per inspection-hour		**Wichita, KS** 22 minutes (1990), 21 minutes (1991)	**Houston, TX** Occupancy Inspection: 33 minutes (1992)	**Scottsdale, AZ** 20 minutes (1999)
Houston, TX 35 minutes (1992)						**Sterling Heights, MI** 20 minutes (1995)
Largo, FL 44 minutes (1990), 67 minutes (1991)						

Table 7.11   Codes Administration Cost Recovery: Selected Cities

Ames, IA
Code enforcement cost recovery by category
  Plumbing/mechanical: 54% (1997)
  Electrical: 128% (1997)
  Structural: 186% (1997)
  Rental housing: 100% (1997)

Eugene, OR
Percentage of construction permit service costs recovered from fees: 125% (1997)

Kansas City, MO
Percentage of codes administration expenditures covered by revenues: 118.2% (1995)

Winston-Salem, NC
Revenues per $1 of budgeted expenditures: $1.14 (1997)

Raleigh, NC
Percentage of construction-related expenses recovered by permit fees: 107% (1996), 90% (1997),
  100% (1998)

Greensboro, NC
Percentage cost recovery: 99% (1997)

Portland, OR
Percentage of Bureau of Buildings program costs covered by applicant fees: 94% (1998)

Macon, GA
Percentage of expenses of Bureau of Inspection & Fees recovered through fees: 85% (1995)

Progress in Savannah is measured in terms of declining percentages of units deemed substandard and, especially, in declining percentages in the "major problem" and "dilapidated" categories.

In Cincinnati, code enforcement efforts directed toward existing housing focus on five performance targets:

- To make initial contact on 80% of complaints within 5 working days
- To service requests for permit inspections within 24 hours of request
- To barricade a building within 45 days of it being found open
- To close 75% of cases of building or zoning code violations within 1 year of issue date
- To raze condemned buildings within 6 months of referral to the Hazard Abatement Program (City of Cincinnati [OH], *City Manager Recommended Benchmarks and Current Operation Descriptions,* p. 4)

As with inspections related to new construction, general code enforcement activities in several cities are judged in part by the promptness of response by inspectors (Table 7.12). Although responses tend to be a bit slower than in the case of new construction inspections, this is understandable given the generally less pressing financial ramifications of modest delay.

An important dimension of code enforcement performance is the rate of compliance by offending parties following notice of code violation. What is a reasonable compliance rate? A city with compliance rates of less than 50% for various categories of code violation can take little comfort in scanning the figures of jurisdictions featured in Table 7.13. Many report compliance rates in the 70%, 80%, and even 90% range.

Range of Activities

The range of development activities undertaken by most cities, whether mostly consolidated in a single department or scattered among multiple departments, is broad. Although many of the more commonly tracked aspects of performance have been tabulated in various sections of this chapter, several dimensions of performance—for instance, responsiveness to general planning inquiries, the collection and reporting of data relevant to development, the scope and currency of comprehensive planning, compliance with public notification requirements, and the staff and city's record when its decisions are appealed—are important but less often reported. The records of selected cities on these dimensions are reported in Table 7.14.

Table 7.12 Prompt Response to Requests for Zoning and Miscellaneous Code Enforcement Inspections: Selected Cities

Zoning	Weed Abatement	Noise	Existing Housing	Abandoned/Junk Vehicles	Miscellaneous/General
Overland Park, KS Target: Respond to alleged violations within 24 hours	**Santa Ana, CA** Percentage of weed abatement notices posted within 24 hours: 90% (1995)	**Alexandria, VA** Responses to noise complaints within 1 day: 66% (1991), 100% (1998)	**Charlotte, NC** Target: Respond to 95% of all housing code complaints within 8 work hours Actual: 97.4% within 8 work hours (1987, 8-month record)	**Winston-Salem, NC** Percentage of responses within 2 days: 95% (1998)	**Duncanville, TX** Target: Respond to complaints within 8 hours Actual: 75% (1996), 95% (1998)
Oklahoma City, OK Responses within 24 hours of complaint: 95% (1999)	**Winston-Salem, NC** Percentage of responses within 2 days: 98% (1998)		**Irving, TX** Target: Respond to and investigate all multifamily complaints within 24 hours Actual: 100% (1998)	**Santa Ana, CA** (Police Department) Percentage of responses within 48 hours: 93% (1995)	**Prairie Village, KS** Target: Respond to code violation complaints within 24 hours Actual: 1-day average (1995)
Winston-Salem, NC Responses within 3 days of zoning enforcement request: 92% (1992), 96% (1998, midyear record)	**Corvallis, OR** (Fire Department) Hazard and weed complaint responses within 3 days: 100% (1991), 100% (1992)		**Ames, IA** Percentage of rental housing code complaints investigated within 2 days: 99% (1997)	**Wilmington, DE** Target: Remove unregistered abandoned vehicles within 4 days following identification; registered abandoned vehicles within 10 days	**Oakland, CA** Percentage of inspections performed by the next business day: 92% (1996)
Orlando, FL 100% of site investigations of public complaints completed within 4 workdays of receipt (1991)			**Palo Alto, CA** Target: Respond to at least 90% of complaints relating to substandard dwellings and improper use or occupancy within 2 working days Actual: 100% (1997)	**Palo Alto, CA** Percentage of abandoned/stored vehicle complaints responded to within 5 working days: 100% (1997) Percentage of abandoned/stored vehicle cases resolved within 10 working days: 99% (1997)	**Plano, TX** Code violation complaints processed within 48 hours: 100% (1997) Average complaint closure: 15 days (1997)

Fayetteville, AR
90% of complaints answered within 1 week (1991)

Cincinnati, OH
Target: Investigate within 5 workdays of complaint
Actual: 84% (1993)

Tacoma, WA
Target: Investigate at least 70% of complaints within 2 weeks of receipt
Average initial response time: 2 days (1997)

San Diego, CA
Percentage of initial contact with alleged violators within 15 days: 80% (1998)

Oklahoma City, OK
Percentage of abandoned housing inspections completed within 4 days of notification: 97% (1999)

Ann Arbor, MI
Target: Investigate at least 99% of complaints within 10 working days

Cincinnati, OH
Target: Respond to at least 80% of housing code compliance complaints within 10 workdays
Actual: 67% (1993)

San Diego, CA
Percentage of initial contact with alleged violators within 15 days: 80% (1998)

Charlotte, NC
Target: Respond to abandoned auto complaints within 7 days
Actual: 100% (1987)

Lubbock, TX
90% of abandoned vehicles towed within 10 days of notification (1991)

Oakland, CA
(Police Department)
Target: Process and remove abandoned vehicles within 10 days
Actual: 8-day average (1990), 10-day average (1991), 10-day average (1992)

Boston, MA
92% of abandoned vehicles removed within 14 days of notification sticker being applied to vehicle (1992)

College Station, TX
Percentage of calls investigated within 48 hours: 92% (1998)

San Clemente, CA
Responses within 48 hours: 75% (1998)

Chesapeake, VA
Average initial response time: 2 days (1997)

Chandler, AZ
Percentage of code enforcement requests responded to within 3 days: 90% (1995)

Cincinnati, OH
Target: Investigate citizen complaints within 3 workdays and take appropriate action within 10 workdays
Actual: 65% within target time frame (1995)

(continued)

85

Table 7.12 Prompt Response to Requests for Zoning and Miscellaneous Code Enforcement Inspections: Selected Cities (continued)

Zoning	Weed Abatement	Noise	Existing Housing	Abandoned/Junk Vehicles	Miscellaneous/General
					Oklahoma City, OK Percentage of environmental code nuisance inspections completed within 4 days of notification: 98% (1999) **Hurst, TX** Target: Respond promptly and supply notice of disposition within 4 workdays Actual: 95% within 4 workdays (1997) **Palo Alto, CA** Percentage of enforcement actions initiated on valid complaints within 5 working days: 93% (1997) Percentage of complaints resolved within 3 months following notification of the responsible party: 98% (1997) **Kansas City, MO** Target: Respond within 5 days Actual: 57% within 5 days (1995) **San Antonio, TX** Percentage of initial inspections within 7 days: 68% (1995), 89% (1997) Percentage within 30 days: 85% (1995), 97% (1997)

Table 7.13 Compliance Rates for Code Enforcement Actions: Selected Cities

			Compliance Following Notice of Violation				
General	Fire Code	Grass and Weeds	Housing	Junk/ Abandoned Vehicles	Trash and Debris	Zoning	Miscellaneous
College Station, TX Percentage of cases resolved within 90 days: 98% (1998)	**Winston-Salem, NC** Rate of compliance: 81.3% (1992)	**Denton, TX** Rate of compliance: 98% (1991)	**Charlotte, NC** Target: 100% of serious violations remedied within 120 days	**Winston-Salem, NC** Corrections as a percentage of abandoned vehicle investigations: 91% (1997)	**Charlotte, NC** Rate of compliance: 92% within 30 days (1994, 5-month record)	**Fort Collins, CO** Resolved voluntarily: 98% (1991)	**Alexandria, VA** Percentage of noise complaints closed within 10 days: 47% (1998)
Denton, TX 90% (1991)	**Reno, NV** Target: At least 75% within 30 days	**Lubbock, TX** 98% (1992, estimated)	**Irving, TX** Target: Secure at least 88% voluntary compliance for all residential code violations Actual: 100% (1998)	**Denton, TX** Rate of compliance: 90% (1991)	**Denton, TX** 90% (1991)	**Scottsdale, AZ** Percentage of compliance achieved within 30 days (prior to issuance of civil zoning citation): 98% (1999)	**Denver, CO** Nuisance abatement compliance: 98.9% (1998)
Anaheim, CA Target: Resolve 90% of all documented violations within 30 days	**Oak Ridge, TN** Compliance within 180 days: 81% (1991), 70% (1992)	**Winston-Salem, NC** 95% (1997), 91% (1998)	**Lubbock, TX** Rate of compliance: 87.4% (1991)	**High Point, NC** 85% (1998)	**Tallahassee, FL** Target: 70% compliance for care-of-premises violations	**Lubbock, TX** 93.9% (1991)	**Decatur, IL** Percentage of nuisance cases brought to compliance: 90% (1991)
Greensboro, NC Percentage of code violations abated within 60 days: 90% (1997)		**Hurst, TX** Corrected within 10 days: 90% (1990)	**St. Petersburg, FL** Target: 85% within 180 days Actual: 67.7% (1990)	**Lubbock, TX** 83.7% (1991)		**Winston-Salem, NC** 88.3% brought into compliance within 6 months (1997)	**Denton, TX** Compliance following notice of sign ordinance violation: 80% (1991)
Greenville, SC Compliance rate within 60 days: 88% (1990)		**Sterling Heights, MI** 85% (1995)				**San Diego, CA** Percentage of zoning code cases resolved within 6 months: 65% (1998)	

Table 7.13 Compliance Rates for Code Enforcement Actions: Selected Cities (continued)

			Compliance Following Notice of Violation	Junk/			
General	Fire Code	Grass and Weeds	Housing	Abandoned Vehicles	Trash and Debris	Zoning	Miscellaneous
Peoria, AZ Violations resolved within 45 days: 85% (1991)							
San Antonio, TX Voluntary compliance: 79% (1995), 85% (1997) Average number of days to close a case: 33 (1997)			**Winston-Salem, NC** Percentage of substandard units brought into compliance: 83% (1998)				
Irving, TX 80% (1995)			**Chapel Hill, NC** Target: At least 80% within 60 days Actual: 79.7% (1999)				
Oakland, CA Percentage of cases resolved within 25 days: 74% (1996)			**San Diego, CA** Percentage of housing code cases resolved within 6 months: 73% (1998)				
Jacksonville, FL Owner compliance: 73.75% (1994)			**Decatur, IL** 61% (1991)				
St. Petersburg, FL 66% (1995)			**Tallahassee, FL** Target: 50% compliance for substandard structures				

Table 7.14   Odds and Ends in Development Administration: Selected Cities

---

RESPONSIVENESS TO PLANNING INQUIRIES

Corpus Christi, TX
Target: Respond to citizen inquiries within 8 working hours (1999)

Ann Arbor, MI
Target: Respond within 24 hours

Alexandria, VA
Percentage of zoning inquiries answered within 24 hours: 98% (1995)

Scottsdale, AZ
Percentage of phone calls responded to within 24 hours: 98% (1999)

Norman, OK
Target: Provide requested information within 2 days (1999)

DATA COLLECTION AND REPORTING

Palo Alto, CA
Target: Provide construction statistics to local, state, and federal agencies within 5 days of month's
end
Actual: 100% (1997)

UP-TO-DATE INFORMATION

Norman, OK
Targets: Update new subdivision, water, and sewer information into database within 3 weeks of receipt of as-built and filing date of the final plat; update zoning database within 1 week of ordinance
adoption (1999)

Palo Alto, CA
Percentage of information material and publications updated within 30 days of relevant code changes:
100% (1997)

UP-TO-DATE ZONING MAP

Charlotte, NC
Target: Update zoning maps within 30 days of rezoning (1999)

Palo Alto, CA
Target: Update zoning maps within 30 days of rezoning
Actual: 50% (1997)

COMPREHENSIVE PLANNING

Corpus Christi, TX
Target: 100% of the city's areas having approved plans not more than 5 years old (1999)

Tucson, AZ
Percentage of city covered by specific neighborhood, area, and subregional land use plans: 83.6%
(1997)

(continued)

Table 7.14   Odds and Ends in Development Administration: Selected Cities
                   (continued)

NOTICE TO PUBLIC

Scottsdale, AZ
  Target: Require applicants for all general plan amendments, rezoning, and major use permits to post
    development sites with red, 4' × 4' public hearing information sign at least 20 days prior to public
    hearing
  Actual: 89% (1998), 80% (1999)
  Target: Post agendas and case fact sheets on the internet 30 days prior to hearing
  Actual: 100% (1999)

TIMELY ENVIRONMENTAL STUDIES

Palo Alto, CA
  Percentage of initial studies and environmental determinations performed within 2 weeks for annu-
    ally scheduled capital improvement projects and within legal timelines for unscheduled projects:
    100% (1997)

Oakland, CA
  Average processing time for environmental impact reviews: 75 days (1996)
  Percentage of reviews completed within 30 days: 75% (1996)

ENVIRONMENTAL SENSITIVITY

Palo Alto, CA
  Target: Convert at least 75% of all proposals for creek flood protection from concrete solutions to
    more natural solutions
  Actual: 67% (1997)

GRADING PERMITS

Winston-Salem, NC
  Percentage of grading permits issued within 9 days: 96% (1996), 100% (1997)

PROMPT DECISION ON DEMOLITION REQUESTS

San Antonio, TX
  Percentage of demolition requests reviewed within 3 days: 89% (1997)

CERTIFICATE OF OCCUPANCY

Cary, NC
  Percentage of certificate of occupancy inspections conducted on the date requested: 95% (1999)

Scottsdale, AZ
  Percentage of certificates of occupancy completed within 24 hours: 100% (1999)

Greensboro, NC
  Percentage of certificates of occupancy issued within 2 business days: 89% (1997)

PROMPT RESOLUTION OF COMPLAINTS

Greensboro, NC
  Average time to resolve complaints: 12 hours (1997)

Palo Alto, CA
Target: Obtain resolution of 75% of complaints within the department's authority and responsibilities within 4 months
Actual: 70% (1997)

PROMPT RESOLUTION OF CASES

Kansas City, MO
Percentage of buildings/zoning code violation cases resolved within 90 days: 42% (1995)

JUDGMENT OF STAFF WORK UPON APPEAL

Eugene, OR
Appeals as a percentage of all land use applications: 1.5% (1996), 2.9% (14 of 488) (1997)
Percentage of appeals to hearing officials resulting in decision to uphold staff recommendation: 71% (1996), 92.9% (13 of 14) (1997)
Percentage of appeals to the planning commission resulting in decision to uphold hearing official/staff recommendation: 100% (1996), 100% (7 of 7) (1997)
Percentage of appeals to the Land Use Board of Appeals in which local jurisdiction prevailed: 100% (1996), 100% (3 of 3) (1997)
Percentage of court appeals in which local jurisdiction prevailed: 100% (1996)
Appeals beyond the Land Use Board of Appeals: 0 (1997)

8

Emergency Communications

The primary function of emergency communications is to receive messages from persons in need of emergency services and to summon assistance on their behalf. In some cases, emergency communications focus on, and are assigned within, a single department—typically, a fire or police department. In other cases, the emergency communications operation covers a combination of police, fire, ambulance, and perhaps other emergency services and may be performed by one of the participating departments on behalf of all or, alternatively, may be established as a separate operational entity.

Standards

Communities seeking standards applicable to emergency communications will be interested in a pair of guidelines addressing police communications. The first pertains to equipment capability and is a standard promoted in Pennsylvania that calls for police radio systems to be "engineered to produce a 12-decibel or greater SINAD ratio to the radio receivers in 95 percent of the service area" (Southwestern Pennsylvania Regional Planning Commission [SPRPC], 1990, p. V-7).

The other is a standard developed by the Commission on Accreditation for Law Enforcement Agencies (CALEA; 1993) that specifies the information to be recorded when a request for service is received: date and time of request, name and address of complainant (if possible), type of incident reported, location of incident reported, time of dispatch, time of officer arrival, time of officer return to service, and disposition or status of reported incident (p. 81.3). Managers of emergency communications units that have met basic equipment capability and data collection requirements may wish to proceed beyond the basics and to focus more directly on dispatcher workload and key dimensions of dispatcher performance.

Table 8.1   Dispatcher Workload: An Example From a Single
                   Municipality

Task	Approximate Daily Workload per Dispatcher
Radio communications	275
Telephone communications	82
Alarms handled	12

SOURCE: Statistics for the city of Santa Ana, California, 1995–96, from City of Santa Ana (CA), *Annual Budget 1996–97*, p. 2.66.

Table 8.2   Production Ratios in Emergency Communications:
                   Selected Cities

Eugene, OR
   Calls received per labor hour: 24 (1996), 23 (1997)

Scottsdale, AZ
   Total phone activity per authorized line personnel: 15,770
   Total 911 calls per authorized line personnel: 4,407 (1999)

Plano, TX
   Number of calls answered per communications specialist (annual): overall, 12,826;
      emergency calls, 4,108; nonemergency calls, 8,718 (1997)
   Number of calls dispatched per communications specialist (annual): 4,038 (1997)

Dispatcher Workload

One municipality's analysis of its emergency communication dispatchers' workload yielded the set of average daily activities identified in Table 8.1. Workload statistics are apt to vary from one community to another, depending on population, dispatcher responsibility for a single function or for multiple emergency functions, and variation among communities in the rate of emergency incidents (Table 8.2). In a study involving 14 North Carolina cities, calls answered per telecommunicator in a year's time ranged from 5,019 to 20,948 among small cities and 9,048 to 25,460 among larger units (Table 8.3).

Even within an individual community, workload will vary by hour of the day and by day of the week. Nevertheless, the figures for Santa Ana, California, and more general statistics from other cities indicate substantial capacity for a single operator.

Table 8.3   Emergency Communications: Statistics from North Carolina's
Comparative Performance Project, 1998

	6 Cities of 63,000–260,000 Population		8 Cities of 16,000–62,000 Population		
	Average	Range	Average	Range	Pacesetters
Calls answered per telecommunicator	14,064	9,048–25,460	10,586	5,019–20,948	Raleigh (25,460 calls per telecommunicator)
Time from initial ring to answer (seconds)	9	5–12			Durham (5 seconds from initial ring)
Percentage answered within 3 rings (18 seconds)	95.2%	90%–99%	88%	80%–98%	Winston-Salem (99% within 3 rings)
Time from call receipt to dispatch (seconds)	72.3	40–134	51	25–96	Hickory (25 seconds)

SOURCES: *Interim Report on Services for Seven Cities* (Chapel Hill, NC: Institute of Government/University of North Carolina, 1998), pp. 90–101; and *Performance and Cost Data: Phase III City Services* (Chapel Hill, NC: Institute of Government/University of North Carolina, 1999), pp. 156–173.
Reprinted with permission of the Institute of Government, The University of North Carolina at Chapel Hill.

Benchmarks

Three performance dimensions are especially critical to emergency communications:

- Speed
- Accuracy
- Good judgment

Response to callers should be prompt, and emergency units should be dispatched quickly and to the proper location—but ideally, only if emergency personnel are really needed.

Speed

To a frantic caller reporting an emergency, precious seconds are critical, and each unanswered ring seems like an eternity. Time *is* critical in every phase of emergency response, and many cities measure the performance of communications personnel in quickly answering 911 telephone calls (Table 8.4). Most set their targets at answering emergency calls with a maximum of 10 seconds, a maximum of 20 seconds, or somewhere in between. The "quick draw"

Table 8.4   Prompt Answer of 911 Calls: Selected Cities

Denver, CO
Average 911 answer time: 2.0 seconds (1998)
Average nonemergency answer time: less than 10 seconds (1998)

Fayetteville, AR
Actual: 4-second average (1991)

Houston, TX
Target: Within 5 seconds
Actual: 99% within 5 seconds (1992)

Lubbock, TX
Target: At least 95% within 5 seconds (1997)

Oakland, CA
Target: Within 10 seconds
Actual: 5-second average (1996)

Tucson, AZ
100% within 10 seconds (1997)

Kansas City, MO
99% within 10 seconds (fire) (1995)

Bellevue, WA
Actual: 95% within 10 seconds (1998)

Reno, NV
Target: At least 95% within 10 seconds (1999)

Scottsdale, AZ
Actual: 83% within 10 seconds (1999)

Raleigh, NC
Target: Within 2 rings (12 seconds)
Actual, within 2 rings: 87.0% (1997), 75.0% (1998)
Actual, within 5 rings: 98.0% (1997), 95.0% (1998)

San Jose, CA
Target: At least 95% within 15 seconds
Actual: 87% of fire emergency calls and 85% of police emergency calls answered within 15 seconds

Durham, NC
96% within 3 rings (1998)

Overland Park, KS
Target: 3 rings or less

Winston-Salem, NC
99% within 19 seconds (1998)

Cincinnati, OH
Target: At least 95.9% within 20 seconds
Actual: 97.5% (1993)

Portland, OR
Target: Within 20 seconds
Actual: 86.3% within 20 seconds (1998)

Charlotte, NC
95% within 36 seconds (police) (1999)

leaders among cities examined for this volume were Denver, Colorado, with a reported average of 2 seconds; Fayetteville, Arkansas, with an average of 4 seconds; and Houston, Texas, reporting response to 99% of its 911 calls within 5 seconds.

How quickly, then, should an operator proceed from answering the call to actually dispatching emergency units? Although some cities report average dispatch times of 3 minutes or longer, most of the ones examined for this volume set targets or report actual experience at 2 minutes or less for emergency dispatches (Table 8.5). Lubbock and Duncanville, Texas; Eugene, Oregon; and Boca Raton, Florida, were pacesetters among the cities examined independently for this volume. Lubbock, for example, reported a 30-second average for fire dispatches, and Duncanville reported an average of 45 seconds for priority calls.

Other comparative studies corroborate the figures reported in Table 8.5 as reasonable benchmarks of dispatching speed—although in a few cases offering a challenge to the pacesetters for the top spot. In the North Carolina study reported in Table 8.3, Hickory, North Carolina, reported an even quicker average of only 25 seconds from call receipt until dispatch. A similar study involving eight South Carolina cities with populations of between 12,000 and 75,000 reported emergency dispatch averages ranging from a 30-second average to an average of slightly more than 5 minutes (*South Carolina Municipal Benchmarking Project: 1998*, 1999, pp. 109–119). Four cities reported averages of 30 seconds. A study by the ICMA (1999a) reported dispatch times for top priority police calls ranging from a 30-second average to a 4-minute average among 13 cities with populations between 10,000 and 100,000, and ranging from a 1-minute average to a 4.1-minute average among 26 larger cities (pp. 27, 77, 308). Average times from call entry to dispatch for medical emergencies requiring advanced life support were even quicker.

Some cities have established priority tiers for promptness of dispatch to various types of calls. For example, a five-tier call prioritization system was developed for handling calls to the Dallas, Texas, police department (City of Dallas [TX], 1989). Priority 1 calls included shootings, knifings, officers in need of assistance, emergency blood transfers, and felonies in progress, and had a performance target of dispatch within 1 minute. Priority 2 calls had a performance target of dispatch within 3 minutes; priority 3, within 8 minutes; priority 4, within 60 minutes. Priority 5 calls were to be handled entirely by telephone without dispatch, if the caller agreed to that approach.

Yet another aspect of emergency communication speed pertains not to call dispatching but instead to promptness in providing requested information to field officers regarding wants and warrants. Overland Park, Kansas, sets its performance target at response within 1 minute, if possible—and within 5 minutes in every case.

Table 8.5 Rapid Dispatch: Selected Cities

Lubbock, TX
Actual: 1.5-minute average for police emergency calls; 30-second average for fire calls; 15-second average for emergency medical service (EMS) transfer calls (1997)

Duncanville, TX
Target: Automatic Number Identification (ANI) to entry response time of less than 90 seconds on priority 1 and 2 calls; less than 5 minutes on priority 3 calls
Actual: Average of 45 seconds on priority 1 and 2 calls; average of 120 seconds on priority 3 calls (1996)

Eugene, OR
Target: Average of 48 seconds on fire/EMS emergency calls
Actual: 56-second average (1995)

Boca Raton, FL
Actual: 53-second average for life-threatening calls (1998)

Alexandria, VA
Actual: 1.5-minute average for police emergency; 6.0-minute average for police nonemergency; 0.9-minute average for fire/EMS (1998)

Palo Alto, CA
Percentage of police emergency calls dispatched within 60 seconds: 99% (1997)
Percentage of police nonemergency calls dispatched within 30 minutes: 100% (1997)
Percentage of fire emergency calls dispatched within 90 seconds: 100% (1997)
Percentage of utility emergency calls dispatched within 10 minutes: 99% (1997)
Percentage of joint department responses to public utility problems coordinated within 10 minutes: 99% (1997)

Savannah, GA
Target 1: Within 2 minutes for police emergency calls and 4 minutes for "immediate" calls
Actual: 2-minute 39-second average dispatch delay for police emergency calls; 3-minute 17-second average for "immediate" calls (1997)
Target 2: Within 1 minute for fire emergency calls
Actual: 99% (1998, estimated)

College Station, TX
Percentage of priority 1 police calls dispatched within 3 minutes: 100% (1998)
Percentage of priority 1 fire calls dispatched within 1 minute: 97% (1998)

Denver, CO
Target 1: Dispatch 98% of police priority 1 calls for service within 2 minutes
Target 2: Dispatch 90% of police priority 2 and 3 calls within 5 minutes
Target 3: Dispatch 99% of fire emergency calls within 1 minute from receipt
Actual: 91% of fire emergency calls dispatched within 1 minute (1998)

Chandler, AZ
88.3% entered into computer-aided dispatch within 1 minute (1998)

San Antonio, TX
1.09-minute average for fire calls; 1.19-minute average for EMS (1997)

(continued)

Table 8.5   Rapid Dispatch: Selected Cities (continued)

Ann Arbor, MI
Target: Average of 1 minute or less for emergency calls
Actual: 1.1-minute average for police emergencies (1998)

Durham, NC
67-second average for priority 1 calls (1998)

Bellevue, WA
1-minute 14-second average for priority 1 fire calls; 2-minute 39-second average for priority 1 police
calls (1997)

San Jose, CA
Target 1: 90-second processing time for at least 90% of priority 1 police calls
Actual: 88% (1995)
Target 2: At least 90% of fire emergency calls dispatched within 2 minutes
Actual: 95% within 2 minutes (1995)

Cincinnati, OH
Actual: 1.57-minute average (1993)

Raleigh, NC
Average time from call answer to dispatch: 1.67 minutes (1997), 2.47 minutes (1998)
Average time from call disconnect to dispatch: 1.0 minutes (1997), 0.82 minutes (1998)

Fort Collins, CO
1-minute 58-second average for emergency police calls (1996)

Little Rock, AR
Actual: 2-minute average (1-minute telephone processing time plus 1-minute dispatch, 1995)

Accuracy

Dispatch errors can be disastrous. Sending a fire or police unit to the wrong location can result in the loss of precious minutes at a time when there are none to spare.

Dispatcher error rates reveal the frequency of accuracy lapses (Table 8.6). They may pertain strictly to the inaccurate radio transmission of critical information or, as in the case of Chandler, Arizona, they may incorporate other aspects of the job, such as improper prioritization of service calls or inaccuracies in communications paperwork. In addition to reporting its overall error rate (a mere 0.4% in 1998), Chandler also reports the percentage of state and national criminal information system reports entered accurately by communications personnel (99.7% in 1998). Kansas City, Missouri, reports the percentage of fire alarms dispatched with the proper first-due company (95% in a recent year) and the percentage of fire incident reports prepared without error (95%). Whether focusing on specific elements of emergency communica-

Table 8.6   Dispatch Accuracy: Error Rates for Three Cities

City	Dispatcher Error Rate
Chandler, AZ	0.4%[a] (1998)
Duncanville, TX	0.4% (1990)
Charlotte, NC	0.5%[b] (1987)

a. Chandler, Arizona, reported an accuracy rate of 99.6% for 1997–1998, a rate "established by validating samples of each employee's work product which includes review of 911 call tape recordings, incident entry computer information, and calls for service priority assignments" (City of Chandler [AZ], *1999–2000 Annual Budget,* p. 225).
b. A reported transmission accuracy rate of 99.5% for Charlotte was based on a sample of 18,484 fire emergency dispatches (City of Charlotte [NC], *FY88–FY89 Objectives,* p. 22).

tions or on the function as a whole, the critical nature of dispatch accuracy demands the consistent achievement of extremely low error rates.

Judgment

Communications personnel must decide quickly whether to dispatch field units. Some cities report impressive contributions to the effective deployment of police, fire, and other emergency resources through the ability of emergency communications officers to deflect calls that can be handled effectively by other means. Alexandria, Virginia, for example, reported an estimated 29% reduction in field officer workload by directing appropriate calls to a telephone reporting unit rather than dispatching field personnel. Phoenix, Arizona, expects half its calls to be handled without dispatch of emergency units; Orlando, Florida, sets its target at two thirds of all calls. Properly trained and supervised dispatchers who consistently make the right decisions can contribute mightily to efficient and effective emergency operations. On the other hand, dispatchers who have inappropriately delayed calls for emergency units have landed their cities in embarrassing legal and public relations fiascoes. Clearly, good judgment is crucial.

9

Emergency Medical Services

Amunicipality's emergency medical service (EMS) operation performs the critical life-or-death function of stabilizing patients—often the victims of serious illness or injury—and transporting them quickly to a hospital. The adequacy of a community's EMS operations can be assessed in a variety of ways.

The Need for Speed

Among the many key aspects of EMS performance, a crucial dimension is speed of response. For seriously stricken patients or critically injured accident victims, delays of even a few minutes can be catastrophic.

Knowing what to do on arrival is also crucial to effective EMS action. In cardiac care emergencies, for instance, the American Heart Association considers early access (i.e., quick recognition of the problem and prompt response), early CPR (cardiopulmonary resuscitation), early defibrillation, and early advanced care to be the "chain of survival" (Cummins, Ornato, Theis, & Pepe, 1991). Defibrillation administered within 6 minutes dramatically improves the odds of survival (Figure 9.1).

For other medical emergencies as well, the chain begins with prompt response. Because even the most proficient stabilization efforts provided too late may be of little value, response speed is the most frequently reported EMS performance indicator. Paraphrasing comedian Woody Allen only slightly, a big part of success in this business is showing up . . . on time.

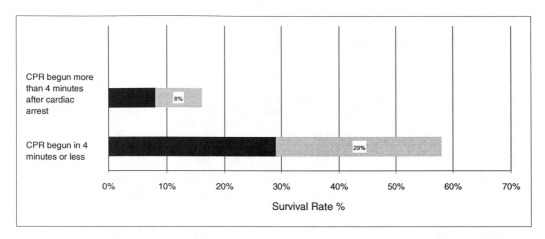

Figure 9.1. Quick EMS Response: A Key to Surviving Ventricular Fibrillation
SOURCE: W.D. Weaver et al., "Factors Influencing Survival After Out-of-Hospital Cardiac Arrest," *Journal of the American College of Cardiology,* 7 (April 1986), p. 754. Reprinted with permission from the American College of Cardiology, *Journal of the American College of Cardiology,* 1986, Vol. 7, p. 754.

Benchmarks

Response Time

A 1989 study of EMS communication services in Washington, D.C. (City of Washington, D.C., 1989b), reported average EMS response times among major cities ranging from 4.8 minutes in Kansas City, Missouri, to 10 minutes in Washington, D.C. (Table 9.1). A broader array of cities examined for this volume reported generally quicker EMS responses (Table 9.2). Several cities reported response times in the 3- to 5-minute range. It should be noted, however, that this second set of cities includes smaller communities with potentially shorter travel distances and less congested routes. Furthermore, self-reporting by these cities might in some cases exclude the time from call receipt until dispatch, a component of total response time explicitly included for all cities in Table 9.1 but mentioned explicitly in reporting documents by only three of the cities in Table 9.2. Even with this possible difference, however, the response times reported by many cities in the second set were quicker even than the time from dispatch until arrival reported for most cities of the first set.

The reasonableness of these EMS response time benchmarks is corroborated by the cities participating in the ICMA (1999a) comparative performance measurement project. Twenty-four cities with populations of 100,000 or greater reported EMS dispatch-to-arrival times for basic life support (BLS) response ranging from an average of 2.1 minutes (Oklahoma City, Oklahoma) to 8.9 minutes in 1997. The median average was 4.7 minutes from dispatch until arrival. For advanced life support (ALS) response, dispatch-to-arrival time averages ranged from 3.1 minutes (Odessa, Texas) to 8.4 minutes among

Table 9.1   Emergency Medical Service Response Times Among Major Cities, as Reported in 1989

City	Time From Receipt of Call Until Dispatch (minutes)	Time From Dispatch Until Arrival (minutes)	Total Response Time (minutes)
Kansas City	0.80	4.00	4.80
Memphis	1.00	5.00	6.00
Philadelphia	1.00	5.00	6.00
Phoenix	1.00	5.00	6.00
Chicago	1.50	4.70	6.20
San Francisco	1.14	5.26	6.40
Cleveland	1.00	5.50	6.50
St. Louis	0.50	6.00	6.50
Los Angeles	0.60	6.00	6.60
Baltimore	1.00	6.00	7.00
Nashville	1.00	6.00	7.00
New Orleans	1.00	6.00	7.00
El Paso	1.00	6.30	7.30
Houston	1.00	6.50	7.50
San Antonio	1.00	6.70	7.70
Honolulu	1.00	7.00	8.00
Oklahoma City	0.70	7.18	8.04
San Jose	1.30	7.00	8.30
Boston	1.50	7.00	8.50
Washington, D.C.	3.10	6.90	10.00

SOURCE: Adapted from City of Washington, D.C., *Improving Emergency Medical Services Communications* (Washington, DC: Office of the City Administrator, Productivity Management Services, June 1989), p. 6.

cities with populations greater than 100,000, and from 3.9 minutes (Santa Monica, California) to 5.7 minutes among smaller cities (pp. 307, 309, 339).

National statistics on EMS response times and the time from arrival on the scene until hospital delivery following serious accidents sheds further light on this topic (Table 9.3). Among urban accidents resulting in fatalities in 1997, EMS response times exceeded 10 minutes in slightly more than 10% of the cases. Delivery to the hospital usually required another 11 to 40 minutes.

Ten local governments participating in the ICMA (1999a) project reported average times from arrival at the scene to delivery of the patient at a medical facility in 1997 for calls requiring an ALS response (p. 310). The median was a

Table 9.2   Emergency Medical Service Response Times Among a Broader Array
of Cities

Chapel Hill, NC
 3.2-minute average (1999)

Duncanville, TX
 3.7-minute average (1996), 4-minute average (1998)

Palo Alto, CA
 Targets: Respond to at least 90% of emergency medical requests with EMT-D-trained personnel[a]
   within 4 minutes in urban response zone, and to at least 90% of paramedic requests within
   6 minutes
 Actual: 100% (1997)

Cambridge, MA
 Target: Within 4 minutes at least 90% of the time

Reno, NV
 Target: At least 75% within 4 minutes following dispatch (1999)

Eugene, OR
 64% within 4 minutes (1995)

Portland, OR
 46% within 4 minutes (turnout and travel time) (1998)

Cincinnati, OH
 4.15-minute average (1995)

Long Beach, CA[b]
 4.4-minute average from call until arrival (1996)

St. Petersburg, FL
 4.4-minute average (1995)

Alexandria, VA
 4.77-minute average (1998)

Irving, TX
 4-minute 50-second average (1998)

Charlottesville, VA
 94% within 5 minutes (1998)

Plano, TX
 5-minute 5-second average (1997)

Greensboro, NC
 5.3-minute average (1997)

Bellevue, WA[b]
 5.8-minute average from initial call until arrival; 22% within 4 minutes (1998)

Orlando, FL
 Percentage within 6 minutes following dispatch: 96% (1996)

(continued)

Table 9.2   Emergency Medical Service Response Times Among a Broader Array of
Cities (continued)

Albuquerque, NM
Targets: At least 90% of BLS responses on scene within 6 minutes; at least 90% of ALS within 8 minutes (1997)[a]

Corpus Christi, TX
Target: At least 80% within 6 minutes following dispatch (1999)

Philadelphia, PA
6-minute average (1998)

Chesapeake, VA
6-minute 30-second average (1997)

Jacksonville, FL
6-minute 33-second average (1994)

Nashville–Davidson County, TN
6.8-minute average (1998)

San Antonio, TX[b]
Targets: Less than 9-minute average (dispatch and travel time) within the city and suburban areas;
travel time average within 6 minutes for city and within 7 minutes for suburbs
Actual: 7.25-minute average (6.06-minute travel time) within city; 8.28-minute average for suburbs
(1997)

Shreveport, LA
7.65-minute average medic response time; 71% of ALS responses within 8 minutes; 58% of BLS
responses within 4 minutes; 4.35-minute average response time by firefighters to EMS calls;
50% within 4 minutes (1994)

New York, NY
7-minute 54-second average (1998)

Tucson, AZ
Percentage of ALS responses within 8 minutes of dispatch: 89% (1997)

Calgary, Alberta
87% within 8 minutes (1996)

Boston, MA
87.9% within 11 minutes (1996)

San Diego, CA
90% within 12 minutes (1998)

a. EMT-D-trained personnel are emergency medical technicians trained in defibrillation.
b These cities specifically defined response time to include dispatch time and travel time—that is, the time from initial
call until arrival on the scene. Many others exclude dispatch time.
c. BLS = basic life suppprt; ALS = advanced life support.

21.5-minute average, with Miami–Dade County's 6.5-minute average leading
the group.

Table 9.3   EMS Response Times and Hospital Delivery Times for Urban Crashes Resulting in Fatalities, 1997

Time (minutes)	Percentage of Response Times Within Specified Range (EMS notification to EMS arrival)	Percentage of Delivery Times Within Specified Range (EMS arrival at scene to hospital arrival)
0 to 10	89.3	7.5
11 to 20	9.4	34.0
21 to 30	0.8	31.2
31 to 40	0.3	14.3
41 to 50	< 0.05	6.4
51 to 60	< 0.05	2.9
61 to 120	0.1	3.8
Total	100	100

SOURCE: U.S. Department of Transportation, *Traffic Safety Facts 1997* (Washington, DC: National Highway Traffic Safety Administration, November 1998), p. 48. Figures are based on records for 7,865 urban accidents with fatalities.

Among cities examined independently for this volume, the city of San Antonio, Texas, reported that for 1995 the average time from receipt of an EMS call until arrival at a hospital by its units was 47.23 minutes for non-life-threatening incidents and 44.91 minutes for life-threatening incidents.

According to the study by the City of Washington, D.C. (1989a), the national medical community and the EMS industry have defined a two-part standard for EMS responsiveness: "90 percent of [Emergency Medical Technician] responses should be within four minutes, and 90 percent of paramedic responses should be within eight minutes" (p. 101). A standard promoted in Pennsylvania calls for an 8-minute response time on at least 90% of all emergency calls (SPRPC, 1990, p. VII-1). Reported performance targets and experience of the cities examined here suggest that an 8-minute standard might be realistic but that a 4-minute standard lies well beyond the reach of many communities.

Workload

Much as in the case of police officers and firefighters, some communities attempt to gauge the adequacy of their EMS services by considering the number of paramedics per 1,000 population. Such indicators can be misleading, because they fail to take into account a given community's distinctive needs for various emergency services. In reality, such measures are not really perfor-

Table 9.4   Quick and Effective Action on the Scene: Selected Cities

Quick Action on the Scene	*Effective Treatment*
Corpus Christi, TX Target: Medic units will initiate transport of critical trauma patients within 15 minutes of arriving on the scene at least 80% of the time (1999)	**Orlando, FL** Patients improved with treatment: 57% (1995)
Orlando, FL Critical patients transported within 20 minutes: 53% (1994), 65% (1995), 43% (1996) Significant trauma patients transported within 10 minutes: 22% (1994), 28% (1995)	**Bellevue, WA** Cardiac arrest survival rate: 33% (1998) **Tucson, AZ** Percentage of patients saved after suffering witnessed cardiac arrest: 15% (1997)
Shreveport, LA Percentage of prehospital time at major trauma less than 36 minutes: 74% (1994)	

mance indicators at all because they do not measure performance. They measure resource inputs—not performance outputs—and can vary greatly from one community to another. For example, at 0.44 paramedics per 1,000 population, the reported paramedic rate in College Station, Texas, in the early 1990s was almost twice that of San Antonio's 0.24 per 1,000 population in the mid-1990s.

More useful indicators of workload may be secured through direct measurement, rather than by population ratio. In that regard, Chesapeake, Virginia, reported 967 responses per paramedic during a recent year, and Sedgwick County, Kansas, reported an average of 810 calls per crew. San Antonio reported 9.6 responses per ambulance per day and 3,516 responses per year. Calgary, Alberta, reported an average of 244 responses per department employee.

Effectiveness

EMS units in College Station, Texas, strive to provide treatment and achieve stabilization within 15 minutes. Corpus Christi, Texas, attempts to initiate transport of most critical trauma patients within 15 minutes following arrival on the scene (Table 9.4). Orlando, Florida, tries to transport critical patients within 20 minutes and "significant trauma patients" within 10 minutes. Shreveport, Louisiana, limited prehospital time at major trauma to less than 36 minutes for most of its EMS patients.

Table 9.5   Fee Collection and Cost Recovery Rates for EMS
Services

COLLECTION RATE

Tucson, AZ
Collection rate for ALS/ambulance service: 74% (1997)

San Antonio, TX
64.5% of net billings collected (1995)

Burbank, CA
63% (1988), 54% (1989)

Shreveport, LA
55% (1994)

Duncanville, TX
54% (1996)

Palo Alto, CA
Paramedic receivables collection rate: 54% (1997)

Norfolk, VA
Medical billing collection rate: 38% (1997)

COST RECOVERY THROUGH FEES

College Station, TX
Target: 40% to 70%

Calgary, Alberta
40% to 45% (1987–1996)

More significantly, Orlando reported in one recent year's statistics that 57% of its EMS patients improved with treatment. Bellevue, Washington, reported a cardiac arrest survival rate of 33%. Among 17 participating local governments in the ICMA (1999a) project, the median unit delivered 20.7% of those patients suffering full cardiac arrest from medical causes to a medical center with a pulse in 1977 (pp. 311, 341). Lake Forest, Illinois, led the group with a 55.9% average.

Yet another gauge of effectiveness is the rate of complaints. In one recent year, complaints about EMS services received by the city of Boston represented only 0.03% of all ambulance responses.

Collection Rates and Cost Recovery

Fees charged to recipients of EMS services often go unpaid. Reported collection rates among examined cities ranged from 38% to 74% (Table 9.5). Among a variety of proposed performance targets for EMS services in Washington, D.C., was an ambulance fee collection rate of 60% (Exhibit 9.1).

Exhibit 9.1.   Performance Targets and EMS Performance Report

EMS Performance Report   Month: _____ Year: _____

Item Number	Indicators	Last Month	This Month	YTD	Target
1.1	Number of incidents handled				—
2.1	Average call-to-scene time for first fire-fighting unit (in minutes)				4.0
2.2	Average call-to-scene time for first Emergency Ambulance Bureau (EAB) unit (in minutes)				n.a.
2.3	Average call-to-scene time for first responding unit (in minutes)				4.0
2.4	Average time from call to EAB unit dispatch (in minutes)				0.5
3.1	Percentage of fire-fighting units responding within 4 minutes				90%
3.2	Percentage of calls receiving a response within 4 minutes				90%
3.3	Percentage of urgent calls receiving an ALS response within 8 minutes				90%
3.4	Percentage of urgent calls for which an ALS unit is not dispatched				0%
4.1	Percentage of incidents upgraded in priority on the scene				10%
4.2	Percentage of incidents downgraded in priority on the scene				10%
5.1	Percentage of EAB runs evaluated in the field				25%
5.2	Percentage of fire unit medical runs evaluated				90%
6.1	Ambulance fee collection rate				60%
6.2	EAB absenteeism rate				5%

SOURCE: City of Washington, D.C., *Improving Ambulance Operations in Washington, D.C.: A Blueprint for Change* (Washington, DC: Office of the City Administrator, Productivity Management Services, March 1989), p. 106.

Recognition of the nonpayment factor and the high cost of service leads most cities to anticipate a municipal subsidy of the EMS operation. College Station, for example, sets as its target the recovery of 40% to 70% of EMS costs through fees. Calgary achieved cost recovery of 40% to 45% from 1987 through 1996.

10

Finance

Financial management in local government is a complex topic. Attempting to assess the performance of this function in an individual community is no easy task. No simpler is the task undertaken here of compiling a relatively comprehensive set of financial management benchmarks and confining those benchmarks to a single chapter.

Assessing the Financial Health of a Municipality

A "quick-and-dirty" way of gauging the quality of financial management in a given municipality is by noting that city's general financial health. One popular method of doing so—perhaps because of its speed and simplicity—is by checking the local bond rating to see how industry analysts judge the municipality's creditworthiness. Other methods, sometimes more revealing because they address specific aspects of financial management more fully controllable by local officials, are available as well.

Bond Ratings

Capital improvements typically are financed by the issuance of bonds—general obligation (GO) bonds, if the funds are used for traditional, non-revenue-generating functions such as street construction and park development and the city is willing to place its full faith and credit (i.e., taxing authority) behind the issuance to ensure repayment of investors; revenue bonds, when revenue from public projects such as airports or water treatment facilities can be used to pay off their own debt; or industrial development bonds. Prior to the issuance of such bonds, municipal credit rating agencies—the three largest being Fitch Investor Service, Moody's Investor Service, and Standard & Poor's—perform a credit risk evaluation of the municipality to advise potential investors of the city's creditworthiness. That rating is a shorthand declaration of the city's financial health.

In rating municipalities for the issuance of general obligation debt, Standard & Poor's (1995) reviews four factors that are considered especially relevant to a government's "capacity and willingness" to repay its debt:

1. The local economic base, including local employment, taxes, and demographics (e.g., age, education, income level, and skills of the local population)
2. Financial performance and flexibility, including accounting and reporting methods, revenue and expenditure structure and patterns, annual operating and budgetary performance, financial leverage and equity position, budget and financial planning, and contingency financial obligations, such as pension liability funding
3. Debt burden
4. Administration, including local autonomy and discretion regarding financial affairs, background and experience of key administrative officials, and frequency of elections (pp. 20–23; used with permission)

Although rating labels differ, general comparability exists across the rating services (Table 10.1). Higher ratings (for example, BBB and above for Standard & Poor's) are considered "investment grade"; lower ratings (BB and below) are considered "speculative grade." Lower ratings for Standard & Poor's include B, CCC, CC, C, and D.

A checklist of "early warning guidelines" developed by Standard & Poor's is provided in Table 10.2. In addition to basic financial criteria, such as those included in the checklist, socioeconomic factors have been increasingly cited for their importance in rating a local government's creditworthiness (Crowell & Sokol, 1993). Median household income and housing values, for example, are among the factors Standard & Poor's analysts consider (Table 10.3).

ICMA's Financial Trend Monitoring System

In 1980, the ICMA published a set of handbooks that described a Financial Trend Monitoring System (FTMS) for tracking local financial conditions (Groves & Godsey, 1980). The FTMS relies on 36 indicators, shown in Table 10.4.

Minimum standards are not declared for most indicators. Instead, potential "warning trends" are identified and suggestions for analysis are offered. In a few cases, however, relevant "credit industry benchmarks" are noted by the FTMS authors:

Table 10.1   Credit Ratings for Municipal Bonds

Rating			
Fitch	Moody's	Standard & Poor's	Description
AAA	Aaa	AAA	Best quality; highest grade; extremely strong capacity to pay principal and interest; payment is secured by a stable revenue source
AA	Aa	AA	High quality; very strong capacity to pay principal and interest; revenue sources are only slightly less secure than for highest grade bonds
A	A	A	Upper-medium quality; strong capacity to pay principal and interest, but revenue sources are considered to be susceptible to fluctuations in relevant economic conditions
BBB	Baa	BBB	Medium-grade quality; adequate capacity to pay principal and interest, but may become unreliable if adverse economic conditions prevail
BB and lower	Ba and lower	BB and lower	Speculative quality; low capacity to pay principal and interest; represent long-term risk whether relevant economic conditions are favorable or not

SOURCES: Moody's Investor Service, Public Finance Department, *An Issuer's Guide to the Rating Process* (New York: Moody's Investor Service, 1993); Standard & Poor's, *Municipal Finance Criteria* (New York: Author, 1994); and Fitch Investor Service, *Fitch Ratings* (New York: Moody's Public Finance Department, 1994).

NOTE: Within groups, Moody's designates those bonds with the strongest attributes with a "1." For instance, A1 or Aa1 would be of slightly higher quality than A or Aa. Both Fitch and Standard & Poor's attach a "+" or "−" to show slight variation within the rating groups. Examples would be AA− or A+ to indicate a credit better than an A but less than an AA.

- Credit rating firms assume that a local government normally will be unable to collect from 2% to 3% of its property taxes within the year that the taxes are due. If uncollected property taxes rise to more than from 5% to 8%, rating firms consider this a negative factor because it signals potential problems in the stability of the property tax base. An increase in the rate of delinquency for two consecutive years is also considered a negative factor.

- A credit rating firm would regard a current-year operating deficit as a minor warning signal; funding practices and the reasons for the deficit would be carefully assessed before it would be considered a negative factor. The following situations, however, would be given considerably more attention and would probably be considered negative factors: (1) two consecutive years of operating fund deficits, and (2) a current operating fund deficit greater than that of the previous year.

Table 10.2   Early Warning Guidelines: Bond Rating Checklist

Warning signs
- Current-year operating deficit
- Two consecutive years of operating fund deficit
- Current general fund deficit (two or more years in last five)
- Short-term debt (other than bond anticipation notes [BAN]) at end of fiscal year greater than 5% of main operating fund revenues
- Short-term interest and current-year debt service greater than 20% of total revenues
- Total property tax collections less than 92% of total levy
- Declining market valuations—two consecutive years on a three-year trend
- Overall net debt ratio 50% higher than four years ago
- Current-year operating deficit larger than previous year's deficit
- General fund deficit in the current year—balance sheet—current position
- Two-year trend of increasing short-term debt outstanding at fiscal year end
- Property taxes greater than 90% of tax limit
- Net debt outstanding greater than 90% of debt limit
- A trend of decreasing tax collections—two consecutive years on a three-year trend
- Overall net debt ratio 20% higher than previous year

SOURCE: Standard & Poor's, *Credit Overview: Municipal Ratings* (New York: Author, 1983), p. 119. Reprinted by permission.

- The credit industry considers the following situations negative factors: (1) short-term debt outstanding at the end of the year exceeding 5% of operating revenues (including tax anticipation notes but excluding bond anticipation notes), and (2) a two-year trend of increasing short-term debt outstanding at the end of the fiscal year, including tax anticipation notes (Groves & Valente, 1986, pp. 38, 63, 77; reprinted by permission).

Financial Planning Standards in Overland Park

Some municipalities establish their own standards for financial planning and management. Overland Park, Kansas, is one such city (Table 10.5). Although some of the standards adopted by that community for capital planning purposes may not be deemed immediately achievable elsewhere, the outstanding bond ratings awarded to Overland Park by rating agencies suggest that its standards could serve as useful benchmarks.

Selected Indicators of Financial Health

Two facets of financial health that deserve special attention, even in a concise overview of financial management benchmarks, are fund balance and municipal debt.

Table 10.3 Typical Ranges for Tax-Backed GO Ratings

Economic

Income levels as a percentage of the national average. These include both per capita and median household figures. Analysts may also compare income levels against local cost of living indexes.

Very low	≤ 15%
Low	85%
Average	100%
High	120%
Very high	≥ 140%

Market value per capita. These may vary greatly by state depending on assessment practices, homeowners' exemptions, cost of living, etc.

Low	$20,000
Moderate	$40,000
High	$60,000

Taxpayer concentration. Percentage of assessed value in the top 10 taxpayers.

Diverse	≤ 15%
Moderately concentrated	25%
Concentrated	≥ 40%

Financial

Ending general fund balances as a percentage of operating revenues. These are only guidelines. What is considered high and low depends on peak cash-flow needs during the year, as well as whether the fiscal year ends in a historically cash poor or cash rich month.

Total general fund balances.

Strong	≥ 15%, plus no cash flow borrowing over the fiscal year
Adequate	5% to 15%
Low	0% to 5%

Unreserved general fund balances.

Strong	≥ 8%
Adequate	2% to 8%
Low	≤ 2%

Property tax burdens. Overlapping tax as a percentage of market value.

Low	1.0% of market value
Moderate	1.5% to 2.0%
Moderately high	2.0% to 2.5%
Very high	≥ 2.5% of market value

Debt

Debt to market value. Not including pension funding debt.

Low debt burden	≤ 3%
Moderate debt burden	3% to 6%
High debt burden	≥ 6%

(continued)

Table 10.3   Continued

Combined general fund/debt service fund debt service to operating expenditures "Carrying Charge." [a]
Not including pension funding debt.

Low	$\leq$ 5%
Moderate carrying charge	10%
High carrying charge	$\geq$ 15%

Overall debt per capita.

Low	$1,000
Moderate	$1,000 to $2,500
High	$\geq$ $2,500

Debt to income. Standard & Poor's index.

Low	0% to 3%
Moderate	3% to 6%
High	$\geq$ 6%

Appropriate debt amortization over 10 years.

25% over 5 years
50% over 10 years

SOURCE: David G. Hitchcock and Hyman C. Grossman, "Benchmark General Obligation Ratios," Standard & Poor's *CreditWeek Municipal,* 9 (February 8, 1999), p. 10. Reprinted by permsision of Standard & Poor's, A Division of the McGraw-Hill Companies.
NOTE: The ratios in this table represent benchmarks that Standard & Poor's analysts usually consider high, low, or moderate, regardless of rating category or point in the national economic cycle.
a. Carrying charges for special service districts may not be a relevant statistic; collecting a debt service levy may be their only operation.

Fund Balance. Fund balance has been defined in general terms as "excess, surplus or unbudgeted money" (Carter & Vogt, 1989, p. 33). More precisely, it is "the cumulative difference of all revenues and expenditures from the government's creation. It can also be considered to be the difference between fund assets and fund liabilities, and can be known as fund equity" (Allan, 1990, p. 1).

A suitable fund balance provides a source of working capital to meet cash flow needs, offers a cushion against revenue shortfalls and unanticipated expenditures, provides an accumulation of funds that some cities use for capital projects to avoid bond sales, helps others meet bond indenture requirements, produces investment income, and contributes to a favorable bond rating (Hembree, Shelton, & Tyer, 1999; Olson, Shuck, Vinyon, & Lilja, 1980). Appropriations from the fund balance of the general fund often are used to smooth peaks and valleys in property and sales tax revenues—with greater appropriations from the fund balance when current revenues are down and smaller appropriations when they are up—to reduce the need for frequent or erratic rate adjustment.

Table 10.4   Financial Trend Monitoring Indicators

Number & Indicator	Formula	Warning Trend
1. Revenues per Capita	$\dfrac{\text{Net Operating Revenues in Constant Dollars}}{\text{Population}}$	Decreasing net operating revenues per capita (constant dollars)
2. Restricted Revenues	$\dfrac{\text{Restricted Operating Revenues}}{\text{Net Operating Revenues}}$	Increasing amount of restricted operating revenues as a percentage of net operating revenues
3. Intergovern- mental Revenues	$\dfrac{\text{Intergovernmental Operating Revenues}}{\text{Gross Operating Revenues}}$	Increasing amount of intergovern- mental revenues as a percentage of gross operating revenues
4. Elastic Tax Revenues	$\dfrac{\text{Elastic Operating Revenues}}{\text{Net Operating Revenues}}$	Decreasing amount of elastic operating revenues as a percentage of net operating revenues
5. One-Time Revenues	$\dfrac{\text{One-Time Operating Revenues}}{\text{Net Operating Revenues}}$	Increasing use of one-time operating revenues as a percentage of net operating revenues
6. Property Tax Revenues	Property Tax Revenues in Constant Dollars	Decline in property tax revenues (constant dollars)
7. Uncollected Property Taxes	$\dfrac{\text{Uncollected Property Taxes}}{\text{Net Property Tax Levy}}$	Increasing amount of uncollected property taxes as a percentage of net property tax levy
8. User Charge Coverage	$\dfrac{\text{Revenues From User Charges}}{\text{Expenditures for Related Services}}$	Decreasing revenues from user charges as a percentage of total expenditures for related services
9. Revenue Shortfalls	$\dfrac{\text{Revenue Shortfalls}^{\text{a}}}{\text{Net Operating Revenues}}$	Increase in revenue shortfalls as a percentage of actual net operating revenues
10. Expenditures per Capita	$\dfrac{\text{Net Operating Expenditures in Constant Dollars}}{\text{Population}}$	Increasing net operating expenditures per capita (constant dollars)
11. Employees per Capita	$\dfrac{\text{Number of Municipal Employees}}{\text{Population}}$	Increasing number of municipal employees per capita
12. Fixed Costs	$\dfrac{\text{Fixed Costs}}{\text{Net Operating Expenditures}}$	Increasing fixed costs as a percentage of net operating expenditures
13. Fringe Benefits	$\dfrac{\text{Fringe Benefit Expenditures}}{\text{Salaries and Wages}}$	Increasing fringe benefit expendi- tures as a percentage of salaries and wages
14. Operating Deficits	$\dfrac{\text{General Fund Operating Deficit}}{\text{Net Operating Revenues}}$	Increasing general fund operating deficits as a percentage of net operating revenues

(continued)

Table 10.4   Financial Trend Monitoring Indicators (continued)

15. Enterprise Losses	Enterprise Profits or Losses in Constant Dollars	Recurring enterprise losses (deficits) (constant dollars)
16. Fund Balances	$$\dfrac{\text{Unreserved Fund Balances}}{\text{Net Operating Revenues}}$$	Declining unreserved fund balances as a percentage of net operating revenues
17. Liquidity	$$\dfrac{\text{Cash and Short-Term Investments}}{\text{Current Liabilities}}$$	Decreasing amount of cash and short-term investments as a percentage of current liabilities
18. Current Liabilities	$$\dfrac{\text{Current Liabilities}}{\text{Net Operating Revenues}}$$	Increasing current liabilities at the end of the year as a percentage of net operating revenues
19. Long-Term Debt	$$\dfrac{\text{Net Direct Bonded Long-Term Debt}}{\text{Assessed Valuation}}$$	Increasing net direct bonded long-term debt as a percentage of assessed valuation
20. Debt Service	$$\dfrac{\text{Net Direct Debt Service}}{\text{Net Operating Revenues}}$$	Increasing net direct debt service as a percentage of net operating revenues
21. Overlapping Debt	$$\dfrac{\text{Long-Term Overlapping Bonded Debt}}{\text{Assessed Valuation}}$$	Increasing long-term overlapping bonded debt as a percentage of assessed valuation
22. Unfunded Pension Liability	$$\dfrac{\text{Unfunded Pension Liability}}{\text{Assessed Valuation}}$$	Increasing unfunded pension liability as a percentage of assessed valuation
23. Pension Assets	$$\dfrac{\text{Pension Plan Assets}}{\text{Annual Pension Benefits Paid}}$$	Decreasing value of pension plan assets as a percentage of benefits paid
24. Accumulated Employee Leave	$$\dfrac{\text{Total Days of Unused Vacation and Sick Leave}}{\text{Number of Municipal Employees}}$$	Increasing number of unused vacation and sick leave days per municipal employee
25. Maintenance Effort	$$\dfrac{\text{Expenditures for Repair and Maintenance of General Fixed Assets (Constant Dollars)}}{\text{Quantity of Assets}}$$	Declining expenditures for maintenance of general fixed assets per unit of asset (constant dollars)
26. Capital Outlay	$$\dfrac{\text{Capital Outlay from Operating Funds}}{\text{Net Operating Expenditures}}$$	A three-year or longer decline in capital outlay from operating funds as a percentage of net operating expenditures
27. Depreciation Expense	$$\dfrac{\text{Depreciation Expense}}{\text{Cost of Depreciable Fixed Assets}}$$	Decreasing depreciation expense as a percentage of total depreciable fixed assets (at cost) for enterprise funds and internal service funds
28. Population	Population	Rapid change in population size

29. Median Age	Median Age of Population	Increasing median age of population
30. Personal Income per Capita	$$\frac{\text{Personal Income in Constant Dollars}}{\text{Population}}$$	Decline in the level, or growth rate, of personal income per capita (constant dollars)
31. Poverty House-holds or Public Assistance Recipients	$$\frac{\text{Poverty Households or Public Assistance Recipients}}{\text{Households in Thousands}}$$	Increasing proportion of poverty households or public assistance recipients
32. Property Value	$$\frac{\text{Constant Dollar Change in Property Value}}{\substack{\text{Constant Dollar Property Value} \\ \text{Prior Year}}}$$	Declining growth or drop in the market value of residential, commercial, or industrial property (constant dollars)
33. Residential Development	$$\frac{\text{Market Value of New Residential Development}}{\substack{\text{Market Value of Total New} \\ \text{Development}}}$$	Increasing market value of residential development as a percentage of market value of total development
34. Vacancy Rates	Vacancy Rates	Increasing vacancy rates in residential, commercial, or industrial buildings
35. Employment Base	Local Unemployment Rate; Number of Jobs Within the Community	Increasing rate of local unemploy-ment or a decline in the number of jobs within the community
36. Business Activity	Retail Sales; Number of Community Businesses; Gross Business Receipts; Number of Acres Devoted to Business; Market or Assessed Value of Business Property	Decline in business activity as measured by retail sales, number of business units, gross business receipts, number of acres devoted to business, and market or assessed value of business property (constant dollars where appropriate)

SOURCE: Adapted from Sanford M. Groves and Maureen Godsey Valente, *Evaluating Financial Condition: A Handbook for Local Government* (3rd ed.) (Washington, DC: International City/County Management Association, 1994).
a. Net operating revenues budgeted less net operating revenues (actual). Reprinted with permission of the International City/County Management Association, 777 North Capital Street NE, Suite 500, Washington, DC 20002. All rights reserved.

A certain mystical quality surrounds the consideration of what constitutes an appropriate amount of fund balance. Some authors have suggested that the credit industry considers a general fund balance to be at a "healthy level" if it equals 10% of annual revenues or expenditures (Olson et al., 1980, p. 11). Others suggest that credit analysts view a fund balance of 5% as the minimum "prudent level," although lower or higher levels may be deemed appropriate in specific instances depending on long-term financial trends and other local factors (Allan, 1990, p. 6). Some local government managers argue for fund balances equal to 1 to 3 months' expenditures—or 8.3% to 25% of the annual budget (Hembree et al., 1999, p. 17). The Local Government Commission of

Table 10.5   Financial Planning Standards Established by the City of Overland Park

Several financial standards have been developed as a basis for evaluating the financial soundness of the 5-year capital plan. These standards serve as guidelines in scheduling and balancing the composite of financing methods for projects included in the plan. The standards are summarized below.

Percentage of Ending Cash to Operating Expenditures   13% to 15%
The General Fund's unreserved undesignated fund balance divided by the General
Fund's operating expenditures (excluding transfers). (Source of standard: staff.)

Percentage of Pay-As-You-Go Funding to Total Program   20% to 30%
The amount of pay-as-you-go funding used to finance capital improvement
projects divided by the total amount of the Capital Improvement Program.
(Source of standard: staff.)

Percentage of Pay-As-You-Go Funding to City Funds   45% to 55%
The amount of pay-as-you-go funding used to finance capital improvement
projects divided by the total amount of city funds financing the Capital
Improvement Program. (Source of standard: staff.)

Percentage of Debt to Total Program   35% to 50%
The amount of city's general obligation debt incurred to fund capital improvement
projects divided by the total amount of the Capital Improvement Program.

Percentage of Debt to City Funds   45% to 55%
The amount of the city's general obligation debt incurred to fund capital
improvement projects divided by the total amount of city funds used to
finance capital improvement projects. (Source of standard: staff.)

Total Direct Debt per Capita $675
The amount of per capita direct bonded debt (debt for which the city has
pledged its full faith and credit) issued by the city. (Source of standard:
Moody's Investor Service.)

**Percentage of Direct and Overlapping Debt to Market Value of
Tangible Property**   4% to 5%
The city's direct bonded debt and overlapping debt as a percentage of
estimated market valuation of property within the city. (Source of standard:
Moody's Investor Service.)

**Percentage of Debt Service Cost to General Fund
Operating Expenditures**   Less than 25%
The city's annual payments to the Bond & Interest Fund for debt service cost
divided by total operating expenditures (excluding transfers) of the General Fund.
(Source of standard: ICMA.)

Mill Levy Equivalent of Bond & Interest Transfer   6 mills
The mill levy required to replace the General Fund's annual transfer to the Bond &
Interest Fund. These monies are used to pay annual principal and interest payments
on the city's maturing general obligation debt. (Source of standard: staff.)

**Percentage of City Funds Financing the Capital Improvement
Program (CIP) to the Operating Budget**   45% to 55%
The amount of city funds used to finance the 5-year CIP divided by the current
General Fund Operating Budget and the total city funds financing the CIP.
(Source of standard: staff.)

SOURCE: City of Overland Park (KS), *1999 Annual Budget*, pp. 7.8–7.9.

the North Carolina Department of State Treasurer recommends a general fund balance of at least 8% of expenditures (Carter & Vogt, 1989).

How much is enough? It is probably impossible to say with certainty for any given community, but a look at the amount of cash reserves held by other cities and a review of the fund balance policies actually in place among respected communities may be instructive to municipalities that are attempting to map their own fund balance and reserve course. Local policies prescribing fund balance accumulations in 56 of the cities examined for this volume are reported in Table 10.6. These cities may be divided fairly evenly between those prescribing fund balances greater than 10% and those prescribing 10% or lesser amounts as adequate minimums. Underrepresented on this table are smaller cities, which often strive to maintain larger fund balances than their more populous counterparts. Because officials in smaller communities may feel especially vulnerable to the financial impact of a major emergency, many of these towns and cities have been found in recent studies to maintain fund balances equal to 50% of general fund expenditures or more (Carter & Vogt, 1989; Massey & Tyer, 1990).

Some cities take special steps to protect the fund balance from raids for purposes other than true emergencies. Consider, for example, Greensboro, North Carolina, which maintains an undesignated fund balance equal to 9% of its General Fund. Appropriations from the undesignated fund balance require seven votes in the affirmative from the nine-member city council (City of Greensboro [NC], p. 11).

Debt. Debt is also a useful indicator of a community's financial health. Although many aggressive and vibrant communities incur debt to secure capital for new infrastructure or public amenities, too much debt may signal impending problems. Among the long-accepted general warning signals for potentially burdensome debt, as cited by different cities, are the following:

- Overall net debt exceeding 10% of assessed valuation
- An increase of 20% over the previous year in overall net debt as a percentage of market valuation
- Overall net debt as a percentage of market valuation increasing 50% over the figure for 4 years earlier
- Overall net debt per capita exceeding 15% of per capita personal income
- Net direct debt exceeding 90% of the amount authorized by state law

Debt Limits. Among the cities examined for this volume, most report state-imposed restrictions limiting the city's general obligation debt to a ceiling expressed as a percentage of the total assessed value of property within the city. Eugene, Oregon, for example, reported one of the tightest restrictions—a ceiling imposed by the State of Oregon at 3% of assessed valuation—but noted that the limitation did not apply to assessment, sewer, and off-street parking debt. At the other end of the spectrum, Fort Smith, Arkansas,

(text continues on page 122)

Table 10.6   Policies of Selected Cities Regarding Fund Balances and Undesignated
                   Reserves

City	Policy
Cary, NC	Maintain a fund balance equal to 30% of operating expenditures
College Park, MD	Maintain a fund balance equal to 25% of General Fund expenditures
Smyrna, GA	Maintain an unreserved, undesignated fund balance equal to at least 25% of the operating budget (3 months' operation)
Tempe, AZ	Maintain a fund balance equal to 25% of General Fund revenues
Fort Smith, AR	Strive to achieve fund balances/working capital at 25% of total expenditures in addition to maintaining a contingency reserve of at least 10% of general operating revenues
Fairfield, CA	Maintain ending cash and current receivable balances (including carryover and vehicle/leave reserves) of at least 20% of the approved budget
San Luis Obispo, CA	Maintain a fund balance of at least 20% of operating expenditures in the General Fund and Enterprise Funds
Southfield, MI	The "desirable" level of unreserved and undesignated fund balance to be maintained in each operating fund is 20% of net operating revenues for the year
Burbank, CA	Maintain a 15% unappropriated working reserve and a 5% unappropriated emergency reserve in the General Fund
Hanover Park, IL	Contingency reserves for miscellaneous purposes and emergencies shall not exceed 20% of the General Fund
College Station, TX	Maintain a fund balance of at least 15% of General Fund expenditures (55 days' operation). An additional contingency of 2.5% should also be maintained
Duncanville, TX	Maintain a fund balance equal to at least 16.67% of budgeted expenditures (equivalent to 60 days' expenditures)
Eugene, OR	Maintain an unappropriated ending fund balance equal to 2 months of General Fund operating expenditures
Hurst, TX	Maintain a fund balance sufficient to cover at least 60 days' expenditures
Lubbock, TX	Maintain an unreserved, undesignated fund balance equal to at least 2 months' expenditures
Denton, TX	In addition to a budgeted reserve of 8% to 10% of the general operating fund available for unanticipated expenditures, maintain a fund balance sufficient to provide 30 to 45 days' working capital
Ames, IA	Maintain a fund balance equal to 15% of General Fund expenditures
Chandler, AZ	Maintain an unrestricted contingency reserve of 15% of General Fund revenues

Danville, KY	Maintain a reserve for economic uncertainties and emergencies equal to 15% of General Fund revenues
Grand Junction, CO	Long-range plan: Maintain a fund balance in the General Fund equal to 15% of expenditures and fund balances in enterprise funds equal to 5% of operating expenditures
Greenville, SC	Maintain a fund balance equal to 15% of the general operating budget
Kalamazoo, MI	Maintain an unreserved fund balance equal to at least 15% of General Fund revenues and transfers from other funds
Grand Prairie, TX	Maintain a 12.5% operating reserve in the General Fund
Durham, NC	Maintain an undesignated fund balance in the General Fund equal to at least 12% of adjusted appropriations (budgeted appropriations minus debt service and transfers to other funds)
Dayton, OH	Maintain sufficient year-end reserves to cover 6 to 10 weeks of expenditures
Longview, TX	Maintain a cash balance of 10% for service delivery funds and a 15% balance in funds with major facilities and equipment
Alexandria, VA	Maintain an unreserved General Fund balance equal to at least 10% of General Fund revenue
Blacksburg, VA	General Fund balance should be targeted at no less than 10% of operating expenditures exclusive of capital improvement and debt service expenditures
Fayetteville, AR	Maintain an undesignated fund balance of at least 10% of current year operating expenditures for the General Fund and Street Fund
Flagstaff, AZ	Maintain a fund balance of at least 10% of General Fund expenditures
Fort Worth, TX	Maintain the General Fund undesignated fund balance at 10% of current year budget expenditures
Glendale, AZ	Maintain a fund balance equal to 10% of annual General and Street Fund revenue
Prairie Village, KS	Maintain a fund balance equal to at least 10% of General Fund operating expenditures
Sterling Heights, MI	Maintain an undesignated, unreserved General Fund balance equal to 10% of the general operating budget
Winston-Salem, NC	Maintain an unrestricted fund balance equal to at least 10% of budgeted expenditures
Gainesville, GA	In addition to a contingency reserve equal to 2% of the General Fund budget, maintain a fund balance equal to 1 month's operating expenditures
Greensboro, NC	Maintain an undesignated fund balance equal to 9% of General Fund expenditures, and accumulate a capital reserve account of at least $10 million. Maintain a fund balance of at least 8% in all other operating funds
Oak Park, MI	Maintain an undesignated fund balance at not less than 10% of revenues and not less than 8.33% (1 month) of expenditures

(continued)

Table 10.6   Policies of Selected Cities Regarding Fund Balances and Undesignated
Reserves (continued)

Plano, TX	Maintain a fund balance equal to at least 30 days of operating expenses
Norman, OK	Maintain in reserve a portion of fund balance equal to 8% of operating expenditures to protect against unexpected revenue loss or expenditure requirement
Redondo Beach, CA	Maintain General Fund "reserves" for future contingencies equal to at least 7.5% of the General Fund budget
Oklahoma City, OK	Maintain an unbudgeted General Fund reserve of 6% to 10% of annual operating expenditures (in addition to a budget contingency of 2% or greater)
Auburn, AL	Achieve an ending fund balance equal to 6% of budgeted General Fund expenditures
Albuquerque, NM	Maintain a fund balance equal to 5% of recurring revenue
Ann Arbor, MI	Maintain a stable undesignated General Fund balance within a range of 5% to 15% of operating expenditures
Avon, CT	Maintain a fund balance at 5% to 10% of budgeted expenditures
Boca Raton, FL	Maintain a fund balance equal to at least 5% of General Fund expenditures
Cincinnati, OH	Maintain a working capital reserve of 5% to 8% of general operating revenues
Irving, TX	Maintain in addition to General Fund reserves an unreserved cash balance in the General Fund at not less than 5% of expenditures
Little Rock, AR	Maintain an emergency or working capital reserve at approximately 5% of current operating budget appropriation
Oakland, CA	Maintain the General Fund reserve balance at a level equal to 5% of General Fund appropriations
San Antonio, TX	Achieve a reserve equal to 5% of the General Fund operating budget
Savannah, GA	Maintain a fund balance between 5% and 10% of General Fund expenditures
Shreveport, LA	Maintain an operating reserve of $5 million or 5% of revenue, whichever is greater
St. Petersburg, FL	Maintain a cash-working fund balance of 5% of current year's General Fund operating appropriations
White Plains, NY	The "carryover fund balance" should be at least 4% of the general fund operating budget, and shall be separate from the reserve for financing

NOTE: The fund balance policies reported here were compiled from documents published during the 1990s by the respective municipalities and depict the policies in effect for these cities at some time during that decade.

reported a general obligation debt limit of 25% of total assessed property value. Most reported debt ceilings at 5% to 15% of assessed value.

Table 10.7   Debt Ratio Policies of the City of Alexandria

	Target	Limit
Debt as a percentage of fair market real property value	1.1%	not greater than 1.6%[a]
Debt per capita as a percentage of per capita income	2.25%	not greater than 3.25%
Debt service as a percentage of general government expenditures	8.0%	not greater than 10%

SOURCE: City of Alexandria (VA), *Fiscal Year 2000 Approved Operating Budget,* pp. 9.50–9.53.
a. The legal limitation, established by the Virginia Constitution, states, "No city or town shall issue any bonds or other interest-bearing obligations which, including existing indebtedness, shall at any time exceed ten percent of the assessed valuation of real estate in the city or town subject to taxation, as shown by the last preceding assessment for taxes." The Alexandria policy is more restrictive than the legal limitation.

Debt limits take a different form in some cities. Avon, Connecticut, for instance, reports that general statutes in Connecticut limit municipal indebtedness to no greater than seven times annual receipts from taxation.

Some cities impose debt restrictions on themselves, beyond those prescribed by their state. Ames, Iowa, for example, reports that the state limits general obligation debt to 5% of assessed valuation and that the Ames city council imposes a further limit "which reserves 25% of that capacity" (City of Ames [IA], p. 431). Locally imposed restrictions in St. Petersburg, Florida, limit general obligation debt to 6% of property valuation—just one half of the amount allowed by state law (City of St. Petersburg [FL], *Approved Program Budget: Fiscal Year 1997,* p. 19). Alexandria, Virginia, is limited by the state constitution to a 10% ceiling, but has both a local policy that restricts debt to 1.6% of total property value and a local target of 1.1% or less (Table 10.7).

Not all local governments have their allowable debt restricted by the state. In sharp contrast to most of its counterparts across the nation, Decatur, Illinois, makes this simple declaration: "The City has no debt restriction or debt limits as it is an Illinois home-rule unit" (City of Decatur [IL], *Annual Performance Budget: 1996–1997,* p. 17).

Warning signals triggered when general obligation debt exceeds a specified percentage of the assessed value of all taxable property within the city are commonly prescribed, although the recommended limits of prudence differ from one authority to another. Another useful reference is the full market value of taxable property, which may differ from assessed value depending on assessment ratio. Many cities report that their direct net debt, excluding short-term operating debt and bonds fully supported by enterprise revenues, constitutes 1% to 3% of the full market value of all taxable property within the community.

Debt Service. High levels of debt impose debt service burdens on municipal budgets, forcing undesirably high percentages of local resources to be devoted to the payment of principal and interest on the debt. Rules of thumb accepted by many cities over the years, and often attributed to the credit rating industry as well, hold that net direct debt service approaching 10% of operating revenues or expenditures is acceptable, but debt service exceeding 20% could be a potential problem (see, for example, City of Portland [OR], 1992, p. 34).

Although cities such as Sioux Falls, South Dakota; Portland, Oregon; and Alexandria, Virginia report debt service ratios below even 3%, other municipalities contend with higher amounts (Table 10.8). Many are reasonably content if no more than 15 cents, 20 cents, or a quarter out of every dollar goes toward debt service. Blacksburg, Virginia, and Winston-Salem, North Carolina, for example, try to keep debt service at 15% of operating expenditures or less. Fort Worth, Texas, attempts to keep it below 20% of operating revenues (Figure 10.1). Iowa City, Iowa, tries to limit debt service to 25% of the general tax levy. Overland Park, Kansas, and Fayetteville, Arkansas, try to hold debt service to within 25% of operating expenditures and operating revenues, respectively.

Accuracy in Revenue Forecasting

The accurate projection of revenues is extremely important to the smooth operation of municipal government. The task of "crystal balling" future revenues often falls to the finance department.

Most cities take a conservative approach to the forecasting of revenues, hoping that any error is on the side of *underestimating* actual revenues. Ideally, the percentage variance between projected and actual revenues will be quite small. A review of performance documents suggests that many revenue forecasters do quite well. Revenue projections within 3% of accuracy appear fairly common (Table 10.9).

Finance departments perform an array of duties related to financial accountability and stewardship. They must account for revenues and expenditures, promptly reconciling accounts and keeping local officials and operating departments informed about financial status on a timely basis; they must collect taxes; they must handle revenues efficiently, pay invoices on a timely basis, and manage cash and investments to the advantage of the municipality. But what constitutes good performance of these duties?

The GFOA has developed a set of criteria for judging the adequacy of financial reporting in local government. Cities and counties whose comprehensive annual financial reports meet these criteria qualify for "certificates of conformance in financial reporting." Few standards have been published for other finance department duties, but individual cities have established their own performance targets.

Table 10.8   Debt Service Ratios: Selected Cities

DEBT SERVICE AS A PERCENTAGE OF OPERATING EXPENDITURES

Sioux Falls, SD
Actual: 0.57% to 1.99% (1982–1990), 0.35% (1991)

Alexandria, VA
2.49% (1998)

Norman, OK
4.2% (1997)

Boca Raton, FL
5.69% (1997)

Boston, MA
Target: 7% or less
Actual: 5.8% (1996)

Sterling Heights, MI
Actual: 6.26% (1995)

Charlottesville, VA
6.64% (1998)

Fort Collins, CO
Actual: 7.6% (1996)

White Plains, NY
Actual: 8.1% (1995)

Winston-Salem, NC
Policy: Annual debt service payments shall not exceed 15% of total general government
  expenditures
Actual: 10% (1998)

Raleigh, NC
Actual: 11.4% (1999)

Blacksburg, VA
Target: Not to exceed 15% of operating expenditures exclusive of capital improvement and
  debt service (1999)
Actual: 11.5% (1997)

Rockville, MD
12.7% (1997)

Overland Park, KS
Target: Less than 25%
Actual: 22.7% (1995)

Duncanville, TX
Actual: 24.4% (1996), 22.6% (1998)

Lubbock, TX
Actual: 28.3% (1997)

(continued)

Table 10.8   Debt Service Ratios: Selected Cities (continued)

DEBT SERVICE AS A PERCENTAGE OF OPERATING REVENUES

Portland, OR
Actual: Less than 2% (1991)

Greeley, CO
7.11% (1997)

Savannah, GA
8.91% (1997)

Jacksonville, FL
Target: Not to exceed 20%
Actual: 12.37% (1998)

Tempe, AZ
Target: 16% or less
Actual: 14.87% (1999)

Iowa City, IA
Target: Limit annual debt service to 25% of general tax levy

Fayetteville, AR
Target: Limit debt service to 25% of total operating revenues or less

Duncanville, TX
Target: Not to exceed 33⅓% of revenue from property tax and sales tax relief

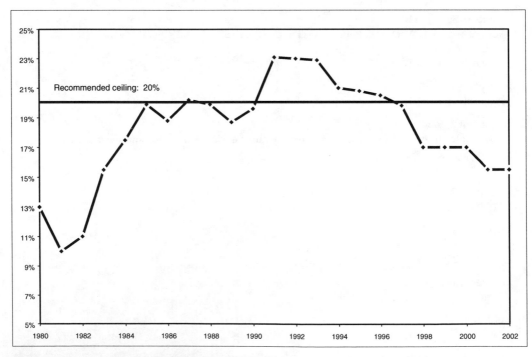

Figure 10.1. Debt Service as a Percentage of Net Operating Revenues in Fort Worth
SOURCE: City of Fort Worth (TX), *Annual Budget and Program Objectives: 1998–99*, p. D.32.

Table 10.9   Accuracy of Revenue Forecasts: Performance Targets and Actual
Experience of Selected Cities

City	Target	Actual Experience
Durham, NC	Within 1%	Actual revenue as percentage of projected: 103.25% (1998)
Plano, TX		Actual revenue as percentage of projected revenue: 100% (1997)
Oklahoma City, OK		Actual variance of 1.2% (1999)
Winston-Salem, NC		Actual revenue as percentage of projected: 101.6% (1998)
Milwaukee, WI	Within 2%	Actual variance of 0.8% (1997)
St. Petersburg, FL	Within 2%	Actual revenue as percentage of projected: 103.0% (1995)
San Clemente, CA	Within 2%	Actual revenue as percentage of projected: 106% (1998)
Duncanville, TX		Actual revenue as percentage of projected: 102.2% (1997)
Oakland, CA	Within 2.5%	Actual variance of 2.3% (1996)
Boston, MA		Actual revenue as percentage of projected: 102.7% (1996)
San Jose, CA	Within 3%	Actual variance of 1% (1995)
Portland, OR	Within 3%	Actual: Variance of 2.3% (1997), 3.0% (1998)
Savannah, GA		Actual General Fund revenue as percentage of projected: 103% (1990), 98% (1991)
Macon, GA		Actual variance of 3% (1995)
Santa Ana, CA	Within 3%	Actual variance of 3.69% (1995)
Rockville, MD		Actual variance of 3.7% (1997)
Nashville–Davidson County, TN	Within 5%	
Overland Park, KS		Sales tax, 2.6% variance; all other revenues, 14.0% variance (1997)

Many cities target the release of monthly financial reports by the middle of
the next month, some even earlier (Table 10.10). Among the cities examined
for this volume, most of those reporting reconciliation statistics attempt to
reconcile bank accounts within 30 days of month's end (Table 10.11). Several
municipalities target the completion of their annual audits and release of their
comprehensive annual financial reports (CAFR) by the fourth or fifth month
of the following fiscal year (Table 10.12). Palo Alto, California, tries to issue
an unaudited version of its annual financial report within 75 days of the end of
the fiscal year.

Table 10.10   Prompt Issuance of Monthly Financial Reports: Selected Cities

City	Performance Target	Actual Experience
College Park, MD	By the 7th workday of the following month	83% (1999)
Savannah, GA	Within 7 days of month's end	7.0 days (1991), 7.6 days (1997)
Flagstaff, AZ	Within 8 workdays of the end of the month at least 95% of the time	92% (1997)
Boston, MA	Within 8 days	84% within 8 days (1996)
Norman, OK	Within 10 workdays of the end of the month at least 95% of the time	8 workdays (1997)
Greenville, SC	By the 10th workday of the following month	
Bellevue, WA	Within 10 workdays of the end of the month	50% (1998)
Denton, TX	Month-end closing within 5 workdays; monthly operations report within 7 workdays of month's end	95% (1997)
Boca Raton, FL	Month-end closing within 10 workdays; monthly report within 2 workdays of closing	
Hurst, TX	Produce expenditure report within 2 weeks of month's close at least 90% of the time	83% within 2 weeks (1997)
Palo Alto, CA	Produce and distribute at least 85% of monthly departmental reports by the 15th of the following month	100% by the 15th (1997)
Oakland, CA	Cash management reports issued by the 15th of the following month	100% by 15th (1992)
Lubbock, TX	Close within 10 working days of month's end; interim financial reports within 5 days of closing	83% of closings within 10 workdays of month's end; 100% of interim reports within 5 days of closing (1997)
College Station, TX	By the 15th business day of the following month	92% (1996, estimated)
San Clemente, CA	Within 15 workdays of month's end	75% (1998)
Victoria, TX	Within 20 days of month's end	98% (1997, estimated)
Cambridge, MA	Interim financial report within 20 days of end of period	
Cary, NC	Within 20 days of month's end	
Overland Park, KS	Monthly closeout within 10 calendar days of month's end; interim financial statements issued within 4 weeks of closeout	100% of closeouts within 10 days of month's end; 100% of interim statements within 4 weeks of closeout (1997)
Shreveport, LA	Monthly closeouts within 10 workdays of month's end; 100% (1994) interim financial statements issued within 4 weeks of closeout	

Table 10.11   Prompt Reconciliation of Accounts: Selected Cities

Boca Raton, FL
Targets: Reconcile month-end balances of outstanding accounts payable within 5 days of
month-end close; bank reconciliation will be completed within 15 days of calendar month end

Corpus Christi, TX
Target: Reconcile bank statements within 2 weeks of receipt (1999)

Denver, CO
Percentage of bank accounts reconciled by last day of subsequent month: 80% (1998)

Ann Arbor, MI
Percentage of bank reconciliations completed within 30 days of month end: 75% (1998)

San Clemente, CA
Target: 100% of bank statements reconciled to general ledger cash balances within 30 days of
month's end
Actual: 50% (1998)
Target: 100% of general ledger reconciliations (accounts receivable, deposits, utility billing)
completed within 30 days of month's end
Actual: 64% (1998)

College Station, TX
Percentage of completed general ledger reconciliations and updates within 30 days of the period
close date: 91% (1998)

Cambridge, MA
Percentage of bank/cash/investment accounts reconciled with 45 days: 99% (1996)

Boston, Massachusetts, reports prompt action on two of these financial management functions and others (Table 10.13). Most monthly reports were distributed within 8 days following the end of the month; most account reconciliations were processed within 5 days; and other functions were handled in a similarly expeditious fashion.

Most of the municipal documents examined for this chapter reported current property tax collection rates in the mid- to upper-90% range (Table 10.14). The collection of other revenues often is more problematic, but leading municipalities pursue these receivables, as well as delinquent taxes, diligently (Table 10.15). For example, the policy on handling receivables in Plano, Texas, stipulates that a second bill is sent when an account becomes 30 days past due (City of Plano [TX], pp. 383–384). At 60 days past due, a third bill is sent along with notification that the account will be turned over to an outside collection agency if it becomes 90 days past due. At 90 days, the account is sent to an outside agency for collection.

Aggressive cash management demands the prompt processing of newly received revenues (Table 10.16) and the timely—or perhaps, *appropriately*

Table 10.12   Timely Audit and Financial Statement: Selected Cities

City	Performance Target
Ann Arbor, MI	Complete CAFR within 90 days of end of fiscal year
Cambridge, MA	Release Annual Report within 92 days of the end of the fiscal year Actual: 176 days (1995), 168 days (1996, estimated)
Lubbock, TX	Issue CAFR within 4 months of the end of the fiscal year Actual: Within 4 months (1997)
College Park, MD	Complete audit and release CAFR within 4 months after the end of the fiscal year
Oak Ridge, TN	Complete audit process within 4 months after the end of the fiscal year; print and distribute CAFR by middle of the 5th month
Chandler, AZ	Complete audit within 4 months after end of fiscal year; release CAFR by the end of the 5th month
San Antonio, TX	Complete annual financial statement within 151 days of the end of the fiscal year Actual: 177 days (1997)
Anaheim, CA	Submit an unqualified financial statement within 5 months of the end of the fiscal year
Corvallis, OR	Publish CAFR within 5 months following the end of the fiscal year
Oakland, CA	Issue CAFR within 5 months of fiscal year close
Portland, OR	Issue CAFR within 175 days after the end of the fiscal year
Norman, OK	Publish CAFR within 180 days of the end of the fiscal year Actual: 170 days (1997), 175 days (1998)

Table 10.13   Promptness Targets for the Processing of Financial Transactions/
Reports: Boston

	Target	Actual (1996)
Distribution of monthly reports	Within 8 days	84%
General Fund transfers	Completed within 3 days	100%
Review weekly payrolls	Within 2 days	100%
Payrolls posted	Within 7 days	96%
Transfers processed	Within 4 days	90%
Vendor invoices processed	Within 5 days	78%
Encumbrances processed	Within 3 days	86%
Account reconciliations	Within 5 days	85%

SOURCE: City of Boston (MA), *Mayor's Management Report: FY 1996* (Vol. 2), pp.45–46.

Table 10.14   Reported Collection Rates for Current Year Taxes: Selected Cities

City	Collection Rate for Current Taxes
Sterling Heights, MI	99.9% of property tax levy (1995)
Blacksburg, VA	99.7% (real estate tax levy, other taxes, licenses, and permits) (1997)
College Park, MD	99.62% of property tax levy (1991)
Cambridge, MA	99.5% of real estate levy; 99.5% of personal property levy; 86.3% of motor excise vehicle tax levy (1995)
Iowa City, IA	Taxes collected, as percentage of property tax budget: 99.4% (1989), 99.1% (1990)
Prairie Village, KS	98.8% of property taxes (1995)
Denton, TX	98.7% of property tax levy (1997)
Grand Junction, CO	96.67% to 99.67% of property tax levy (1982–1990), 98.6% (1991)
Duncanville, TX	Target: At least 98% Actual: 98.4% (1996), 97.5% (1997), 97.5% (1998)
Avon, CT	98.3% of levy (1997)
White Plains, NY	98.25% (1995)
Saginaw, TX	97.9% (1990), 98.2% (1991)
San Antonio, TX	98.2% of property taxes (1997)
Boston, MA	98% of real estate and personal property taxes (1996)
Danville, KY	98% of property taxes (1997)
Durham, NC	98% of property tax levy (1998)
Oak Ridge, TN	98% (1997)
Plano, TX	98.0% of ad valorem tax levy (1998)
Lubbock, TX	97.5% (1998)
Shreveport, LA	97% of property taxes (1994)
Albuquerque, NM	94.4% of property taxes (1995)
Oak Park, MI	93.93% of property tax levy (1997)
Eugene, OR	93.0% of property tax levy (1997)

timed—payment of invoices (Table 10.17). Some municipalities consider timely payment to mean prompt payment—for instance, within 2 or 3 days of receipt at accounts payable or within 2 or 3 days of authorization by the receiving department. Others regard timely payment to mean the timing of payments to retain cash in the possession of the municipality as long as possible for investment purposes but not so long as to annoy vendors or result in penalties for late payment. Lubbock, Texas, among others, tries not only to minimize late payment penalties but also to maximize vendor discounts for

Table 10.15   Collections Success: Selected Cities

TARGETS

Palo Alto, CA
Target 1: Collect at least 70% of paramedic receivables
Actual: 54% (1997)

Target 2: Collect at least 60% of DUI receivables
Actual: 82% (1997)

Target 3: Collect at least 90% of false alarm receivables
Actual: 100% (1997)

Target 4: Collect at least 75% of property damage receivables
Actual: 79% (1997)

PROMPT RECEIPT

Eugene, OR
Percentage of notes receivable that are over 31 days past due: 5.6% (1996)

Cary, NC
Target: No more than 10% of receivables more than 30 days old
Actual: 12% more than 30 days old (1999)

Portland, OR
Percentage of invoices collected within 60 days: 85% (1998)

San Jose, CA
Targets: (a) No more than 35% of active, invoiced accounts receivable more than 90 days old;
(b) no more than 20% of business tax accounts with an outstanding balance more than 90 days old
Actual: (a) 40.5%, (b) 13.7% (1995)

Ann Arbor, MI
Percentage of delinquent invoices older than 90 days: 25% (1998)

RECOVERY OF DELINQUENCIES

Glendale, AZ
Percentage of delinquent funds recovered: 86% (1997)

---

prompt payment when offered. Lubbock's target is to receive at least 98% of available vendor discounts. Its success rate in the mid-1990s was 60%.

Cash management targets and performance indicators suggest two other major dimensions of performance as well: (a) maximum prudent investment of idle funds and (b) modest costs for performing the cash management function (Table 10.18).

Wichita, Kansas, judges its municipal investments by comparing its rate of return with that of U.S. Treasury bills (Figure 10.2). Many other cities follow a similar strategy of comparing their investment returns with a fluctuating standard—typically U.S. Treasury bills or, in some cases, their own state's investment pool (Table 10.19). By pegging their own performance to a standard that

Table 10.16   Prompt Deposit of Revenues: Performance Targets and Actual
Experience of Selected Cities

Duncanville, TX
Percentage of bank deposits made for same day business: 100% (1998)

Santa Ana, CA
Percentage of payments received and processed the same day: 100% (1995)

Tucson, AZ
Percentage of utility payments processed the same day as received: 73% (1997)
Percentage of tax, license, and other payments processed the same day as received: 100% (1997)

Orlando, FL
Percentage of revenue received by the treasury bureau that is deposited and reported within
1 workday of receipt: 100% (1996)

Alexandria, VA
Percentage of checks deposited within 1 day of receipt by the treasury division: 99.5% (1995)

Reno, NV
Percentage of central cashier receipts deposited within 24 hours: 99% (1990)

Savannah, GA
Target: Process general fund and water and sewer receipts within 1 day at least 90% of the time
Actual: 90% within 1 day (1988–1991)

floats upward or downward with market trends, their performance may be judged not simply by a raw interest rate but instead by how their interest rate compares with another avenue of investment available to the city.

Within the central tasks of financial management are dozens of odds and ends that many proficient finance operations track to be sure they are getting the quality and consistency they expect. Several are listed in Table 10.20. Top performers pay attention to the details.

Table 10.17   Timely Payment of Invoices: Selected Cities

Upper Arlington, OH
Turnaround time on payment vouchers: 2 days (1991)

Oakland, CA
Target: Process vendor payments within 3 days of receipt of payment request
Actual: 2.75-day average (1996)

Fort Worth, TX
Target: Process payment documents within 5 days of receipt
Actual: 8-day average (1991)

Shreveport, LA
Target: Process within 5 days
Actual: 85% within 5 days (1994)

Charlottesville, VA
Percentage of vendor bills paid accurately and within 1 week of receipt in the finance office: 99%
   (1998)

Tucson, AZ
Percentage of vendor invoices paid within 7 calendar days of receipt: 100% (1997)

Boston, MA
Percentage of vendor invoices processed within 5 days of receipt: 78% (1996)
Percentage of vendor payments within 20 days: 44% (1996)

Portland, OR
Turnaround time from invoice to payment: 17 to 20 days (1991)

Savannah, GA
Invoices paid within 21 days of receipt by payables: 86% (1997)

Bellevue, WA
Average number of days from receipt of goods or services to payment of vendor: 24 (1998)

Boca Raton, FL
Target: Issue vendor checks within 30 days of invoice date

Duncanville, TX
Target: Within 30 days of invoice receipt

Plano, TX
Policy: Invoices shall be paid within 30 days of receipt (1999)

Corpus Christi, TX
Target: Process invoices within 30 days (1999)

Orlando, FL
Invoices paid within 30 days: 99.8% (1991), 95% (1998, estimated)

Sterling Heights, MI
Percentage of invoices processed within 30 days: 98% (1995)

College Station, TX
Target: Mail payment by 3 p.m. on the Friday prior to the due date
Percentage of vendor invoices paid within 30 days of invoice date: 90% (1998)

Denton, TX
Percentage of vendor invoices processed within 30 days: 90% (1997)

Oklahoma City, OK
Percentage of claims paid within 30 days of receipt of correct invoice by accounts payable: 89% (1999)

Hurst, TX
Target: Pay at least 90% of invoices within 30 days of receipt
Actual: 86% (1997)

Palo Alto, CA
Percentage of invoices paid within 30 days of invoice date: 73% (1997)

Albuquerque, NM
Percentage of invoices paid within 30 days: 66% (1998)

Table 10.18   Cash Management: Selected Cities

INVESTMENT OF IDLE FUNDS

Rochester, NY
Average daily invested cash as percentage of average daily cash ledger balance: 99.87% (1991)

Savannah, GA
Target: Less than 0.5% of portfolio's book value in idle cash
Actual: 0.08% (1997)

Long Beach, CA
Target: Maintain a fully invested position of at least 99.5% of all funds

Sunnyvale, CA
Target: 99.5% of available funds placed in interest-bearing investments

Albuquerque, NM
Target: Average cash balance less than 1.0%
Actual: 0.8% (1998)

College Station, TX
Percentage of available cash invested: 99% (1998)

San Jose, CA
Target: Invest 99% of available funds
Actual: 99% (1995)

Chapel Hill, NC
Target: Invest at least 98% of all funds daily

Lubbock, TX
Target: Invest at least 98% of available funds

Victoria, TX
Percentage of available cash invested in securities: 96% (1996)

(continued)

Table 10.18   Cash Management: Selected Cities (continued)

Irving, TX
  Target: No more than 4.75% of available funds left idle
  Actual: 4.89% idle (1991)

New York, NY
  Total average daily resources invested: 93.4% (1991), 89.1% (1992)

CASH MANAGEMENT COSTS

Corpus Christi, TX
  Management cost per $1,000 of investment income: $0.024 (1991)

Fort Collins, CO
  Investment program cost as percentage of total portfolio: 0.06% (1991)

Gresham, OR
  Target: Cash management costs (e.g., personnel and bank service charges) should not exceed 0.25%
    of assets managed

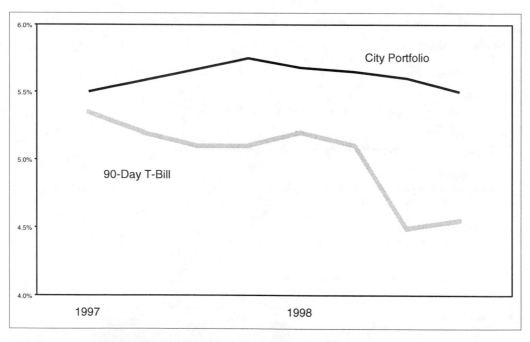

Figure 10.2. Yield on City of Wichita's Investments Compared to Federal Funds
SOURCE: City of Wichita (KS), *2000/2001 Annual Budget*, p. 23.

Table 10.19   Favorable Return on Investments: Selected Cities

COMPARED WITH FEDERAL FUNDS

Charlotte, NC
118.1% of the Public Investor Ten Bill Index rate (1999)

Shreveport, LA
Average return on investments: 10% greater than 3-month treasury bill rate (1994)

College Station, TX
Portfolio rate of return as a percentage of similarly weighted average maturity treasury notes:
   108% (1998)

Boston, MA
City return as percentage of federal funds rate: 107.7% (1996)

Denver, CO
Actual: 93 basis points above the 90-day treasury bill index (1998)

Alexandria, VA
Actual: 86 basis points above the average federal funds rate (1997)

Tucson, AZ
Target: Return on investment at least 0.25% greater than return on a 3-month treasury bill
Actual: 0.71% greater (1997)

Albuquerque, NM
Target: Investment return exceeds 1-year constant maturity treasury note
Actual: +40 basis points (1998)

Kalamazoo, MI
Actual: 40 basis points above the 3-month treasury bill average (1997)

Boca Raton, FL
Target: Investment earnings will exceed the 3-month treasury bill moving average by at least 20 basis
   points
Actual: +35 basis points (5.72% to 5.37%) (1997)

Chandler, AZ
Target: Rate of return equal to or greater than 90-day treasury bill
Actual: +30 basis points (5.55% to 5.25%) (1995)

Fort Collins, CO
Actual: +30 basis points (6.1% to 5.8%) (1996)

Rockville, MD
Actual: +30 basis points (5.5% to 5.2%) (1997)

Oklahoma City, OK
Target: Exceed the rate of return on 1-year U.S. Treasury bills
Actual: +28 basis points (5.46% to 5.18%) (1999)

Hurst, TX
Target: Rate of return exceeding the 3-month treasury bill rate (1999)

(continued)

Table 10.19   Favorable Return on Investments: Selected Cities (continued)

COMPARED WITH INVESTMENT POOLS

Milwaukee, WI
Target: Rate of return greater than Wisconsin Local Government Investment Pool
Actual: +64 basis points (1997)

Corvallis, OR
Target: Yield at least 0.5% above the state pool
Actual: 0.5% above the state pool (1991)

St. Petersburg, FL
Target: Meet or exceed the rate of the State of Florida Investment Pool
Actual: +25 basis points (1995)

San Clemente, CA
Actual: +9 basis points (5.75% to 5.66%) (1998)

Eugene, OR
Return on investments as a percentage of Local Government Investment Pool return: 100% (1996)

Nashville–Davidson County, TN
Target: Exceed rates of 90-day treasury bill index and state investment pool

Decatur, IL
Target: Average rate of return at least equal to that of the state investment pool

COMPARED WITH OTHER INVESTMENTS

Oakland, CA
Rate of return relative to market rates: +0.87% (1991), +0.20% (1996)

Savannah, GA
Difference between investment yields and highest yields offered by large money center banks:
+0.30% (1991), +0.25% (1992, estimated)

Auburn, AL
Target: Average rate of return meeting or exceeding jumbo certificates of deposit, as quoted in Wall
Street Journal

NOTE: A basis point is $\frac{1}{100}$ of 1%.

Table 10.20   Financial Administration Odds and Ends: Selected Cities

TAX ADMINISTRATION/CUSTOMER WAITING TIME

Philadelphia, PA
Average waiting time for walk-in taxpayers seeking assistance with tax matters: 17 minutes 21 seconds (1996), 14 minutes 15 seconds (1998)
Average waiting time for call-ins: 3 minutes 14 seconds (1996), 3 minutes 51 seconds (1998)

DAILY RECEIPTS BALANCING

San Jose, CA
Target: Balance at least 95% of daily receipts by 5 p.m.
Actual: 94% (1995)

CLOSING OF MONTH

Bellevue, WA
Average number of days to close month in financial system: 4.2 (1998)

San Antonio, TX
Target: Close within 9 working days of month's end
Actual: 10 workdays (1997)

BILLINGS

San Clemente, CA
Target: Process at least 85% of all accounts receivable within 5 days of receipt
Actual: 100% (1998)

Palo Alto, CA
Target: Issue at least 95% of billings within 7 working days of billing request from departments
Actual: 100% (1997)

PROMPT ISSUANCE OF MUNICIPAL LIEN CERTIFICATES

Cambridge, MA
99% issued within 10 business days of request (1996)

FIXED ASSETS INVENTORY

Palo Alto, CA
Target: Post new fixed assets within 60 days of payment and record retirements within 60 days of disposal
Actual: 100% (1997)

ACCURATELY LOCATED TAXPAYERS

Philadelphia, PA
Percentage of outgoing mail returned for lack of current address: 6.5% (1996), 6.4% (1998)

JOURNAL ENTRIES

College Station, TX
Ratio of correcting journal entry transactions to total journal entry transactions: 0.5% (1998)

Corpus Christi, TX
Target: Journal entries should have an error rate no greater than 5% (1999)

Palo Alto, CA
Target: Post no more than 25 adjusting journal entries during the preparation of the city's annual financial statements (1999)

CANCELED PAYMENTS

Calgary, Alberta
Canceled payments, as a percentage of total: 0.38% (1996)

MANUAL CHECKS

Calgary, Alberta
Manual checks, as a percentage of total: 1.75% (1996)

11

Fire Service

M ore than most other municipal service units, fire departments revere tradition. And understandably so. More than 300 years after the establishment of America's first public fire department, the roots of fire service tradition run deep (Paulsgrove, 1997, p. 10.4).

Some fire department traditions are charming and contribute to an aura that fosters good will. A dalmatian living in the station house—or perhaps borrowed to ride on a gleaming fire truck in local parades—helps link the modern fire service with its colorful and heroic past. The Saturday morning ritual of firefighters washing their bright red fire engines—most *still are red* despite scientific studies suggesting better visibility of other colors—and polishing their ample chrome is a scene the public enjoys. Where traditions collide with managerial pursuit of greater efficiency and effectiveness, however, the resulting confrontations often are less pleasurable.

Given their traditional bent, it is perhaps not surprising that many fire departments place considerable stock in long-standing rules of thumb that address everything from fireground tactics to prescribed levels of staffing. More than a few city executives with only a general background in fire services have met their match in the person of a fire chief able to cite with ease a battery of statistics, standards, and rules of thumb to bolster the department's position on a budgetary or operational issue. Understandably, nonspecialists feel at a disadvantage in exchanges of this type. So what is the best strategy? Mayors and city managers should enter such a fracas armed with a reasonable familiarity with relevant facts and figures themselves.

Overview

Various fire statistics offer potentially useful insights, but often they must be placed in proper context. Popular images of fire departments staffed with career firefighters square with reality in most cities with a population of 25,000

Table 11.1   Department Type by Population Protected, 1997

| | Type of Department | | | | |
Population Protected	All Career (%)	Mostly Career (%)	Mostly Volunteer (%)	All Volunteer (%)	Total (%)
1,000,000 or more	83.3	16.7	0.0	0.0	100.0
500,000 to 999,999	71.4	19.0	9.5[a]	0.0	100.0
250,000 to 499,999	71.8	20.5	7.7	0.0	100.0
100,000 to 249,999	72.3	13.9	11.9	2.0	100.0
50,000 to 99,999	68.3	17.0	11.9	2.8	100.0
25,000 to 49,999	46.8	23.4	23.4	6.3	100.0
10,000 to 24,999	17.1	21.0	38.3	23.6	100.0
5,000 to 9,999	2.3	6.7	38.4	52.6	100.0
2,500 to 4,999	2.1	1.3	14.6	82.0	100.0
Under 2,500	0.5	0.1	4.8	94.6	100.0
All Departments	6.4	4.9	15.7	73.0	100.0

SOURCE: National Fire Protection Association (NFPA) Survey of Fire Departments for U.S. Fire Experience, 1997, as reported in Michael J. Karter, Jr., *U.S. Fire Department Profile Through 1997* (Quincy, MA: National Fire Protection Association, November 1998), p. 17. Reprinted with permission from the National Fire Protection Association, Quincy, MA 02269.

NOTE: Type of department is broken into four categories. *All-career* departments are composed of 100% career fire-fighters. The *mostly career* category is composed of 51% to 99% career firefighters, whereas *mostly volunteer* departments are composed of 1% to 50% career firefighters. *All-volunteer* departments are composed of 100% volunteer firefighters.

a. This statistic reflects the inclusion of several county fire departments that have a considerable number of volunteer firefighters on their staffs.

or more, but less accurately depict the reality of small-town America, where volunteer firefighters still play a major role (Table 11.1). Moreover, fighting fires is not the only thing—or even by some measures the *main* thing—that fire departments do. More than half of all calls to the typical fire department are for medical aid, approximately one tenth are false alarms, and less than one tenth report actual fires (Figure 11.1). Time on the fireground fighting a real blaze typically consumes only a small portion of a firefighter's working day.

Among fire department personnel, the fire chief, a few other executive officers, and the department's clerical employees may have work schedules similar to other municipal personnel, but most firefighters do not. Workweeks for paid firefighters typically range from 40 to 56 hours (Paulsgrove, 1997, p. 10.24). Workweeks of 56 hours usually indicate the use of staggered schedules calling for 24 hours on duty by a given firefighter followed by 48 hours off

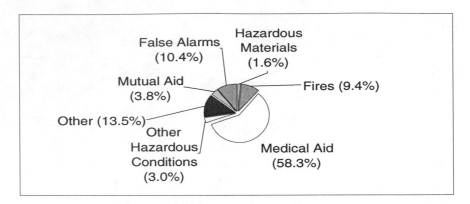

Figure 11.1. Nature of Fire Department Calls, 1998
SOURCE: National Fire Protection Association, *NFPA Survey*, September 1999 update. Reprinted with permission from the *Fire Journal* (Vol. 93 No. 5) Copyright © 1999, National Fire Protection Association. All rights reserved.

duty. Among departments with workweeks of 48 hours or less, a common schedule calls for a 10-hour day shift and a 14-hour night shift.

National statistics for fire departments often differentiate between departments with career firefighters and those that rely primarily on volunteers and, where staffing ratios are cited, between departments with 56-hour workweeks and those using other schedules. The prudent manager will look for those distinctions.

Typical Topics of Budgetary Debate

Three aspects of fire service typically draw considerable attention at budget time:

1. Appropriate staffing level
2. The hazardous nature of fire fighting and contentions that firefighter compensation should recognize that characteristic
3. The appropriate number and types of fire-fighting equipment

Although each of these topics deals more directly with inputs than with performance, the frequency with which each becomes an issue in most communities prompts their brief discussion here before proceeding to actual benchmarks of performance.

Staffing Levels

Staffing statistics compiled by the National Fire Protection Association (NFPA) reveal different lengths of workweek and ratios of career firefighters per 1,000 population for various sizes of communities (Table 11.2). Among

Table 11.2   Career Firefighters per 1,000 People by Workweek and Population
Protected, 1995–1997

Population Protected	Career Firefighters per 1,000 People					
	Lowest Reported	Median	Highest Reported	40–45 hour workweek	46–51 hour workweek	52–60 hour workweek
1,000,000 or more	0.71	1.45	1.76	1.60	1.47	0.98
500,000 to 999,999	0.71	1.52	2.78	2.12[a]	1.82[a]	1.33[a]
250,000 to 499,999	0.54	1.11	2.93	—	—	—
100,000 to 249,999	0.00	1.33	3.37	2.44	1.68	1.38
50,000 to 99,999	0.00	1.33	3.33	2.16	1.76	1.36
25,000 to 49,999	0.00	1.17	3.93	2.20	1.62	1.60

SOURCE: NFPA Fire Service Survey, 1995–1997, as reported in Michael J. Karter, Jr., *U.S. Fire Department Profile Through 1997* (Quincy, MA: National Fire Protection Association, November 1998), pp. 5, 8. Reprinted with permission from the National Fire Protection Association, Quincy, MA 02269.

NOTES: *The rates listed above are based on data reported to the NFPA and do not reflect recommended rates or some defined fire protection standard.*

The rates of a particular size of community may vary widely because departments face great variation in their specific circumstances and policies, including unusual structural conditions, type of service provided to the community, geographic dispersion of the community, and other factors.

Career rates are shown only for communities with more than 25,000 population in which departments are composed mostly of career firefighters.

a. This figure is for the combined population category 250,000 to 999,999.

cities of 100,000 to 249,999 persons, for example, at least one department was all volunteer (reporting no career firefighters), the highest ratio was 3.37 career firefighters per 1,000 persons, and the median was 1.33. Among communities in that population range, the average ratio for fire departments with a 52- to 60-hour workweek was 1.38 career firefighters per 1,000 people.

Staffing norms differ by region. Northeastern municipalities tend to employ higher ratios of career firefighters than do other regions, with western municipalities employing the lowest ratios (Table 11.3).

From time to time, various ratios of firefighters to population are offered as prescriptions, as rules of thumb, or even as standards. Unfortunately, such ratios sometimes are applied without any thought given to important local factors, such as the use of volunteers in a given community, the region in which the community is located, its population and density, and various other local characteristics that influence fire risk. Statistics such as those presented here that reveal low, median, and high ratios for various population categories, as well as variations by workweek and region, can be of greater value to managers and other public officials than can a "one size fits all" rule of thumb. By comparing the ratio in one's community with the relevant cluster, a more

Table 11.3   Median Rates of Career Firefighters per 1,000 People by Region and Population Protected, 1997

Population Protected	Northeast	North Central	South	West
250,000 or more	1.74	1.64	1.52	0.91
100,000 to 249,999	2.46	1.35	1.46	0.82
50,000 to 99,999	2.07	1.28	1.58	0.86
25,000 to 49,999	1.76	0.99	1.50	0.89

SOURCE: NFPA Survey of Fire Departments for U.S. Fire Experience, 1997, as reported in Michael J. Karter, Jr., *U.S. Fire Department Profile Through 1997* (Quincy, MA: National Fire Protection Association, November 1998), p. 9. Reprinted with permission from the National Fire Protection Association, Quincy, MA 02269.

NOTES: *The rates listed above are based on data reported to the NFPA and do not reflect recommended rates or some defined fire protection standard.*

The rates of a particular size of community may vary widely because departments face great variation in their specific circumstances and policies, including unusual structural conditions, type of service provided to the community, geographic dispersion of the community, and other factors.

Career rates are shown only for communities with more than 25,000 population, where departments are composed mostly of career firefighters.

appropriate determination can be made as to whether local staffing is in the ballpark.

Perhaps more controversial than attempts to use overall staffing ratios as prescriptive devices have been attempts to prescribe minimum staffing on individual fire apparatus. The U.S. Department of Labor's Occupational Safety and Health Administration (OSHA) now requires departments to have at least four firefighters on the scene before starting interior structural fire-fighting operations, except in rare instances. The International Association of Fire Fighters (IAFF, 1993), insists that these firefighters should arrive at the fire scene together on their engine—all members of the same engine company. The NFPA (1992) also has long encouraged a minimum of four firefighters per engine prior to commencing fire-fighting operations inside a structure; however, neither NFPA nor OSHA insists that they all arrive on the same vehicle. NFPA's position is that,

while members can be assigned and arrive at the scene of an incident in many different ways, it is strongly recommended that interior fire fighting operations not be conducted without an adequate number of qualified firefighters operating in companies under the supervision of company officers.

It is recommended that a minimum acceptable fire company staffing level should be 4 members responding on *or arriving with* each engine and each ladder company responding to any type of fire. The minimum accept-

Table 11.4   Death Rates in Fire Fighting and Other Hazardous Occupations

Occupation	Deaths per 100,000 Employees (per year)
Mining and quarrying (includes oil and gas extraction)	23.7 (1997)
Agriculture (includes forestry and fishing)	20.3 (1997)
Construction	13.5 (1997)
Transportation and public utilities	11.9 (1997)
Fire fighting (career-municipal)	11.5 (1998)
Manufacturing	3.2 (1997)

SOURCES: *Accident Facts,* 1998 (Itasca, IL: National Safety Council) and A. E. Washburn, P. R. LeBlanc, and R. F. Fahy, "Firefighter Fatalities," *NFPA Journal,* 93 (July/August 1999), p. 60. Reprinted with permission from the *Fire Journal* (Vol. 93 No. 4) Copyright © 1999, National Fire Protection Association. All rights reserved.

able staffing level for companies responding in high-risk areas should be 5 members responding *or arriving with* each engine company and 6 members responding *or arriving with* each ladder company. (NFPA, 1992, p. 40, italics added; reprinted with permission from *NFPA 1500: Fire Department Occupational Safety and Health Program,* Copyright 1992, National Fire Protection Association, Quincy, MA 02269. This reprinted material is not the complete and official position of the National Fire Protection Association on the referenced subject, which is represented only by the standard in its entirety.)

Although the difference between the IAFF and NFPA positions may seem subtle, it is not. The NFPA position allows much greater flexibility in the deployment of firefighters when not actually engaged in fire fighting.

Hazardous Duties

Spokespersons for proposals to increase the pay of firefighters often mention the hazardous nature of that occupation, sometimes contending that it is America's most dangerous work. Although a manager or mayor might choose to dispute the "most dangerous" label (Table 11.4), there is no disputing that fire fighting is hazardous work with considerable risk of injury or even death. Nevertheless, a reasonable command of relevant statistics on firefighter deaths and injuries will equip city officials not only to rebut any exaggerated claims, but also to place local experience in a national context. If local firefighters' deaths and injuries exceed national norms, a careful review of the community's training program, firefighter fitness levels, and fireground procedures might be in order.

Only about half of all firefighter deaths occur while actually fighting fires (Figure 11.2). Others occur en route to the fire scene or on the way back to the

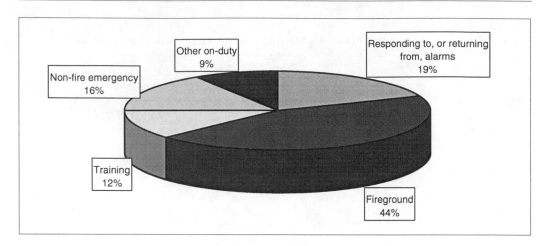

Figure 11.2. Firefighter Deaths by Type of Duty, 1998
SOURCE: Arthur E. Washburn, Paul R. LeBlanc, and Rita F. Fahy, "Firefighter Fatalities," *NFPA Journal, 93* (July/August 1999) p. 56. Reprinted with permission from the *Fire Journal* (Vol. 93, No. 4), Copyright © 1999, National Fire Protection Association. All rights reserved.

station, during training, or while performing other duties. Not surprisingly, death rates tend to be greater among older firefighters—reaching rates of 1.8 per 10,000 firefighters among those in their 50s and 2.6 per 10,000 among firefighters 60 years of age or older, compared with fewer than 1 death per 10,000 among firefighters less than 50 years of age (Washburn, LeBlanc, & Fahy, 1999, p. 58). Heart attacks claim far more firefighters' lives than any single type of injury (Figure 11.3).

Fire departments serving larger communities understandably experience greater numbers of fires and greater numbers of firefighter injuries in absolute terms, but is the job of fire fighting really more hazardous in large cities than in smaller communities? Statistics indicate that the answer is "yes." When measured as number of fireground injuries per 100 fires, larger communities tend to experience higher rates of injury (Table 11.5). Even that statistic, however, tends to understate the true differential. Because big-city firefighters respond to more fires, they are more exposed to the risk of injury than are their small-town counterparts. Each year, 16 out of every 100 big-city firefighters experience a fireground injury, compared with fewer than 2 out of every 100 firefighters in communities with populations less than 5,000.

Equipment Levels

As every city official who has budgeted for major fire equipment already knows, fire engines are expensive. Unquestionably, fire-fighting efforts can be severely handicapped—sometimes with tragic consequences—by inadequate equipment. Nevertheless, few communities possess the financial resources to meet all their fire department's desires for the latest and most sophisticated fire-fighting equipment. City officials struggling to find the appropriate

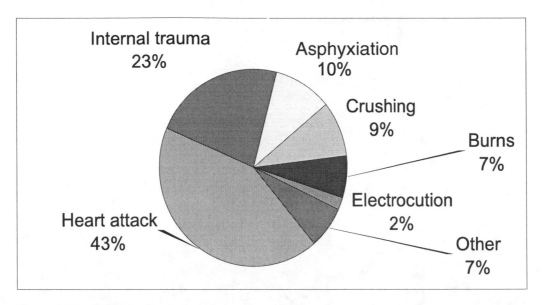

Figure 11.3. Firefighter Deaths by Nature of Injury, 1998
SOURCE: Arthur E. Washburn, Paul R. LeBlanc, and Rita F. Fahy, "Firefighter Fatalities," *NFPA Journal, 93* (July/August 1999), p. 58. Reprinted with permission from the *Fire Journal* (Vol. 93 No. 4) Copyright © 1999, National Fire Protection Association. All rights reserved.

balance between their responsibility for ensuring public safety and their obligation for financial prudence may find helpful information in the statistical norms for fire equipment inventories in communities of various sizes.

Obviously, larger communities own more pumpers and aerial equipment and operate more fire stations than do smaller communities, but their *rates* of pumpers, aerial apparatus, and stations per 1,000 population are not as great (Table 11.6). Although considerable variation is evident, the national average is approximately 1 pumper per 3,800 population, 1 aerial apparatus per 41,700 population, and 1 fire station per 5,400 population.

Most cities with more than 100,000 people possess at least 10 pumpers (Table 11.7). Cities with populations of 10,000 to 50,000 tend to have 3 or 4 pumpers, whereas cities of less than 5,000 people are lucky to have more than 2. Aerial equipment is common among cities with populations greater than 100,000, fairly common among cities with more than 25,000 people, and relatively rare among cities of less than 10,000 people (Table 11.8).

Fire Service Performance Statistics

Statistics on staffing levels, firefighter deaths and injuries, and numbers of stations and heavy fire-fighting equipment can be useful to city officials as they grapple with budgetary and compensation issues, but those statistics are not

Table 11.5 National Statistics on Reported Fires and Fireground Injury Rates, 1998

Population of Community Protected	Average Number of Reported Fires					Number of Fireground Injuries per 100 Fires					Number of Fireground Injuries per 100 Firefighters
	National Average	Northeast	North Central	South	West	National Average	Northeast	North Central	South	West	National Average
500,000 to 999,999	3,872.8	*	*	4,044.5	2,475.5	4.0	*	*	3.0	1.8	16.1
250,000 to 499,999	2,329.8	*	2,593.8	2,637.3	1,399.3	2.1	*	3.5	1.6	1.6	11.8
100,000 to 249,999	795.9	1,044.6	767.1	932.3	542.3	2.3	6.2	1.9	1.3	1.9	7.5
50,000 to 99,999	344.0	406.9	289.8	447.7	278.5	2.1	4.1	2.3	1.3	1.4	6.4
25,000 to 49,999	160.0	146.2	140.8	188.0	175.3	1.9	4.5	1.9	1.1	1.1	5.1
10,000 to 24,999	88.0	80.9	75.7	107.6	91.3	1.8	2.8	2.0	1.6	0.5	3.9
5,000 to 9,999	47.1	40.2	38.6	59.0	59.3	2.1	3.5	1.8	1.5	1.2	3.0
2,500 to 4,999	27.6	24.6	22.8	35.6	31.6	1.8	2.4	1.8	1.4	0.9	1.7
Under 2,500	12.6	12.5	10.0	19.7	11.7	1.6	3.2	1.0	2.0	1.7	1.2

SOURCE: NFPA's Survey of Fire Departments for U.S. Fire Experience (1998), as reported in Michael J, Karter, Jr., and Paul R. LeBlanc, "Firefighter Injuries," *NFPA Journal*, 93 (November/December 1999), pp. 47–56. Reprinted with permission of the National Fire Protection Association, Quincy, MA 02269.

NOTE: *denotes insufficient data.

Table 11.6   Average Apparatus and Station Rates, 1995–1997

Population Protected	Pumpers per 1,000 People	Aerial Apparatus per 1,000 People	Stations per 1,000 People
1,000,000 or more	0.033	0.016	0.034
500,000 to 999,999	0.062	0.020	0.054
250,000 to 499,999	0.059	0.015	0.061
100,000 to 249,999	0.077	0.016	0.071
50,000 to 99,999	0.086	0.018	0.080
25,000 to 49,999	0.116	0.023	0.091
10,000 to 24,999	0.195	0.034	0.126
5,000 to 9,999	0.343	0.037	0.196
2,500 to 4,999	0.568	0.027	0.344
Under 2,500	1.265	0.033	0.955
National average	0.260	0.024	0.186

SOURCE: NFPA Fire Service Survey, 1995–1997, as reported in Michael J. Karter, Jr., *U.S. Fire Department Profile Through 1997* (Quincy, MA: National Fire Protection Association, November 1998), p. 19. Reprinted with permission from the National Fire Protection Association, Quincy, MA 02269.

NOTES: These results reflect average apparatus and station rates per 1,000 people by population protected among reporting jurisdictions. The NFPA emphasizes that they do not reflect recommended rates or some defined fire protection standard.

Note that pumpers reported above had a capability of 500 gallons per minute (gpm) or greater. Many departments reported other suppression vehicles, including apparatus with pumps less than 500 gpm, hose wagons, brushfire vehicles, tankers, and so forth.

benchmarks of performance. National or regional norms, perhaps categorized by community size or length of workweek, can tell officials whether local practices and characteristics lie within or outside normal bounds, but they do not address performance.

Benchmarks of performance provide external pegs against which local service measures may be compared. Ideally, the performance measures of a fire department include more than simple workload indicators, such as the number of alarms, number of inspections, and hours of training. Although valuable, workload indicators only report what was done. They fail to reveal how efficiently the work was performed, its quality, or its effectiveness in achieving desired results. A good set of measures will address these key aspects of performance, and in so doing will often provide a better means of comparison with national norms or selected counterparts (for suggested fire service measures, see Hatry et al., 1992, pp. 94–95; Parry, Sharp, Vreeland, & Wallace, 1991, pp. 36–43).

Table 11.7   Pumpers (750 gpm or greater) by Community Size, 1995–1997

	Percentage of U.S. Fire Departments With				
Population Protected	1–5 Pumpers (%)	6–9 Pumpers (%)	10–19 Pumpers (%)	20–39 Pumpers (%)	40 or More Pumpers (%)
1,000,000 or more	0.0	0.0	0.0	7.7	92.3
500,000 to 999,999	0.0	0.0	0.0	60.0	40.0
250,000 to 499,999	0.0	3.4	39.0	54.2	3.4
100,000 to 249,999	9.9	33.1	49.4	5.8	1.7
50,000 to 99,999	57.5	31.1	9.5	1.8	0.0
Population Protected	No Pumpers (%)	1 Pumper (%)	2 Pumpers (%)	3–4 Pumpers (%)	5 or More Pumpers (%)
25,000 to 49,999	0.9	3.4	15.6	51.0	29.0
10,000 to 24,999	0.7	6.2	30.0	53.4	9.8
5,000 to 9,999	1.3	12.1	40.9	41.6	4.1
2,500 to 4,999	2.2	21.9	49.8	25.1	1.0
Under 2,500	9.4	42.9	38.4	8.9	0.4

SOURCE: 1995–1997 NFPA Fire Service Survey, as reported in Michael J. Karter, Jr., *U.S. Fire Department Profile Through 1997* (Quincy, MA: National Fire Protection Association, November 1998), p. 20. Reprinted with permission from the National Fire Protection Association, Quincy, MA 02269.
NOTE: Pumpers reported above had a capability of 750 gpm or greater. Many departments reported other fire suppression vehicles including apparatus with pumps less than 750 gpm, hose wagons, brushfire vehicles, tankers, and so forth.

Intercity Comparison

Progress in a municipal fire department's performance may be judged by comparing performance indicators from one period to another. Although such a review could reveal improvements, it would not address the adequacy of performance in a broader sense. Only by comparing local performance with some external benchmark can the adequacy of local operations truly be assessed. As noted by Parry et al. (1991) in the GASB's volume on Service Efforts and Accomplishments (SEA) indicators for fire department programs, the consideration of intercity norms can be useful:

It would seem that recommended disclosures should focus on some intercity performance norms. A tie-in to average response time and average fire loss per capita experienced by peer cities, certification programs, and insurance grading systems provide benchmarks that place performance in context. (p. 24; reprinted by permission)

| Table 11.8 | Aerial Apparatus by Community Size, 1995–1997 |

Population Protected	Percentage of U.S. Fire Departments With				
	No Aerial Apparatus (%)	1–5 Aerial Apparatus (%)	6–9 Aerial Apparatus (%)	10–19 Aerial Apparatus (%)	20 or More Aerial Apparatus (%)
1,000,000 or more	0.0	15.4	0.0	15.4	69.2
500,000 to 999,999	0.0	4.0	16.0	64.0	16.0
250,000 to 499,999	10.0	53.3	18.3	18.3	0.0

Population Protected	No Aerial Apparatus (%)	1 Aerial Apparatus (%)	2 Aerial Apparatus (%)	3–4 Aerial Apparatus (%)	5 or more Aerial Apparatus (%)
100,000 to 249,999	16.2	22.5	22.0	26.2	13.1
50,000 to 99,999	22.0	43.0	25.0	9.5	0.7
25,000 to 49,999	34.2	51.8	12.7	1.3	0.0
10,000 to 24,999	49.8	46.5	3.3	0.4	0.0
5,000 to 9,999	73.7	25.4	0.9	0.0	0.0
2,500 to 4,999	90.3	9.3	0.4	0.0	0.0
Under 2,500	96.6	3.0	0.4	0.0	0.0

SOURCE: 1995–1997 NFPA Fire Service Survey, as reported in Michael J. Karter, Jr., *U.S. Fire Department Profile Through 1997* (Quincy, MA: National Fire Protection Association, November 1998), p. 21. Reprinted with permission from the National Fire Protection Association, Quincy, MA 02269.

Workload measures from other municipal fire departments cannot perform that benchmarking function. Knowledge that the fire department in Gainesville, Georgia, responded to 3,641 emergency calls in 1999 is not especially useful to the fire chief or other municipal officials in Muleshoe, Texas, or Oliver Springs, Tennessee—or to officials in most other communities across the nation. On the other hand, indicators that are standardized or measure service quality—for instance, indicators reporting that Gainesville's average response time from dispatch until arrival at the scene was 3 minutes and 21 seconds or that the fire department in Wichita, Kansas, was able to confine 91% of all fires to the room of origin—may usefully serve as benchmarks.

Fire Insurance Ratings

In almost all of the nation's 50 states, the Insurance Services Office (ISO) rates the fire readiness of individual communities as an informational service to potential insurers. Alternate systems performing the same function may be found in the few states that are exceptions.

ISO rates are based on the adequacy of a community's water supply, its fire department, and its fire alarm system. Public protection classifications range from the most desirable rate of 1 to the least desirable of 10. Water supply considerations focus on hydrant availability, hydrant condition, and the availability of adequate supplies of water at sufficient pressure. Fire department considerations focus on staffing levels and geographic distribution of stations, training, and equipment. Fire alarm considerations—weighted at 10% compared with 40% for water supply and 50% for fire department—focus primarily on dispatch operations and the availability of sufficient staff and equipment (Granito, 1997). Selected standards for ISO ratings may be found in Table 11.9.

A fire insurance rate has the advantage of conveying a great deal of information by means of a single indicator. Often, it is useful to compare staffing levels, costs, and deployment patterns with other communities possessing the same rating. If officials attempt to judge different fire departments on the strengths on their fire ratings, however, it is important to note two important facts. First, 40% of a community's rating is based on the adequacy of its water supply, something typically beyond the control of the fire department. Second, there appears to be a fairly strong correlation between favorable fire insurance rates and community size. Because large cities are more likely to possess favorable fire insurance ratings than are small towns, it might be most useful to select as benchmarking partners other communities of similar population that have equal or better fire insurance rates.

Several municipalities supplying documents used in the development of this book reported their fire insurance rate as an indicator of performance in fire services. Some, including a few relatively small communities that defy the general trend, reported favorable public protection classifications. For example, the most favorable rating of 1 was reported by Plano, Texas. A rating of 2 was reported by Gainesville, Georgia; High Point, North Carolina; and Little Rock, Arkansas. Class 3 ratings were reported by Cary, North Carolina; Kalamazoo, Michigan; and Kingsport and Oak Ridge, Tennessee.

Fire Incidence Rate

A fire department with a narrow view of its mission might expect to suppress fires within a defined geographic area—and to do nothing more. A broader view of fire department mission includes responsibility for minimizing loss of lives and property, and incorporates a fire *prevention* role. Departments that subscribe to this broader view typically undertake more ambitious inspection programs in an effort to influence both the rate and severity of fire incidents. The effectiveness of such a strategy has been confirmed by NFPA studies indicating that average fire incidence rates in communities in which almost all public buildings are inspected annually are approximately one half the rate of

Table 11.9   Selected Standards for Fire Insurance Ratings

Basic Equipment for Class 9
- At least one apparatus with pump capacity of 50 gpm or more at 150 psi (pounds per square inch), and at least a 300-gallon water tank

Basic Equipment for Class 8 or Better
- Apparatus with pump capacity of 250 gpm or more at 150 psi
- Water system capable of delivering 250 gpm for 2 hours, plus consumption at the maximum daily rate at a fire location

Fire Flow at Residential Structures
- For 1- and 2-family dwellings not exceeding 2 stories in height, the following Needed Fire Flows shall be used:

Distance Between Buildings	Needed Fire Flow
Over 100'	500 gpm
30'–100'	750 gpm
11'–30'	1000 gpm
10' or less	1500 gpm

- Other habitational buildings, up to 3500 gpm maximum

Reserved Telephone Lines at Communication Center

Population Served	Number of Reserved Lines	
	Fire	Business
Up to 40,000	1	1
40,001–125,000	2	2
125,001–300,000	3	3
Over 300,000	4	3

- When emergency calls for other than fire are received over the fire number, double the number of needed reserved fire lines indicated above. Automatic telephone dialing equipment used to report alarms from private fire detection systems should have an emergency line separate from the normal fire and business numbers.

Number of Engine Companies

Needed Fire Flow (gpm)	Number of Engine Companies Needed	Number of Ladder Companies Needed
500–1,000	1	—
1,250–2,500	2	—
3,000–3,500	3	—
4,000–4,500	4	1
5,000–5,500	5	2
6,000–6,500	6	2
7,000–7,500	7	3
8,000–8,500	8	3
9,000–9,500	9	4
10,000	10	4
11,000	11	4
12,000	12	5

Hydrant Distribution
- Full credit awarded if hydrant is no more than 300 feet away; partial credit for 301 to 600 feet away; less credit for 601 to 1,000 feet away
- Built-on areas not within 1,000 feet of a recognized water system may still be eligible for class 9

SOURCE: Includes copyrighted material of Insurance Services Office, Inc., with its permission. Excerpted from *Fire Suppression Rating Schedule,* Edition 6-80. Copyright © Insurance Services Office, Inc., 1980.

communities with less ambitious inspection programs (Hall, Koss, Schainblatt, Karter, & McNerney, 1979).

An aggressive fire department with a broad view of its mission, then, hopes not only to suppress promptly those fires that do occur but also to prevent fires. Fire incidence rates record the success—or more accurately, the lack of success—in achieving that latter objective. City officials may wish to track their progress in reducing the rate of fire incidents from year to year; they may furthermore wish to compare their own community's figures with the fire incidence rates of other cities. Although both comparisons provide useful information, special care is required to interpret intercity comparisons properly. Gloating over possession of a more favorable fire incidence rate than cities with substantially older housing stock and higher numbers of abandoned or poorly maintained manufacturing facilities not only may be unseemly but also may reflect an inaccurate assessment of relative fire department performance. With that note of caution in mind, a review of fire incidence rates in other communities may nevertheless provide local officials a context in which to judge their own numbers—giving them a sense of where they stand.

Among the documents examined for this volume, fire incidence rates were reported in three ways: (a) fire incidents per 1,000 structures, (b) structural fires per 1,000 population, and (c) fire incidents per 1,000 population (Table 11.10). Cities reporting those statistics managed to hold the rate of residential fires to less than 10 per 1,000 occupancies. They managed to hold the rate of structural fires to less than 5 per 1,000 population. And fire incidents of all types occurred at a rate of less than 9 per 1,000 population. As a further reference point, a panel considering municipal service standards for the LCC (1994) considers fire incidence rates below 9.8 per 1,000 population to be characteristic of high-service-level cities, rates above 11.8 to reflect low-service levels, and rates in between to indicate medium fire protection service (p. 24).

Statistics among cities of 100,000 or greater participating in the ICMA comparative performance measurement project ranged from 0.27 *residential* structure fire incidents per 1,000 population in 1998 (Bellevue, Washington) to 3.56 per 1,000 population (ICMA, 1999b, p. 169). Total fire incidents per 1,000 population ranged from 1.06 (Gresham, Oregon) to 31.5 (p. 174).

False Alarms

Not every alarm summons firefighters to an actual fire incident. Nationally, about one tenth of all emergency calls received by fire departments, including calls for medical aid, are false alarms (NFPA, 1999). Malfunctions causing activation of automatic detection systems accounted for 43.5% of all false alarms in 1998 (Karter, 1999, p. 1).

Some cities are fortunate to have false alarm rates well below the national average (Table 11.11), but even those communities devote considerable

Table 11.10   Fire Incidence Rates: Selected Cities

FIRE INCIDENTS PER 1,000 STRUCTURES

Long Beach, CA
2.0 single-family residential structure fires per 1,000 occupancies (1996)

San Antonio, TX
Structural fires per 1,000 occupancies (single-family residential): 3.66 (1995), 5.31 (1997)

Gresham, OR
Targets: Not more than 6.5 residential fire incidents per 1,000 occupancies (0.65%); not more than
5.5 commercial fire incidents per 1,000 occupancies (0.55%)

Duncanville, TX
7 residential fire incidents per 1,000 structures (0.7%); 36.3 nonresidential fire incidents per 1,000
structures (3.63%) (1996)

Savannah, GA
9.0 dwelling fires per 1,000 structures (0.90%) (1993)

STRUCTURAL FIRES PER 1,000 POPULATION

Charlotte, NC
1.15 (1999)

Seattle, WA
1.2 (1998)[a]

Sunnyvale, CA
Target: 1.43 (1994)

Denver, CO
1.5 (1997)[a]

Portland, OR
1.7 (1998)

Sacramento, CA
1.7 (1998)[a]

Kansas City, MO
3.2 (1998)[a]

Cincinnati, OH
4.2 (1997)[a]

FIRE INCIDENTS PER 1,000 POPULATION

Chandler, AZ
Fires per 1,000 population: 4.70 (1995), 1.93 (1998)

Plano, TX
3.17 (1997)

Iowa City, IA
Fires per 1,000 population: 4.1 (1991)

Fayetteville, AR
4.7 working fires per 1,000 population (1991)

Portland, OR
Fire incidents per 1,000 population: 5.0 (1998)

Cincinnati, OH
Fire incidents per 1,000 population: 7.8 (1991)

Corpus Christi, TX
8.1 fires per 1,000 population (1991)

a. Indicates figures reported in City of Portland (OR) Office of the City of Portland Auditor, *Service Efforts and Accomplishments: 1997–98* (December 1998), pp. C2–C3.

Table 11.11   False Alarms

National Statistics
- NFPA survey[a]—10.4% of all fire department emergency calls (including medical aid) in 1998 were false alarms

Local Statistics

False Alarms as a Percentage of All Alarms Received	Ratio of Actual Fires to False Alarms	False Alarms per 1,000 Population
Tallahassee, FL	**Savannah, GA**	**Loveland, CO**
4.0% (1991)	1.5:1 (1991), 1.6:1 (1992, estimated)	3.08 (1990), 0.77 (1991)
Southfield, MI	**Southfield, MI**	**Tallahassee, FL**
6% (1991)	1.2:1 (1991)	2.2 (1991)
Longview, TX	**Ames, IA**	**Ames, IA**
11.0% (1991)	0.7:1 (1997)	5.3 (1997)
Milwaukee, WI		**9 SC cities[b]**
12% (1997)		7.2 (1998)

a. National false alarm statistics from the NFPA survey, updated September 1999.
b. False alarms in selected South Carolina cities reported in *South Carolina Municipal Benchmarking Project: 1998* (Columbia, SC: University of South Carolina, Institute of Public Affairs, 1999), pp. 175–221.

resources to needless fire runs and impose unnecessary risks on firefighters and others sharing city streets. Every city that experiences *any* false alarms has a false alarm problem; comparison with other communities and the national average permits it to know how big that problem is.

Response Time

Quick response is instrumental to fire rescue and effective fire suppression. Although prompt arrival is no more important than competent performance at the scene, studies have shown a positive correlation between quicker response times and lower loss of life and property (Gordon, Drozda, & Stacey, 1969; Rider, 1979). Quick response counts. Often in fire emergencies, it counts a great deal.

Fire service professionals sometimes subdivide what they call the "response time sequence" into five parts:

1. *Dispatch Time,* during which the call is received, the message interpreted, and appropriate units selected and dispatched

2. *Turn-Out Time,* during which crews receive the information and prepare to leave the station

3. *Travel Time,* from the station to the scene (i.e., from "wheel start to wheel stop")

4. *Access Time,* the interval required to climb stairs, reach the interior of a mall, and so forth

5. *Set-Up Time,* involving crew deployment and apparatus placement (Barr & Caputo, 1997, p. 10.251)

More commonly, however, response time is thought to consist of two major components: (a) the time between a citizen's call for help and the dispatch of an emergency unit and (b) the time between dispatch and arrival.[1] Excessive speed by heavy fire equipment is a poor way to reduce response time, because it imposes undue risk on the lives of firefighters, motorists, and pedestrians. Far better avenues for reducing response time lie in minimizing the time consumed in the first component (dispatch time), in reducing the time required to get out of the station (turn-out time), and also in selecting station locations to minimize travel distances.

A panel of local government officials appointed by the LCC (1994) prescribed a 5-minute response for high-service-level departments in 1994 (p. 23). A more forgiving regional standard in Pennsylvania calls for first alarm responses within 8 minutes (SPRPC, 1990, p. VI-5). Among the cities examined for this volume, those that reported comprehensive response time—that is, dispatch time *and* travel time—were competitive with the 5-minute standard and comfortably within the 8-minute mark (Table 11.12). The median among cities reporting only the time from dispatch until arrival was an average response time to fire emergencies of 4 minutes and 16 seconds. Other quick action elements and factors contributing to prompt response are reported by a few cities in Table 11.13.

Many fire departments respond not only to fire emergencies but to other calls as well. Some maintain separate time statistics for different categories of emergency, often reporting somewhat slower response time averages for specialty categories other than medical emergencies. For example, Winston-Salem, North Carolina, reported an average response time of 12.35 minutes by its hazardous materials unit (Table 11.14). The fire department of San Jose, California, responded to all hazardous material incidents within 15 minutes during one recent year.

Controlling the Fire

The prompt arrival of a highly trained and well-equipped fire company is the desire of every fire chief—and the hope of every emergency caller. Proper equipment and staffing levels remain a topic of debate, with fire service organizations and fire unions calling for numbers that exceed the equipment and

(text continues on page 161)

Table 11.12   Response Times to Fire Emergencies: Selected Cities

TIME FROM CALL RECEIPT AT 911 UNTIL ARRIVAL AT SCENE

Long Beach, CA
3-minute 42-second average (1996)

San Antonio, TX
4-minute 58-second average (1997)

Milwaukee, WI
93.7% within 5 minutes of call (1997)

Fort Worth, TX
5-minute 4-second average (1999)

Plano, TX
5-minute 17-second average (1999)

Irving, TX
5-minute 24-second average (1998)

Bellevue, WA
6.0-minute average; 64% within 6 minutes (1998)

TIME FROM DISPATCH UNTIL ARRIVAL AT SCENE
(EXCLUDING DISPATCH TIME)

Wilmington, DE
2-minute 7-second average (1999)

Winston-Salem, NC
2-minute 27-second average (1998); 2-minute 52-second average (1999)
Percentage of responses within 4 minutes: 84% (1997), 86% (1998)

Danville, KY
2-minute 47-second average[a] (1997)

Chapel Hill, NC
3-minute 18-second average; 95% within 5 minutes (1999)

Gainesville, GA
3-minute 21-second average[a] (1999)

Indianapolis, IN
3-minute 26-second average (1998)

Smyrna, GA
3-minute 30-second average (1999)

Blacksburg, VA
3-minute 46-second average, in town; 6-minute 42-second average, outlying county; 4-minute
10-second average, Virginia Tech campus[a] (1997)

Fayetteville, AR
3-minute 46-second average[a] (1998)

San Antonio, TX
3-minute 52-second average (1997)
Percentage within 4 minutes: 66.5% (1997)
Percentage within 5 minutes: 79.5% (1997)

New York, NY
4-minute 19-second average (Manhattan); 4-minute 21-second average (Bronx); 4-minute
    47-second average (Staten Island); 3-minute 57-second average (Brooklyn); 4-minute 53-second
    average (Queens)[a] (1998)

Corpus Christi, TX
Target: At least 90% within 4 minutes (1999)

Palo Alto, CA
Among calls within urban response zone, percentage from front ramp of station to scene within
    4 minutes: 100% (1997)

Denver, CO
95% within 4 minutes;[a] 100% within 6 minutes (1998)

Cambridge, MA
Target: 4 minutes or less for the first arriving unit at least 95% of the time; 8 minutes or less for
    the entire fire alarm assignment at least 90% of the time
Actual: 95% within 4 minutes (1999)

High Point, NC
4-minute average (1998)

Philadelphia, PA
4-minute average[a] (1998)

Greeley, CO
Percentage of responses within 4 minutes:[a] 71%
Percentage within 5 minutes: 83% (1997)

Portland, OR
43% within 4 minutes (turnout and travel time); 77% within 5 minutes 20 seconds; 4-minute
    24-second average (1998)

Germantown, TN
4-minute 2-second average[a] (1995)

Wichita, KS
4-minute 6-second average (1999, estimated)

Norfolk, NE
4-minute 8-second average (1997)

Tempe, AZ
4-minute 11-second average; 33% within 3.5 minutes[a] (1997)

Hurst, TX
4-minute 12-second average (1999)

Richmond, VA
4-minute 12-second average (1999)

(continued)

Table 11.12   Response Times to Fire Emergencies: Selected Cities (continued)

Chandler, AZ
4-minute 13-second average[a] (1998)

Oklahoma City, OK
4-minute 15-second average (1999)

Ames, IA
4-minute 16-second average[a] (1997)

College Station, TX
4-minute 16-second average (1999)

Glendale, AZ
4-minute 16-second average[a] (1997)

Greensboro, NC
4-minute 18-second average[a] (1997)

Raleigh, NC
4-minute 24-second average[a] (1998)

Irving, TX
4-minute 26-second average (1998)

Sacramento, CA
4-minute 30-second average (1998)

Norman, OK
4-minute 32-second average (urban area); 81% within 6 minutes[a] (1998)

Phoenix, AZ
4-minute 40-second average (1999)

Duncanville, TX
4-minute 41-second average[a] (1998)

Alexandria, VA
4-minute 47-second average[a] (1998)

Nashville–Davidson County, TN
4.9-minute average inside urban service district; 10.7-minute average outside urban service district[a] (1998)

Tucson, AZ
Percentage of responses within 5 minutes of dispatch: 100% (1997)

Durham, NC
95% within 5 minutes[a] (1998)

Oakland, CA
95% of truck and engine companies responding within 5 minutes[a] (1996)

Charlottesville, VA
94% within 5 minutes[a] (1998)

Cary, NC
Target: At least 90% within 5 minutes[a]
Actual: 83% (1999)

Denton, TX
80% within 5 minutes[a] (1997)

Athens–Clarke County, GA
72% within 5 minutes[a] (1998)

Reno, NV
Target: At least 72% within 5 minutes (1999)

Grand Prairie, TX
5-minute 5-second average (1999, estimated)

Brentwood, TN
5-minute 12-second average[a] (1995)

Eugene, OR
Target: 80% within 4 minutes
Actual: 64% within 4 minutes (1995); 5-minute 17-second average (1997)

San Clemente, CA
5-minute 17-second average;[a] 84% within 7 minutes (1998)

Houston, TX
5-minute 35-second average (1999)

Orlando, FL
93% within 6 minutes from dispatch (1996)

San Diego, CA
6-minute average[a] (1998)

Tallahassee, FL
8-minute average (1999, estimated)

---

a. Response time is undefined in the reporting document for this city and is presumed conservatively to indicate the time from dispatch until arrival of the first reporting unit.

staffing levels found in many—and perhaps even most—cities. One such prescription, appearing in an NFPA publication (Granito, 1997), is provided in Table 11.15.

Firefighters and equipment are important suppression resources, but in and of themselves, tallies showing the number of firefighters in a community and the number of fire trucks offer a poor gauge of fire service performance. Suppression effectiveness is a matter of how those firefighters and their equipment perform in the heat of an emergency. Some cities use better and more direct gauges, reporting fire loss statistics (Table 11.16) or—perhaps better

Table 11.13   Other "Quick Action" Characteristics: Selected Cities

TURNOUT TIME

Chandler, AZ
Target: Achieve a 45-second turnout time (from notification to en route) for at least 75% of all
emergency calls
Actual: 55-second average (1995), 53-second average (1998)

San Jose, CA
72% within 90 seconds (1995)

SIGNAL PREEMPTION INSTALLATIONS

Eugene, OR
Target: 85% of signalized intersections equipped with signal preemption
Actual: 25% (1995)

QUICK ACTION ON SCENE

Orlando, FL
Water on fire within 2 minutes of arrival: 69% (1994), 43% (1996)
Search 1- and 2-family residences within 4 minutes of arrival: 72% (1994), 67% (1996)

Table 11.14   Hazardous Materials Incident Preparation and Response: Selected
Cities

PLANNING AND PREVENTION

Palo Alto, CA
Target 1: Review 90% of plans for hazardous materials projects and notify applicant of deficiencies
within 10 working days
Actual: 98% (1997)

Target 2: Review plans and grant approval or denial within 15 working days of receiving complete
document at least 90% of time
Actual: 100% (1997)

Target 3: Inspect at least 90% of facilities with permits designating annual inspection
Actual: 86% (1997)

RESPONSE TIMES OF HAZARDOUS MATERIALS UNITS

Winston-Salem, NC
12.35-minute average (1998)

San Jose, CA
Responses within 15 minutes: 100% (1995)

Table 11.15   Recommended Equipment and Fire Force for Interior Attack

High-Hazard Occupancies (Schools, hospitals, nursing homes, explosives plants, refineries, high-rise buildings, and other high life hazard or large fire potential occupancies):

At least 4 pumpers, 2 ladder trucks (or combination apparatus with equivalent capabilities), 2 chief officers, and other specialized apparatus as may be needed to cope with the combustible involved; not fewer than 24 firefighters and 2 chief officers.

Medium-Hazard Occupancies (Apartments, offices, mercantile and industrial occupancies not normally requiring extensive rescue or fire-fighting forces):

At least 3 pumpers, 1 ladder truck (or combination apparatus with equivalent capabilities), 1 chief officer, and other specialized apparatus as may be needed or available; not fewer than 16 firefighters and 1 chief officer.

Low-Hazard Occupancies (One-, two-, or three-family dwellings and scattered small businesses and industrial occupancies):

At least 2 pumpers, 1 ladder truck (or combination apparatus with equivalent capabilities), 1 chief officer, and other specialized apparatus as may be needed or available; not fewer than 12 firefighters and 1 chief officer.

Rural Operations (Scattered dwellings, small businesses, and farm buildings):

At least 1 pumper with a large water tank (500 gal [1.9m^3] or more), one mobile water supply apparatus (1000 gal [3.78 m^3] or larger), and such other specialized apparatus as may be necessary to perform effective initial fire-fighting operations; at least 12 firefighters and 1 chief officer.

Additional Alarms:

At least the equivalent of that required for Rural Operations for second alarms; equipment as may be needed according to the type of emergency and capabilities of the fire department. This may involve the immediate use of mutual aid companies until local forces can be supplemented with additional off-duty personnel. In some communities, single units are "special called" when needed, without always resorting to a multiple alarm. Additional units also may be needed to fill at least some empty fire stations.

SOURCE: John Granito, "Evaluation and Planning of Public Fire Protection." In Arthur E. Cote and Jim L. Linville (Eds.), *Fire Protection Handbook,* 18th edition (Quincy, MA: National Fire Protection Association, 1997), p. 10.34. Reprinted with permission from *Fire Protection Handbook,* 18th edition, Copyright © 1997, National Fire Protection Association, Quincy, MA 02269.

still—measuring and reporting suppression effectiveness through indicators of fire spread and control time (Table 11.17). National statistics suggest that almost three fourths of all structural fires are contained through fire suppression efforts to the room of the fire's origin (Hall & Cote, 1997, p. 1.13). Many of the examined documents that report fire spread indicated even higher percentages contained at that level. Statistics on control time exhibited considerable variation, with most reported averages falling in the 15- to 35-minute range.

Table 11.16 Fire Loss Ratios: Selected Cities

FIRE LOSS AS A PERCENTAGE OF TOTAL PROPERTY VALUE

Danville, KY
Fire loss as a percentage of total property value: 0.03% (1997)

Ames, IA
Fire loss as a percentage of total valuation of all property: 0.05% (1997)

Raleigh, NC
Fire loss as a percentage of total property value in the city: 0.32% (1998), 0.18% (1999)

Sterling Heights, MI
Fire loss as a percentage of total property value: 0.26% (1995)

Portland, OR
Fire loss as a percentage of value of property: 0.48% (1998)

LOSS PER $1,000 VALUATION

Plano, TX
Property loss per $1,000 valuation: $0.21 (1997)

PROPERTY SAVED

Lubbock, TX
Percentage of value at risk saved in structural fires: 99% (1995), 97.7% (1997)

Table 11.17

Arson Investigation

Regrettably, some fires are not the result of faulty equipment or accidents. Some are set deliberately. High-performance fire departments are proficient at spotting such instances and, through skillful investigations and aggressive follow-up actions, often report impressive clearance and conviction rates (Table 11.18).

Fire Safety Education, Inspections, and Code Enforcement

Departments that take seriously their role in the *prevention* of fires take full advantage of fire safety education opportunities, and establish aggressive inspection and fire code enforcement programs (Table 11.19). Hurst, Texas, for example, takes its fire safety program to every local elementary school. High rates of annual inspection are reported by Palo Alto, California, and Chesapeake, Virginia. And several cities report success rates of greater than 90% in correcting fire code violations.

Under Control: Performance Targets and Actual Experience of Selected Cities

CONTROLLING FIRE SPREAD

Palo Alto, CA Structural fires confined to the room or area of origin: 100% (1997)

Oak Ridge, TN Percentage of building fires confined to room of origin: 98% (1992), 90% (1997)

Percentage of fires confined to level of involvement found on arrival of first fire unit: 100% (1997)

Corvallis, OR Target: Control structural fires and fires involving real property within the level of involvement found on arrival

Actual: 95% (1991), 100% (1992)

Wichita, KS Percentage of fires confined to room of origin: 91% (1998)

Bellevue, WA Percentage of fires confined to room of origin: 89% (1998)

Winston-Salem, NC Percentage of fires contained within room of origin: 84% (1998)

Lubbock, TX Percentage of structural fires confined to room of origin: 79.4% (1995), 79.2% (1997)

Denver, CO Percentage of fires confined to room of origin: 76% (1998)

Chandler, AZ Percentage of structural fires contained to room of origin: 74% (1995), 40% (1998)

Wilmington, DE Target: At least 92% of fires confined to room of origin

Actual: 74% (1992)

Greenville, SC 98% of fires confined to one increment of flame spread (1990)

Orlando, FL Fire contained to less than 80% of building: 86% (1996)

Reno, NV Targets: At least 77% of structural fires confined to room of origin; 98% confined to structure; 2% spread beyond structure (1999)

National Statistics 71.4% of structural fires confined to room of origin; 4.6% confined to floor but not room of origin; 24% spread beyond floor of origin (1989–1993)[a]

CONTROL TIME

Oak Ridge, TN Control time following arrival: 2.4 minute average (1992); 36-second average (1997)

Fayetteville, AR Structural fire control time average: 10.43 minutes (1998)

Alexandria, VA Average time spent at scene of incident: 16.7 minutes (1992), 16.5 minutes (1993)

San Jose, CA Target: Control at least 75% of emergency fire incidents within 15 minutes of arrival; at least 95% within 1 hour

Actual: 79% controlled within 15 minutes; 97% within 1 hour (1989)

Shreveport, LA Average time to control (all incidents): 33.31 minutes (1994)

a. National statistics on fire spread reported in John R. Hall, Jr., and Arthur E. Cote, "America's Fire Problem and Fire Protection." In Arthur E. Cote and Jim L. Linville (Eds.), *Fire Protection Handbook,* 18th edition (Quincy, MA: National Fire Protection Association, 1997), p. 1.13, citing National Fire Incident Reporting System (NFIRS) and NFPA National Fire Experience Survey.

Table 11.18   Arson Clearance and Conviction Rates: Selected Cities

Norfolk, VA
Percentage of arson cases cleared: 81% (1997)

Oklahoma City, OK
Arson clearance rate: 60% (1998), 37% (1999)

San Jose, CA
Percentage of arson or malicious mischief fires in which suspects were arrested: 49% (1995)
Percentage of court cases resulting in conviction: 85% (1995)

Charlotte, NC
Percentage of incendiary fire cases cleared: 45% (1999)

Ann Arbor, MI
Target: Close 40% of arson cases by arrest of perpetrator
Actual: 33% (1998)

Shreveport, LA
Percentage of suspected arson cases solved: 30% (1994)

Portland, OR
Arrests and exceptional clearances as a percentage of all arson fires: 25% (1998)

Oakland, CA
Percentage of cases cleared by arrest or other means: 22% (1996)

Winston-Salem, NC
Apprehensions/arrests as percentage of incendiary/suspicious fires: 18.3% (1997)
Conviction rate: 35.3% (1997), 71% (1998)

Boston, MA
Conviction rate for fires resulting from arson: 11% (1996)

Do inspections prevent fires? Officials in Calgary, Alberta, have ample reason to think that they do. Almost 29% of commercial properties in that city were inspected in 1996 (Table 11.20). Subsequent fire losses suffered by these properties were only 17% of the city's total commercial fire losses that year. The uninspected 71% incurred disproportionately greater losses.

Fire Incident Records

Perhaps every fire department keeps some type of records of fire incidents. In some cases, however, these records fall short of even the bare essentials needed for minimal analysis. Basic fire incidence data should include eight fundamental elements:

Table 11.19 Fire Safety Education, Inspections, and Fire Code Enforcement: Selected Cities

FIRE SAFETY EDUCATION

Hurst, TX
Percentage of elementary school classes taught a 30-minute fire safety lesson: 100% (1997)

Raleigh, NC
Percentage of citizens reached by public fire education: 25% (1998)

Cary, NC
Percentage of population contacted by fire/life safety education effort: 23% (1999)

INSPECTIONS

Palo Alto, CA
Target: Inspect all State Fire Marshal Office–regulated occupancies annually
Target: Inspect 90% of applicable ordinary hazard occupancies annually
Actual: 86% (1997)
Target: Inspect all light hazard occupancies every 36 months
Actual: 99% (1997)

Chesapeake, VA
Percentage of businesses inspected: 80% (1997)

Tucson, AZ
Percentage of commercial buildings inspected: 45% (1997)

Hurst, TX
Percentage of initial apartment inspections completed within 30 days: 90% (1997)

FIRE CODE ENFORCEMENT

Charlotte, NC
Percentage of code violations corrected: 99.5% (1999)

Germantown, TN
Percentage of violations corrected: 99.15% (1995)

Greensboro, NC
Percentage of fire code violations corrected: 98% (1997)

Chesapeake, VA
Percentage of fire code violations corrected: 96% (of 3,782) (1997)

Winston-Salem, NC
Percentage of fire code violations corrected: 92% (1997)

Calgary, Alberta
Percentage of hazards corrected in year found: 88% (1996)

Charlottesville, VA
Compliance rate in correcting housing violations: 85% (1998)

Oak Ridge, TN
Percentage of fire code violations corrected within 180 days: 74% (1997)

Lubbock, TX
Percentage of critical life safety hazards corrected: 65% (1997)

Table 11.20 Fire Inspection Effectiveness in Calgary, 1996

	Commercial Properties	Percentage of Commercial Fire Loss—Total
Inspected	28.61%	17.17%
Uninspected	71.39%	82.83%

SOURCE: City of Calgary (Alberta), *1998 Operating Budget: Volume V—Performance Measures.*

1. Time of dispatch
2. Address
3. Identifications and times of responding units
4. Arrival times at the scene
5. Departure times from the scene
6. Radio traffic tape logs
7. Telephone tape logs
8. Times of significant events at the scene (Stauffer, 1997, p. 10.198)

Equipment Maintenance

Key focal points of maintenance efforts in most fire departments are fire vehicles, pumps, and hydrants. Although principal maintenance of fire vehicles often is performed by mechanics outside the fire department, fire officials can monitor routine preventive maintenance and the increasing age of fire vehicles with an eye toward eventual replacement. If properly maintained, most fire trucks can sustain a lengthy life of service. Generally, however, fire officials prefer their inventory to include at least a few fairly new units of utmost reliability. Although the fire department in Portland, Oregon, projects the useful life of an engine to be 20 years, followed by 5 years in a reserve capacity, the average age of the fleet should be considerably less than 20 years in order to avoid high maintenance costs and problems of low reliability. The average age of Portland's fire engines climbed to more than 10 years at various times in the early and mid-1980s, before declining to a more satisfactory average age of about 6 years in 1991. In New York City, the average age of pumpers was 92 months (7.7 years) in 1997, and the average age of ladder trucks was also 92 months. In Boston, where the city attempts to hold its average age of fire equipment below 7 years, the average age of frontline apparatus was 5 years in 1996.

Periodic testing of pumps to ensure their readiness is an important precaution. Such testing should include all the prescribed steps in Table 11.21. The standards noted within those steps may be considered benchmarks.

Table 11.21   Service Tests of Fire Department Pumps

NFPA 1911, *Standard for Service Tests of Pumps on Fire Department Apparatus,* outlines the procedures for the service test. The test should cover the following items:

1. A test pumping from draft for 20 minutes at 100% of rated capacity, at 150 psi (1039 kPa) net pump pressure; 10 minutes at 70% of rated capacity at 200 psi (1379 kPa) net pump pressure; and 10 minutes at 50% of rated capacity at 250 psi (1724 kPa) net pump pressure.
2. An engine speed test to determine if the engine is capable of reaching its no-load governed speed.
3. A vacuum test to ensure that the pump and the attached piping are still tight.
4. A pressure-control test to ensure that the pressure-control device can control the discharge within the specified limits.
5. A check of the accuracy of the gauges and flow meters.

The purpose of the test is to ensure that the pump is generally in good condition, that the pump casing and various fittings are tight, and that the transfer valve is operating properly, if the pump is a series parallel type. If rated capacity cannot be obtained at 150 psi (1034 kPa) and the pump is not cavitating, or it appears that the pump, engine, pump accessories, or other parts of the power train and pumping equipment are not in good condition, the apparatus manufacturer, an authorized representative, or a competent mechanic should be contacted for advice so the condition can be corrected.

SOURCE: Excerpted from Ralph Craven, "Fire Department Apparatus and Equipment." In Arthur E. Cote and Jim L. Linville (Eds.), *Fire Protection Handbook,* 18th edition (Quincy, MA: National Fire Protection Association, 1997), p. 10.209. Reprinted with permission from NFPA 1911, *Service Tests of Pumps on Fire Department Apparatus.* National Fire Protection Association, Quincy, MA 02269. This reprinted material is not the complete and official position of the National Fire Protection Association on the referenced subject, which is represented only by the standard in its entirety. NOTE: psi = pounds per square inch; kPa = kilopascals.

The middle of an emergency is no time to discover a faulty fire hydrant. Periodic inspections at reasonable intervals can minimize that occurrence, and such inspection need not consume a great amount of time. Fort Collins, Colorado, and Orlando, Florida, attempt to test and service hydrants at least once every 4 years; Oak Ridge, Tennessee, and Palo Alto, California, every 3 years. Durham, North Carolina, tries to check all hydrants annually for operational readiness and appearance. Raleigh, North Carolina, reports that an average hydrant inspection requires 0.14 hours (8.4 minutes). Properly marked hydrants reveal their water flow by the color of their bonnet and nozzle cap (Table 11.22).

Odds and Ends

Little-known facts, figures, and miscellaneous tidbits of information sometimes come in handy. Although most of the performance dimensions mentioned in Table 11.23 were reported by too few cities to depict any trends or norms, they may nevertheless prove helpful to municipal officials.

Table 11.22   Uniform Marking of Fire Hydrants

Hydrant Classification	Hydrant Flow	Color of Bonnets and Nozzle Caps
AA	1500 gpm or greater	Light blue
A	1000 to 1499 gpm	Green
B	500 to 999 gpm	Orange
C	Less than 500 gpm	Red

SOURCE: Gerald R. Schultz, "Water Distribution Systems." In Arthur E. Cote and Jim L. Linville (Eds.), *Fire Protection Handbook,* 18th edition (Quincy, MA: National Fire Protection Association, 1997), p. 6.42. Reprinted with permission from *Fire Protection Handbook,* 18th edition, Copyright © 1997, National Fire Protection Association, Quincy, MA 02269.

One item that *can* be reported as a norm is the national average of residences with smoke detectors, which climbed from 81% in 1988 to 94% by 1997 (Ahrens, 2000, p. i; Hall & Cote, 1991, p. 1.16). The importance of smoke detectors is evident in fire data showing that the small portion of U.S. homes without smoke alarms in 1997 accounted for 38% of all reported home fires and 51% of all home fire deaths (Ahrens, 2000, p. i). Communities that have made the installation of residential smoke detectors a major thrust of their fire safety programs will find the national average to be a useful benchmark. Individual cities' performance targets and experience in other aspects of fire service are similarly instructive.

Production Ratios

Balancing the goals of fire service effectiveness and efficiency can be a complicated endeavor. Good fire safety results are important, but good results at an exorbitant cost bring little satisfaction. Good services, good results, and a reasonable cost are the blend that the most productive departments strive to achieve.

Because so much of the typical fire department's budget is committed to wages, finding and maintaining an appropriate staffing level is a key to achieving optimum productivity. Understaffing is likely to stretch personnel resources too thin and lead to poor performance; overstaffing is inefficient and wasteful. Although some fire service proponents would urge that staffing be based simply on ratios of firefighters per 1,000 population, a more appropriate standard would relate the staffing level to the work performed.

Several cities examined for this volume report production ratios for various elements of their fire services (Table 11.24). Comparative statistics on inspections per inspector, time per inspection, investigations per investigator, incidents per employee, and responses per fire company offer a more relevant

Table 11.23   Fire Service Odds and Ends: Selected Cities

RESIDENCES WITH SMOKE DETECTORS

Oak Ridge, TN
95% (1997)

Philadelphia, PA
95% (1998 survey)

National Average
94% (1997)[a]

BUILT-IN PROTECTION IN COMMERCIAL STRUCTURES

Oak Ridge, TN
95% of all newly constructed or substantially renovated commercial structures are equipped with
automatic fire suppression or alarm systems (1997)

PROMPT RESPONSE TO FIRE SAFETY COMPLAINTS

Hurst, TX
Percentage of code violations addressed within 24 hours: 97% (1997)

New York, NY
Percentage of hazard complaints resolved within 1 day of receipt: 88% (1996), 87% (1997)

Norfolk, VA
Percentage of reported fire code violations inspected within 48 hours of identification: 100% (1997)

Tucson, AZ
Percentage of code complaints addressed within 5 days: 100% (1997)

PROMPT INVESTIGATIONS

College Station, TX
Percentage of investigations of all major fire, hazardous materials, and code compliance cases within
24 hours: 100% (1998)

EXTRICATION

Corvallis, OR
Target: Rescue and extricate persons trapped by motor vehicle accidents, industrial mishaps, or
other emergencies within 15 minutes of arrival
Actual: 100% (1991), 100% (1992)

ADEQUATE WATER PRESSURE

Bellevue, WA
Percentage of customers with at least 1,000 gpm fire flow protection: 92.0% (1997)

HYDRANT TESTING

Palo Alto, CA
Percentage of hydrants tested/serviced per year: 33.3% (1997)

(continued)

Table 11.23   Fire Service Odds and Ends: Selected Cities (continued)

FIRE HYDRANT CONDITION/AVAILABILITY

Philadelphia, PA
Available: 97.3% (1996), 98.6% (1998)

New York, NY
Percentage of hydrants broken/inoperative: 0.54% (1997)

FIREFIGHTER FITNESS

San Luis Obispo, CA
Percentage of employees in "fit" or "athlete" fitness range: 85% (1994)

EMERGENCY MEDICAL CERTIFICATION

San Jose, CA
Percentage of active, sworn personnel with EMT certification: 100% (1995); with EMT-D certification: 95% (1995); with CPR certification: 100% (1995)

EQUIPMENT CAPABILITY

Lubbock, TX
Target: At least 95% of frontline snorkels/aerials and pumpers passing scheduled annual NFPA test
Actual: 80% (1994)

a. National statistics on smoke alarms reported in Marty Ahrens, *U.S. Experience With Smoke Alarms and Other Fire Alarms* (Quincy, MA: National Fire Protection Association, January 2000).

basis for judging the adequacy of fire department staffing in an individual community than does population alone.

Harried municipal executives, besieged by local fire personnel claiming to be overworked, might be especially interested in a statistic that may set the standard for workload. Incredible but true, engine companies in New York City responded to an average of 2,849 calls in 1998. Now, that is busy![2]

Notes

1. Although the call-to-dispatch and dispatch-to-arrival components both should be included to get a complete picture of response time (both are clearly part of response time from the perspective of a citizen awaiting arrival of an emergency unit), some departments report only the second component—the time between dispatch of an emergency unit and its arrival at the scene. If a municipal official suspects that the reported response time of a comparison city is unrealistically quick, further investigation might reveal a different definition of "response time" in that community.

Table 11.24   Fire Service Production Ratios

INSPECTIONS PER INSPECTOR

New York, NY
Inspections per person day (Fire Prevention staff): 5.7 (1996), 5.6 (1997), 5.6 (1998)

Greensboro, NC
Average number of fire code inspections per inspector: 964 (1997)

Bellevue, WA
Average number of fire code inspections per day per inspector: 3.0 (1997)

Victoria, TX
Average number of fire code inspections per inspector: 542 (1996)

College Station, TX
Number of inspections per month per inspector: 38 (1998)

Tempe, AZ
Fire code inspections per inspector: 385 (1997)

TIME PER INSPECTION

Winston-Salem, NC
Person hours per nonresidential building inspection (fire code): 0.75 (1998)

College Station, TX
Time per inspection: 1 hour (1998)

FIRE/ARSON INVESTIGATIONS PER INVESTIGATOR

San Antonio, TX
Arson cases per investigator: 107 (1995)

Shreveport, LA
Fires investigated per investigator: 106 (1994)

INCIDENTS PER ON-DUTY EMERGENCY STAFF

Sacramento, CA
407 (1998)[a]

Portland, OR
353 (1998)

Seattle, WA
347 (1997)[a]

Cincinnati, OH
341 (1997)[a]

Charlotte, NC
308 (1998)[a]

Denver, CO
307 (1997)[a]

Kansas City, MO
254 (1998)[a]

(continued)

Table 11.24   Fire Service Production Ratios (continued)

RESPONSES PER COMPANY

New York, NY
  Average annual responses per engine company: 2,849 (1998)
  Average annual responses per ladder company: 2,577 (1998)
  Average annual responses per unit overall: 2,739 (1998)

Tempe, AZ
  Average responses per company: 2,074 (1997)

San Antonio, TX
  Average annual responses per pumper company: 935 (1992, estimated)

a. Statistics on "incidents per on-duty staff" for designated cities were reported in City of Portland (OR) Office of the City Auditor, *City of Portland Service Efforts and Accomplishments: 1997–98,* (December 1998), pp. C.2–C.3.

2. As impressive as the New York City figure is, some other cities may be able to challenge the heavy workload of engine companies in that city. Phoenix, Arizona, for example, reported that during one year, 56 of its units responded to more than 2,000 calls, 28 responded to more than 3,000, and 5 units responded to more than 4,000 calls (City of Phoenix [AZ], Auditor Department, 1991, p. 10).

12

Fleet Maintenance

Fleet maintenance, including repair and preventive maintenance of city vehicles and other equipment, occasionally is a service contracted with a private shop, but often is the responsibility of a central, municipally operated garage. Because the availability of reliable equipment is crucial to many municipal departments, the demand for quality service is great. With municipal resources stretched and with the availability of private sector alternatives, the pressure for efficient service is also great.

Well-managed fleet maintenance operations track their own performance carefully with at least three purposes in mind:

1. To identify aspects of performance that are slipping toward unacceptable levels before they become serious problems
2. To see if operating procedures or any newly implemented corrective measures are achieving the desired results
3. To document the quality and efficiency of fleet maintenance services, thereby reducing their vulnerability to potential critics and competitors

Typically, performance measures focus on equipment availability, equipment reliability, promptness and quality of repairs, mechanic efficiency, and the success of maintenance efforts to extend the life of the city's fleet.

Benchmarks

Benchmarks that address several dimensions of fleet maintenance quality and efficiency have been offered by professional associations, consultants, and individual cities. A 1993 study of public service fleets, sponsored by the National Association of Fleet Administrators (NAFA), established norms and "upper performance thresholds" for success in minimizing downtime and

175

Table 12.1   Selected Benchmarks for Fleet Maintenance: NAFA Study

Performance Element	Number of Reporting Jurisdictions	Median	Upper Performance Threshold (80th Percentile)
Downtime Percentage			
Police cars	22	5%	2%
Other sedans	46	2%	1%
Light duty trucks	42	2%	1%
Miles Between Breakdowns			
Police cars	25	13,000	36,000
Other sedans	49	11,000	19,161
Light duty trucks	44	7,752	18,321
Vehicle Miles per Mechanic Hour			
Police cars	25	898	1,818
Other sedans	50	747	1,286
Light duty trucks	52	499	950
Mechanic Utilization Rate[a]	48	66%	75%

SOURCE: Excerpted from NAFA Foundation and National Association of Fleet Administrators, *Benchmarking for Quality in Public Service Fleets* (Iselin, NJ: 1993). Copyright © 1993, National Association of Fleet Administrators, Inc. Reprinted with permission.

a. Mechanic Utilization Rate is the percentage of mechanic hours available during a year that are spent doing maintenance and repair work (i.e., "wrench turning") as opposed to other work (e.g., cleaning, running parts, training, etc.). A relatively low rate could reflect overstaffing or inefficient uses of mechanic time.

maximizing miles between breakdowns, vehicle miles per mechanic hour, and mechanic utilization rates (Table 12.1). Winston-Salem, North Carolina, assembled a set of benchmarks from other sources addressing different service dimensions in 1998 (Table 12.2).

Equipment Availability

The NAFA study suggests that top performing fleet maintenance operations will have downtime percentages of 2% or less—or, expressed more positively, equipment availability rates of 98% or higher. Most of the individual cities whose reporting documents were examined for this volume expected more than 90% of their vehicles and equipment to be available for use at any given time (Table 12.3). Only a handful reached the level prescribed for top performers by NAFA. A more typical level of achievement among these respected municipalities was 95% availability, a mark met, for instance, by Fort Worth, Texas, and San Diego, California (Figure 12.1).

Some cities differentiate between downtime for maintenance and downtime attributable to accidents or other causes. Orlando, Florida, for example,

Table 12.2   Winston-Salem's Review of Fleet Maintenance Performance

	Winston-Salem	Average Among Selected Cities[a]	Industry Norm[b]
Number of light equipment vehicles per mechanic	78	—	45
Number of heavy equipment vehicles per mechanic	48	—	25
Total vehicles per mechanic	68	62	—
Employees per supervisor	5.2	6.5	—
Percentage of mechanics' time spent servicing equipment	64%	76%	75%
Percentage of repairs completed in less than 24 hours	55%	81%	70%
Percentage of parts filled from mechanic inventory	68%	71%	70%
Percentage of operators rating service as satisfactory or better	91%	96%	95%

SOURCE: City of Winston-Salem (NC), *Fleet Services Project: Phase I Report* (Winston-Salem, NC: Office of Organizational Effectiveness, 11 May 1998).

a. Based on survey responses from the cities of Asheville, NC; Cary, NC; Des Moines, IA; Fayetteville, NC; Glendale, AZ; High Point, NC; Indianapolis, IN; Raleigh, NC; Richmond, VA; and Rocky Mount, NC.

b. NAFA and David M. Griffith and Associates, *Benchmarking for Quality in Public Service Fleets* (Iselin, NJ: 1993).

reported downtime of 7.5% for police vehicles during a recent year—4.6% for maintenance and 2.9% because of accidents or for other reasons.

Equipment Reliability

The NAFA study suggests that the effective maintenance practices of top performing operations will yield 36,000 miles between police car breakdowns, 19,000 miles for other sedans, and 18,000 miles for light-duty trucks. Individual cities sometimes report their performance statistics regarding mechanic performance and equipment reliability in a different manner. Cambridge, Massachusetts, for example, reported 3.9 breakdowns per public works vehicle one recent year. Scottsdale, Arizona, recently reported its emergency field repairs at 3.2%.

Promptness and Quality of Repairs

Many cities monitor the quality of vehicle and equipment maintenance services by tracking the promptness of repairs and the extent to which initial repairs are done properly and need no rework (Table 12.4). Some cities target 24-hour turnaround for vehicle repair and report high percentages of achievement. Among the examined cities, the prompt-repair leader—at least

Table 12.3   Equipment Availability Rates: Selected Cities

OVERALL AVAILABILITY RATE

Denver, CO
99% (1998)

San Clemente, CA
99% (1998)

Eugene, OR
98.3% (1997)

Scottsdale, AZ
98.3% (1999)

Bellevue, WA
98% (1998)

Palo Alto, CA
96.5% (1997)

College Station, TX
96.3% (1998)

Rockville, MD
96% (1997)

San Jose, CA
96% (1995)

Santa Ana, CA
96% (1995)

Savannah, GA
96.0% (1997)

Fort Worth, TX
Target: 95%
Actual: 95% (1997)

San Diego, CA
95% (1999)

Portland, OR
94.9% (1998)

Tucson, AZ
93% (1997)

Oklahoma City, OK
Target: 95%
Actual: 92% (1999)

Philadelphia, PA
90% (1998)

Raleigh, NC
89% (1997), 90% (1998)

Oakland, CA
Target: Fleet availability of at least 88%
Actual: 88% (1996)

POLICE VEHICLES

Boston, MA
Percentage of marked vehicles available each
day: 92% (1992), 96% (1996)

San Jose, CA
Police fleet availability: 94% (1995)

Orlando, FL
93.5% (1994), 92.5% (1996)

New York, NY
91% (1998)

Philadelphia, PA
89% (1998)

FIRE VEHICLES

High Point, NC
98% (1998)

San Diego, CA
First-line apparatus: 98% (1998)

Winston-Salem, NC
98% (1998)

Raleigh, NC
Availability of first-line apparatus: 96.0% (1997),
96.5 % (1998)

Boston, MA
Percentage of fire vehicle uptime (average):
96% (1996)

Orlando, FL
92.5% (1994), 93.8% (1996)

Denver, CO
92% (1998)

Tucson, AZ
Frontline apparatus, 90%; support equipment,
90% (1997)

New York, NY
Engines: 89%; ladders: 94%; support vehicles:
78% (1998)

PUBLIC WORKS VEHICLES

Boston, MA
Target: 95%
Public works fleet (actual): 91% (1996)

REFUSE COLLECTION VEHICLES

Denver, CO
Trash trucks: 95.1% (1998)

Orlando, FL
87.5% (1996)

Philadelphia, PA
77% (1998)

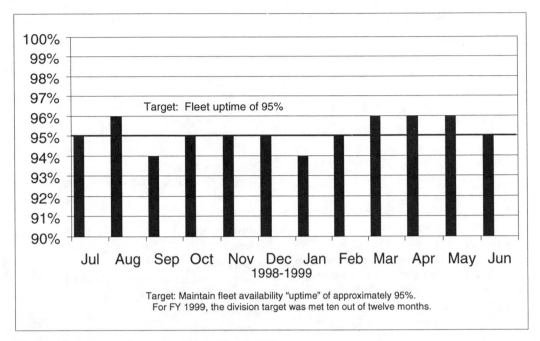

Figure 12.1. Uptime for the San Diego Municipal Fleet
SOURCE: City of San Diego (CA), *Fiscal Year 1999 Semi-Annual Performance Report,* p. 13.

for minor problems—is San Clemente, California, which reports 100% of "quick fix" repairs completed within 1 hour. Perhaps coincidentally, San Clemente secures its fleet maintenance services by contract. Several cities report having less than 3% of repairs returned for subsequent rework.

The promptness of repairs, as well as preventive maintenance activities, may also be gauged by the average downtime imposed on various types of vehicles. In Oklahoma City, for instance, repair downtime for fire apparatus is targeted at 2 days or less, and downtime for preventive maintenance activities at 8 hours or less (Table 12.5).

Mechanic Efficiency

A sophisticated approach to assessing the performance of the municipal garage draws on the engineered standards developed for private shops. Commercially available rate manuals identify performance standards for mechanics, declaring that a competent mechanic should be able to perform a given repair job on a specified type of vehicle in a particular amount of time.[1] A city garage that can usually meet or beat these or other engineered standards may justifiably declare itself to be an efficient shop.

Several of the performance documents examined for this volume declared rates of mechanic efficiency, presumably tied to engineered standards. Tucson,

Table 12.4   Promptness and Quality of Vehicle and Equipment Maintenance Service:
Selected Cities

Prompt Repairs	*Returns for Rework*
San Clemente, CA Percentage of after-hour repair responses within 20 minutes: 100% (1998) Percentage of "quick fix" repairs within 60 minutes: 100% (1998)	**Fort Collins, CO** Percentage returned for rework: 0.13% (1996)
San Diego, CA Percentage of unscheduled repairs completed within 1 day: 74% (1998)	**Oklahoma City, OK** Target: Fewer than 2% of repairs returned for rework Actual: 0.2% (1992)
Knoxville, TN Percentage of equipment repairs completed within 24 hours: 90% —heavy equipment; 98%—light equipment (1991)	**Durham, NC** Actual: 0.35% (1998) **Hurst, TX** Target: Repeat repair rate of 2% or less Actual: 3% (1990), 0.4% (1992)
Long Beach, CA Target: 24-hour turnaround for at least 80% of vehicles in for repair Actual: 80% (1991), 80% (1992)	**College Station, TX** 0.85% (1998)
Raleigh, NC Repair orders completed within 24 hours: 48% (1998)	**Grand Prairie, TX** Actual: Less than 1% (1996) **Cincinnati, OH** Target: Less than 1%
Reno, NV Percentage of service and minor repair jobs completed within 2 workings days: 95% (1990)	**Denton, TX** Actual: 1% (1997) **Greensboro, NC** Target: 3% or less Actual: 1% (1997)
Flagstaff, AZ Percentage of service and minor repair jobs completed within 2 workdays: 70% (1997)	**Cary, NC** Target: Less than 2% Actual: 1.5% (1999)
Fort Worth, TX Percentage completed within 3 working days: 70% (1997)	**Flagstaff, AZ** Target: Less than 1% Actual: Less than 2% (1997)
Boston, MA Target: 95% of repairs completed within 4 days	**Fort Worth, TX** Target: Repeat repair rate of 4% or less Actual: 4% (1997)
	San Clemente, CA Actual: 5% (1998)
	Reno, NV Target: 5% (1999)
	Glendale, AZ Target: 5% or less Actual: 8% (1997)

Table 12.5   Downtime for Fire Vehicle Maintenance: Oklahoma City

	Downtime per Preventive Maintenance Activity		Downtime per Repair	
	Target	Actual FY98–99	Target	Actual FY98–99
Passenger vehicles and suburbans	2 hours	1.94 hours	2 days	1.98 days
Fire apparatus	8 hours	4.44 hours	2 days	2.69 days

SOURCE: City of Oklahoma City (OK), *Annual Budget: Fiscal Year 1999–2000*, p. 226.

Arizona, reported that 85% of the repairs performed by its garage during a recent year were completed within established industry labor standards. Flagstaff, Arizona, similarly reported mechanic productivity of 85%. Reno, Nevada, has established as targets performing oil changes, safety inspections, and break repairs at 98% of the standard book rate cost. Cary, North Carolina, has as its target completing at least 75% of all repairs within industry labor standards.

The quest to provide quality maintenance services at a reasonable cost hinges in part on achieving a proper staffing level. Quality services with a lean staff are possible only if the mechanics are well trained and competent. Decisions regarding appropriate staff size should follow consideration of the types of vehicles and equipment to be maintained, the nature of their use, and their age. A study conducted in Washington, D.C., in the early 1980s offered the following rules of thumb for staffing ratios: one mechanic for every 50 automobiles and light trucks; one mechanic for every 40 heavy trucks; one mechanic for every 30 pieces of heavy equipment; and one mechanic for every 120 mowers and rollers (City of Washington, D.C., n.d., p. 48).

Among cities examined for this volume, vehicle/equipment-to-mechanic ratios ranged from 29:1 to 90:1 (Table 12.6). The average given in the Winston-Salem survey cited previously was 62:1.

Extended Life of City's Fleet

Many city governments establish age and mileage guidelines for the replacement of various types of vehicles. Most, however, tend to exceed their own guidelines (Figure 12.2). Although extending the life of a passenger car, for example, from 80,000 to 90,000 miles may be wise from a capital expenditure standpoint and may be a credit to the excellent equipment maintenance that allows such an extension, local officials should recognize that the

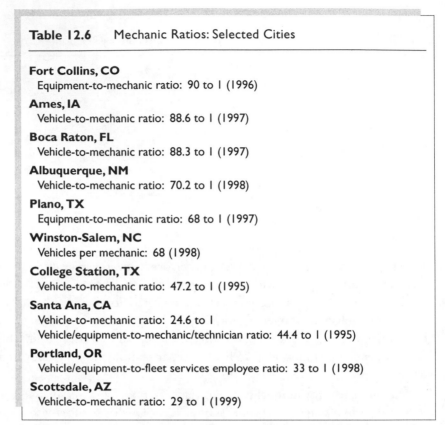

Table 12.6   Mechanic Ratios: Selected Cities

Fort Collins, CO
Equipment-to-mechanic ratio: 90 to 1 (1996)

Ames, IA
Vehicle-to-mechanic ratio: 88.6 to 1 (1997)

Boca Raton, FL
Vehicle-to-mechanic ratio: 88.3 to 1 (1997)

Albuquerque, NM
Vehicle-to-mechanic ratio: 70.2 to 1 (1998)

Plano, TX
Equipment-to-mechanic ratio: 68 to 1 (1997)

Winston-Salem, NC
Vehicles per mechanic: 68 (1998)

College Station, TX
Vehicle-to-mechanic ratio: 47.2 to 1 (1995)

Santa Ana, CA
Vehicle-to-mechanic ratio: 24.6 to 1
Vehicle/equipment-to-mechanic/technician ratio: 44.4 to 1 (1995)

Portland, OR
Vehicle/equipment-to-fleet services employee ratio: 33 to 1 (1998)

Scottsdale, AZ
Vehicle-to-mechanic ratio: 29 to 1 (1999)

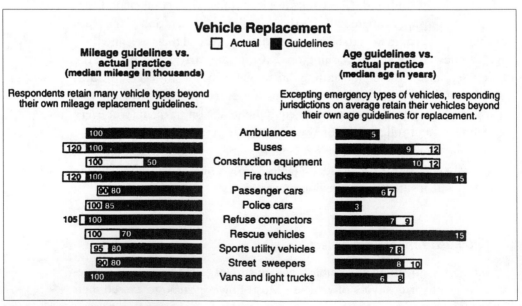

Figure 12.2. Vehicle Replacement Guidelines Versus Actual Practice
SOURCE: *City & State* and Stone Webster Management Consultants, Inc. Adapted from an original graphic by Jerry Parks appearing in Julie Bennett's "Budget Woes Seize Fleet Buying Cycles," *City & State, 10* (2 August 1993), p. 16. Reprinted by permission of *Governing,* successor to *City & State,* and Stone and Webster Management Consultants, Inc.

mechanic workforce needed for maintaining an aging fleet is likely to be greater than that needed for an inventory of newer vehicles and equipment.

Note

1. Two of the most popular rate manuals are *Chilton's Labor Guide and Parts Manual* (Chilton Book Company) and the *Mechanical Labor Estimating Guide,* also known as the *Mitchell Manual* (Mitchell International). Each is revised annually.

13

Gas and Electric Services

In some communities natural gas and electricity are utilities operated by the city government. In such cases, the city typically functions as a distributor of these products rather than as a direct producer. Relatively few cities generate electricity. Most municipalities that provide electrical services purchase electricity and resell it to their customers. Natural gas services typically follow a similar pattern.

Federal and state regulations pertaining to natural gas and electricity form a set of basic standards governing operations and product quality. In addition, the operating norms of the utility industry, as well as the performance targets and actual experience of utility counterparts, can serve as the basis for benchmarks.

Electricity and Natural Gas

Compared with the number of municipalities that operate water and sewer systems, relatively few own electric or gas utility systems. For those that do, relevant benchmarks in the form of industry norms are fairly abundant.

Financial and operating statistics reveal, for example, that the average publicly owned electric utility has a debt equal to approximately one fourth of its total assets, that it nets six cents of every revenue dollar, and that it has between 200 and 300 retail customers for every non-power-generation employee and about 5,000 retail customers per meter reader (Table 13.1). Median values for these and other operating statistics differ by utility size and region (Table 13.2).

A survey of more than 100 electric and gas utilities in the early 1990s revealed performance norms for a variety of important indicators (Mercer Management Consulting, 1992; used by permission). Lead time from an initial customer request for new electric or gas service until completion of installation averaged 15 days for residential customers, 22 days for commercial

Table 13.1 Publicly Owned Electric Utilities Operating Statistics, by Customer Size (1998)

	Total	Median Values, by Customer Size Categories					
		2,000–5,000 Customers	5,000–10,000 Customers	10,000–20,000 Customers	20,000–50,000 Customers	50,000–100,000 Customers	More Than 100,000 Customers
Debt to total assets	0.248 (396)	0.245	0.226	0.236	0.233	0.337	0.597
Operating ratio	0.840 (399)	0.890	0.844	0.857	0.825	0.751	0.669
Net income per revenue dollar	$0.064 (399)	$0.056	$0.069	$0.060	$0.063	$0.107	$0.056
Uncollectible accounts per revenue dollar	$0.0020 (167)	*	$0.0013	$0.0019	$0.0024	$0.0033	$0.0026
Retail customers per non-power-generation employee	261 (181)	268	246	284	288	228	202
Retail customers per meter reader	4,875 (176)	3,650	4,541	4,736	5,461	7,483	7,479
OSHA incidence rate (per 100 employees)	3.0 (168)	0.0	0.8	4.0	3.0	3.7	3.3
Energy loss percentage	4.28% (395)	3.10%	3.95%	4.84%	4.23%	5.20%	3.52%

SOURCE: Adapted from American Public Power Association, *Selected Financial and Operating Ratios of Public Power Systems, 1998* (Washington, DC: Author, February 1996), pp. 2, 8. Used by permission.
NOTE: The number of responses for each category is shown in parentheses.
*Median not calculated fror less than 5 responses.

Table 13.2   Publicly Owned Electric Utilities Operating Statistics, by Region (1998)

	Median Values by Region				
	Northeast	Southeast	North Central/Plains	Southwest	West
Debt to total assets	0.186	0.260	0.237	0.267	0.301
Operating ratio	0.872	0.895	0.822	0.780	0.776
Net income per revenue dollar	$0.060	$0.043	$0.071	$0.116	$0.091
Uncollectible accounts per revenue dollar	$0.0027	$0.0024	$0.0009	$0.0031	$0.0022
Retail customers per non-power-generation employee	244	275	244	228	287
Retail customers per meter reader	4,667	4,749	5,065	3,750	5,913
OSHA incidence rate (per 100 employees)	3.6	2.8	2.4	3.4	3.5
Energy loss percentage	3.54%	4.47%	3.94%	4.92%	4.01%

SOURCE: Adapted from American Public Power Association, *Selected Financial and Operating Ratios of Public Power Systems, 1998* (Washington, DC: Author, February 2000), p. 9. Used by permission.
NOTE: This table is based on data from 399 utilities; however, not all of these utilities were able to provide data for some of the ratios listed.

customers, and 43 days for industrial customers. The average customer of the responding utilities experienced approximately 1.5 interruptions in electric service in 1990, with the average interruption lasting 92 minutes. Altogether, the average customer was without power for approximately 123 minutes during the year. When power outages occurred, the typical utility had a trouble-shooter or electric crew on the scene in approximately 35 minutes. Most utilities reported average response times between 10 minutes and 1 hour. Similarly, the average response time to gas leak calls was 30 minutes, with top-performing utilities responding in less than 20 minutes.[1]

Performance targets and results achieved in various aspects of electric service in Palo Alto, California, and Denton, Texas, are shown in Table 13.3. Gas service performance targets and results in Palo Alto are shown in Table 13.4. In addition to its focus on prompt customer service and effective maintenance, Palo Alto also recognizes the importance of accurate forecasts of utility usage, striving to get its forecasts within 5% of actual consumption. Forecasters met this target easily in 1997, missing actual consumption of electricity by 2.5% and natural gas by only 1%.

Table 13.3   Electric Service Odds and Ends: Palo Alto and Denton

NEW ELECTRIC SERVICE EXTENSION

Palo Alto, CA
Target 1: Respond to customer inquiries within 1 working day
Actual: 1 working day (1997)
Target 2: Review plans for new service requests within 10 working days
Actual: 10 working days (1997)

ELECTRIC SERVICE TURN-ON/TURN-OFF

Palo Alto, CA
Percentage of applicants receiving service by the date requested: 97% (1997)

RESPONSIVENESS TO ELECTRIC SERVICE PROBLEMS/INQUIRIES

Denton, TX
Target: Achieve an average response time of 14 minutes to outages
Actual: 14.5-minute average (1997)

Palo Alto, CA
Target: Respond to customer inquiries or problems within 2 hours
Actual: 0.72-hour average (1997)

ELECTRIC SERVICE RELIABILITY

Denton, TX
Average duration of outage: 43 minutes (1997)

Palo Alto, CA
Target 1: Limit power outages to a systemwide average of 60 minutes per customer or less
Actual: 49.6-minute average (1997)
Target 2: Limit service interrupts to an average of less than 2 per customer

ELECTRIC SERVICE REPAIRS

Palo Alto, CA
Target: Begin field restoration and repairs for unscheduled outages within 1 hour
Actual: 100%; 0.5-hour average (1997)

All public utilities are concerned about *line loss*. In the case of gas and electrical distribution systems, line loss is the difference between the amount of natural gas or electricity metered at the head of the system and the amount actually received by (i.e., sold to) customers. Assuming accurate metering, the tightest gas systems and most efficient electric distribution systems will have the lowest rates of line loss. The gas utility in Tallahassee, Florida, for example, strives to keep its losses at less than 2%, and its counterpart in Corpus Christi, Texas, reported line losses of only 0.6% (Table 13.5). The typical line loss for electrical distribution systems tends to be higher. At least one association of municipalities engaged in the distribution of electricity, however, suggests that average annual system loss should be 6% or less (ElectriCities of North Carolina, 1997, p. 18).

Table 13.4   Gas System Performance Targets and Results: Palo Alto

Palo Alto, CA

Target 1: Procure gas supplies for a delivered cost of less than 110% of the midweek market price (1999)

Target 2: Respond to at least 95% of reported gas leaks within 30 minutes
Actual: 96.7% (1997)

Target 3: Process at least 80% of service orders (turn-ons, turn-offs, and reads) within 2 working days of scheduled date
Actual: 82% (1997)

Target 4: Read at least 98% of all gas meters every 27 to 33 days
Actual: 97% (1997)

Target 5: Maintain a meter reading accuracy of at least 99%
Actual: 98.8% (1997)

Target 6: Limit at least 95% of customer service disruptions due to mainline leak repairs to 4 hours or less
Actual: 96.4% (1997)

Target 7: Review at least 95% of service extension plans within 10 working days
Actual: 100% (1997)

Target 8: Install approved new customer services within 38 days of payment
Actual: 100% (1997)

SOURCE: City of Palo Alto (CA), *1998–99 Adopted Budget* (Vol. 2), pp. 97–106.

A few aspects of meter reading and the business office operations connected with gas and electric utilities have been noted in this chapter. These functions are addressed more extensively in Chapter 31, "Utilities Business Office."

Note

1. Corroborating the statistic for prompt responses, Charlottesville, Virginia, set a target of responding to requests for natural gas emergency service within 20 minutes. It met that target in 74% of all calls in 1998.

Table 13.5   Gas and Electric Line Loss: Selected Cities

GAS LINE LOSS

Corpus Christi, TX
0.6% (1991)

Tallahassee, FL
 Target: Less than 2% unaccounted
 Actual: 2% (1992, estimated)

ELECTRIC LINE LOSS

Apex, NC
 0.17% (1996)

Tarboro, NC
 1.74% (1996)

Elizabeth City, NC
 2.38% (1996)

Lincolnton, NC
 3.21% (1996)

College Station, TX
 4.1% (1998)

Monroe, NC
 4.16% (1996)

Greenville, NC
 4.71% (1996)

Denton, TX
 4.84% (5/92)

Lubbock, TX
 5.29% (1995), 6.46% (1996), 6.72% (1997)

Oak Ridge, TN
 Difference between kilowatt-hour (kWh) purchased and kWh sold:
  5.8% (1991), 5.96% (1992)

Wilson, NC
 5.81% (1996)

14

Human Resource Administration

Professionalism in local government human resource (HR) administration implies, at minimum, competence, diligence, and integrity. More formally, professional human resource administration is associated with five fundamentals:

1. The use of merit systems, or merit principles, in employment decisions and personnel rules
2. The administration of the personnel function by a personnel department
3. The introduction of performance appraisals and provision for merit pay
4. Formal disciplinary and grievance procedures
5. An official affirmative action policy and a formal policy against sexual harassment (Fox, 1993, p. 7)

One basis for judging human resource administration in a given municipality, then, is the extent to which it complies with these five elements of professionalism. A more highly professional HR department is presumed to be superior to one that is less professional.

For the most part, the standards of professional HR administration prescribe structures, policies, and processes, rather than outputs or outcomes. As such, they are *means* to desired ends, rather than ends in themselves. Officials wishing to judge the adequacy of local operations may find far more useful gauges for that purpose among the performance statistics and targets of other municipal HR or personnel departments.

Objective Indicators of Employee Morale _____

As with financial administration, in which performance is sometimes assessed by examining the overall financial condition of a community rather than by

Table 14.1   Employee Turnover Rates in Selected Cities

City	Employee Turnover Rate
Danville, KY	1% (1997)
Santa Ana, CA	3.6% attrition rate (1995)
Fort Collins, CO	5.9% (1991)
Fayetteville, AR	6.3% voluntary turnover rate (1992); 12.2% (1997)
Irving, TX	6.6% (1995)
Scottsdale, AZ	Target: 5% to 8% Actual: 6.9% (1999)
Durham, NC	7% (1998)
Shreveport, LA	8% (1994)
Raleigh, NC	9.2% (1997), 9.5% (1998)
Reno, NV	Target: 10% (1999)
San Antonio, TX	10.2% among civilian employees; 2.0%, uniformed (1997)
Savannah, GA	10.6% (1997)
Chapel Hill, NC	10.8% (1999)
Duncanville, TX	15% (1998)

judging the performance of various financial activities, HR administration is sometimes assessed by examining the general health of an organization from a human resources perspective. Although assessments based on broad indicators offer the advantage of relative simplicity, they can be misleading and, therefore, should be used with care.

Two general indicators of employee morale are turnover and absenteeism. If the HR department is involved in the recruitment of job candidates; the selection, placement, training, and nurturing of new employees; and the adoption of good supervisory practices throughout the organization, then rates of employee turnover and absenteeism may be reasonable indicators of its performance. Good performance should lead to the hiring of employees who are well suited to and enjoy their work. Such employees are less likely to seek new job opportunities elsewhere, to be involuntarily terminated from employment, or to experience high rates of absenteeism. Careful selection, proper training, suitably challenging work, and good supervision all contribute to individual job satisfaction that *ceteris paribus* should be reflected in long tenure and low rates of turnover and absenteeism.

Among cities examined for this volume, most units that reported turnover statistics indicated annual turnover rates of 10% or less (Table 14.1). Danville, Kentucky, led the group with a turnover rate of only 1% in 1997. As a basis of comparison, local governments participating in the ICMA's comparative performance measurement project reported a turnover range of 0.7 to 15.6 turnovers per 100 employees in 1997, and a range of 2.3 to 38.3 turnovers per 100 employees in 1998 (ICMA, 1999a, pp. 393, 456; ICMA, 1999b, pp. 249, 291).

Municipalities reporting rates of sick leave use typically do so either as days or hours of sick leave used per employee annually or as the percentage of total hours consumed by sick leave (Table 14.2). Cities that reported sick leave use rates in the form of "days used" generally reported 5 to 8 days per employee—sometimes higher for workers exposed to inclement weather and other adverse field conditions and lower for office employees. For cities reporting use rates as a percentage of total time for the average employee, rates generally fell within the 2% to 4% range. Participants in the ICMA (1999a) project reported an average of 28.2 sick leave hours used in 1997 per 1,000 hours scheduled among local governments of 100,000 or greater, and 26.9 among smaller units (pp. 403, 461).

Judgments based on turnover and absenteeism should be rendered cautiously. As in the case of judgments on financial management based on overall measures of financial health, important factors other than the quality of human resource administration may influence these numbers. Low rates of turnover and absenteeism often accompany good supervision and high morale, but low rates could also reflect a tight local economy and few job options. Furthermore, it is important to recognize that some degree of turnover—preferably a modest rate—is healthy. An organization that hires good employees should expect to lose some of them to opportunities outside the organization. A moderate rate of turnover opens opportunities for promotion, without which the organization's rising stars probably could not be retained.

Hiring New Employees

In some cities, duties associated with the recruitment, selection, and hiring of new employees are decentralized and handled almost entirely by operating departments. In other cases, a central HR department plays an important role and appropriately may be judged on the municipality's overall proficiency in handling those duties.

In the absence of nationally accepted standards for the performance of employment activities, municipal officials are left to establish their own local standards. Some cities judge their recruitment success in part by the size of the candidate pools for their jobs. Rockville, Maryland, for example, received an average of 31 applications per vacancy in 1997 (Table 14.3). Others set targets for the number of *well qualified* candidates for each recruitment. In its effort to generate strong candidate pools, Norman, Oklahoma, strives to secure three or more qualified applicants for at least 90% of its vacancies.

Some HR departments establish local standards for promptness in conducting qualifying examinations. Dayton, Ohio, for example, tries to sched-

Table 14.2   Rates of Sick Leave Use: Selected Cities

City	Target	Annual Sick Leave Use Rates	
		Employees in General	Specific Departments
Boston, MA		9.28 days per employee (1996)	Parks & recreation: 9.80 days Public works: 14.50 days Human resources dept.: 7.77 days (1996)
Cambridge, MA			Public works: 9.84 days per employee (1995)
Fort Collins, CO		40 hours per employee (1991)	
New York, NY		8.6 days per employee (3.48%) (1998)	Fire: 5.6 days Police: 5.6 days Sanitation: 11.0 days Parks & recreation: 7.2 days (1998)
Orlando, FL	(a) Not more than 2.5% for public works and parks & recreation employees (b) Not more than 2% for human resources dept. employees (c) Not more than 2% for city employees as a whole		Public works: 1.74% Parks & recreation: 2.24% Human resources dept.: 2% (1991)
Savannah, GA	(a) 6 days or less per facilities maintenance employee (b) 5.8 days or less per public development employee (c) 5.03 days or less per police employee (d) 7.4 days or less per fire department employee (e) 4.5 days or less per leisure services employee (f) 5.1 days or less per sanitation employee		Facilities maintenance: 6.10 days Public development: 5.8 days Police: 5.03 days Fire: 7.8 days Leisure services: 4.5 days Sanitation: 5.4 days (1991)
Scottsdale, AZ	Maintain unscheduled leave at or below 3.0% of total hours paid	Unscheduled leave: 2.8% (1999)	
Tempe, AZ	45 hours or less per employee (1999)	47.8 hours per employee (1997)	
Tucson, AZ		6.76 days per employee (1990)	

Table 14.3   Recruitment Yields: Selected Cities

Rockville, MD
  Applications received per vacancy: 31 (1997)

Victoria, TX
  Average number of applications received per posted position: 23 (1997, estimated)

College Station, TX
  Applications received per posted position: 17 (1995)

College Park, MD
  Target: Secure at least 15 applicants for 90% of all external recruitments

Norman, OK
  Target: Secure 3 or more highly qualified candidates for at least 90% of vacant positions
  Actual: 94% (1998)

ule at least 80% of all exams within 120 days from the time they were requested. Oakland, California, considers 135 days to be a reasonable lead time for conducting an exam for sworn positions and 90 days to be reasonable for an exam for non-sworn positions.

Outstanding HR departments waste little time in referring qualified applicants to hiring departments. Otherwise, delays in the process could lead to the loss of top candidates who decide not to wait and instead accept offers from more expedient employers. Quick turnaround is influenced by the role HR is expected to play. When an eligibility list is not required and HR does little screening of applicants, applications often may be forwarded to the hiring department within a few days. When an eligibility list is required and HR is expected carefully to screen candidate qualifications, periods ranging from 5 weeks to 3 months are not uncommon due primarily to the testing process (Table 14.4).

Some cities record the time consumed from beginning to end in the recruitment and hiring process. Others report only a major component—for example, the time from the application deadline to the hiring of the selected candidate. Glendale, Arizona, claims that its Employment Selection Division has the quickest average job recruitment turnaround time among major Arizona cities at 22.5 workdays (City of Glendale [AZ], *1998–1999 Annual Budget,* p. 167; Table 14.5). Some cities establish different timetables depending on the type of recruitment—for example, external or internal, nonprofessional or professional/managerial position. For example, Oklahoma City completes external recruitments in 23 days and internal recruitments in 21 days. Nearby Norman, Oklahoma, tries to extend conditional offers of employment within

Table 14.4   Prompt Referral of Job Candidates: Selected Cities

TIME FROM RECRUITMENT TO REFERRAL

Savannah, GA
  Average time to certify professional applicants: 40 days (1997)
  Average time to certify other applicants: 14 days (1997)

Nashville–Davidson County, TN
  Average number of weeks to produce list: 8 (1997), 6 (1998)

Phoenix, AZ
  Target: Establish eligibility lists within an average of 50 workdays from the recruitment order
  Actual: Average of 42 workdays (1993, estimated)

Anaheim, CA
  Target: Recruit and provide lists of eligible applicants within 50 days after recruitment begins

San Jose, CA
  Target: Create eligible list in an average of 66 days (except police and fire recruitments, and promotional and continuous lists)
  Actual: 56-day average (1995)

Santa Ana, CA
  Average number of days to establish eligibility list: 62 (1995)

Portland, OR
  Target: 2 months to establish eligibility lists
  Actual: 2.25 months (1992, estimated)

San Luis Obispo, CA
  Actual: 75 days to establish eligibility list (1994)

Cincinnati, OH
  Targets: (a) 120 days to issue eligibility lists from structured exams; (b) 60 days from unstructured exams
  Actual: (a) 75-day average with structured exam; (b) 50-day average with unstructured exam (1991)

Palo Alto, CA
  Target: Move 90% of vacancies from job posting to submission of qualified candidates within 90 days
  Actual: 69% (1997)

Reno, NV
  Target: From request to certification of list within 90 days at least 50% of the time; within 120 days 75% of the time (1999)

Philadelphia, PA
  Median time from exam request to creation of eligible list: 120 days (1998)

TIME FROM AD CLOSING TO REFERRAL

San Antonio, TX
  Target: Average turnaround time of 4 workdays from closing advertisement until establishment of eligibility list (1999)
  Actual: 6.0-day average (1995)

(continued)

Table 14.4   Prompt Referral of Job Candidates: Selected Cities (continued)

TIME FROM APPLICATION DEADLINE TO REFERRAL

Greenville, SC
  Target: Referral of applications to departments within 1 day following cutoff date
  Actual: 100% (1990)

Norman, OK
  Target: Referral of qualified candidates to departments within 3 days

Oak Ridge, TN
  Target: Forward applications to hiring manager within 3 days following application deadline

Raleigh, NC
  Target: Referral of applications within 5 workdays of application closing date
  Actual: 91% (1991)

Alexandria, VA
  Average number of working days between application closing date and referral of qualified applicants
    to departments: 11.5 (1998)

90 days for managerial/supervisory positions at least 90% of the time; within 60 days for public safety positions; and within 30 days for all others.

Among local governments reporting to ICMA (1999b) the number of days required to complete a competitive recruitment process, the median among units of 100,000 population or greater was 40 days for external recruitments in 1998, and 24 days for internal recruitments (p. 247). Among smaller local governments, the 1998 medians were 40 days for external recruitments and 18 days for internal recruitments (p. 289).

Classification and Compensation Analysis

Centralized HR departments typically have responsibility for conducting studies or audits to determine the appropriateness of a given position classification and associated compensation for a particular employee or set of employees. Often, such studies are triggered by requests from supervisors or department heads who suspect that a classification or pay level is out of line. How quickly should the HR department be expected to respond? Several cities attempt to complete position audits within 30 days of the request (Table 14.6). Some, such as Grand Prairie, Texas, and San Diego, California, are even quicker, with targets of 15 and 21 days, respectively.

Table 14.5   Prompt Filling of Job Vacancies

Glendale, AZ
Average job recruitment turnaround: 22.5 workdays (1997)

Oklahoma City, OK
Number of days to complete external recruitment process: 23 days (1999)
Number of days to complete internal recruitment process: 21 days (1999)

Norman, OK
Targets: Conditional offers of employment made for 90% of vacancies within 90 days for managerial/
  supervisory positions; within 60 days for public safety positions; within 30 days for all others

Decatur, IL
Average turnaround time for vacancies: 37.8 days (1991), 35 days (1995)

Tempe, AZ
Average time to complete an open, external, competitive recruitment and selection process: 36 days
  (1997)
Average time to complete an open, internal, competitive recruitment and selection process: 24 days
  (1997)

Denton, TX
Targets: Average open-to-fill time of 30 days for nonprofessional positions; 80 days for professional
  and managerial positions
Actual: 40-day average overall (1997)

Winston-Salem, NC
Percentage of vacancies in full-time positions filled within 45 days: 77.4% (1998)

Reno, NV
Percentage of recruitments completed within 60 days: 65%; within 120 days: 100% (1990)

Bellevue, WA
Average length of recruitment process from vacancy notification to job offer: 9.0 weeks (1998)

Corvallis, OR
Percentage of positions filled within 90 days of vacancy: 79% (1992)

Employee Grievances

Austin, Texas, had a miniscule 0.14 grievances per 100 employees in 1998—
and still did not set the pace among ICMA (1999b) project participants!
Thirty-nine participating local governments serving populations of 100,000
or greater reported grievance rates ranging from 0.03 grievances per 100 employees to 4.93 per 100 employees (p. 249). Long Beach, California (0.03
grievances per 100 employees), and Tempe, Arizona (0.10 per 100) set the
pace. Figures such as these can serve as useful benchmarks of employee rela-

Table 14.6   Prompt Position Audits for Reclassifications and Compensation
Requests

Grand Prairie, TX
Target: Complete at least 95% of all reclassification studies within 15 workdays
Projected: 94% (1997)

San Diego, CA
Percentage of classification studies completed within 21 days: 85% (1998)

Oklahoma City, OK
Target: Completed within 30 days
Actual: 100% (1992)

St. Petersburg, FL
Target: Completed within 30 days

Lubbock, TX
Percentage of classification recommendations completed within 30 days of receipt of questionnaire:
80% (1995), 80% (1996), 93% (1997)

Portland, OR
Target: 90% completed within 30 days

Oakland, CA
Percentage completed within 30 days: 80% (1991), 74% (1992)

Boston, MA
100% completed within 48 days of request by department head or union official (1992)

Reno, NV
70% of classification studies completed within 90 days; 80% within 120 days; 85% within 210 days
(1990)

Alexandria, VA
Percentage of classification reviews analyzed and implemented within 6 months of request: 100%
(1998)

tions. Several other measures permit a more direct assessment of HR administration performance in dealing with employee grievances.

Some cities schedule hearings and resolve grievances more promptly than others. Boston, for example, schedules most grievance hearings within 3 weeks, and usually has decisions within 2 weeks of the hearing (Table 14.7). Raleigh, North Carolina, attempts to resolve grievances involving dismissals, demotions, or suspensions within 33 workdays and all others within 75 workdays.

Although promptness in dealing with grievances is important, even more important is the successful resolution of grievances, appeals, and complaints. Palo Alto, California, attempts to resolve at least 90% of its grievances prior to arbitration, and did so 82% of the time in 1997 (Table 14.8). San Jose and Ra-

Table 14.7   Prompt Hearing and Resolution of Employee Grievances: Selected
Cities

City	Time Frame for Hearing Appeals	Time Frame for Resolution of Grievances
Boston, MA	93.1% of grievance hearings scheduled within 3 weeks of receipt (1992)	71% of decisions issued within 2 weeks of hearing (1992)
Raleigh, NC		Targets: Grievances involving dismissals, demotions, or suspensions to be resolved within 33 workdays; all others within 75 workdays Actual: 67% of first category within 33 workdays; 87% of all others within 75 (1991)
Oakland, CA	Target: Administrative hearing and opinion within 30 days	Target: Process appeals within a 60-day maximum
Dayton, OH	Target: Schedule hearings within 45 days of receipt of request	
Cincinnati, OH	Target: Schedule disciplinary appeals within 2 months of request	

leigh were even more successful at 90% and 96%, respectively, in recent years.
A further measure of success is the track record of a city's grievance decisions
as they proceed in some cases to arbitration or other rounds of appeal. In Calgary, Alberta, 76% of the city's grievance decisions were upheld in subsequent
appeals during one recent year.

Employee Training

Many other functions are performed by the typical HR department, but a duty
of particular interest here is employee training. Although many municipal employees receive the bulk of their training in formal college or trade school programs, in training offered by consultants, or in training opportunities offered
in conjunction with professional conferences, HR departments often provide
oversight and frequently arrange on-site training opportunities.

Of particular relevance to the training function is a 1993 recommendation
of the National Commission on the State and Local Public Service (1993),
popularly known as the Winter Commission. That commission encouraged
states and localities to appropriate a "learning budget" equal to at least 3% of
total personnel costs (p. 42). Although the recommended learning budget prescribes resource inputs and, as such, is not truly a performance benchmark,

Table 14.8   Successful Resolution of Employee Grievances: Selected Cities

RESOLVED ADMINISTRATIVELY

Raleigh, NC
Percentage of grievances resolved administratively: 96% (1997)

San Jose, CA
Target: Resolve at least 90% of grievances before arbitration
Actual: 90% (1995)

Palo Alto, CA
Target: Resolve at least 90% of grievances before arbitration
Actual: 82% (1997)

Oklahoma City, OK
Percentage of formal grievances resolved prior to mediation/arbitration: 73%
Percentage resolved through mediation: 0%
Percentage resolved through arbitration: 27% (1998)

Calgary, Alberta
Percentage of grievances resolved prior to reaching the board of commissioners: 72% (1996)

Oakland, CA
Percentage of grievances resolved at the employee relations level: 70% (1996)

FAVORABLE RESOLUTION

Calgary, Alberta
Percentage of grievance decisions upheld: 76% (1996)

Reno, NV
Target: At least 75% of arbitration decisions upholding the city's position (1999)

Tucson, AZ
Percentage of grievance resolutions decided in favor of the city's position: 58% (1997)
Percentage of appeal resolutions decided in favor of the city's position: 81% (1997)

some municipalities may nevertheless wish to compare their own employee development appropriations with the Winter Commission's 3% standard.

Helping Employees Succeed

Much of the work in human resource administration is directed toward recruiting and hiring good employees and creating conditions that help them succeed on the job. New employee orientation programs, job-specific training, good working conditions, appropriate classification and compensation, and the development of capable supervisors—all of the steps taken to ensure a proper working environment, suitable skills, and capable mentoring for new

---

Table 14.9   Successful Completion of Probationary Period: Selected Cities

---

Bellevue, WA
Target: At least 97% successfully completing probation
Actual: 97.4% (1997)

Chandler, AZ
Percentage of new employees successfully passing probation: 96.8% (1995)

Palo Alto, CA
Target: 90% or more successfully completing probation
Actual: 95% (1997)

Santa Ana, CA
Percentage of new employees passing probation: 93% (1995)

Scottsdale, AZ
Percentage passing probation: 90.1% (1999)

College Station, TX
Turnover among new hires in tested positions: 0.03% (1998)
In nontested positions: 12% (1998)

Norman, OK
Target: At least 75% of new hires/promotions successfully completing probation
Actual: 83% (1998)

Irving, TX
Percentage of new hires retained after 6 months: 81.4% (1998)

Milwaukee, WI
Civil service employees first-year retention rate: 66.0% (1997)

Cary, NC
Target: At least 90% successfully completing probation
Actual: 64% (1999)

---

employees—should increase the odds of success during the probationary period for new hires. High rates of probationary success often are considered points of pride by HR departments, an indication that the various parts of the recruitment, hiring, and development process are in place and working properly. Bellevue, Washington, and Chandler, Arizona, each with approximately 97% of new hires successfully completing probation, should be proud of their records—as long as they are confident that the probationary period is being used for its intended purpose (Table 14.9). Cities with lower percentages of new hires completing probation are likely to be envious of Bellevue and Chandler's ability to recruit, hire, and develop capable employees with few apparent errors along the way, but they can take some consolation in the realization that removing an unsatisfactory employee is also a probationary success. An extremely low rate of probationary losses might be reason for almost as much managerial scrutiny as an extremely high rate—just to be sure that a sufficiently rigorous standard for passing probation is being applied.

Odds and Ends

Beyond the major elements of human resource administration already mentioned, many cities establish targets and report results in aspects of HR that are tracked less often by their counterparts. Many of these aspects are noted in Table 14.10.

Table 14.10   Odds and Ends in Human Resource Administration: Selected Cities

PROMPT REVIEW OF REQUESTS TO FILL POSITIONS

Philadelphia, PA
Percentage of routine requests approved within 5 days: 92%
Average approval time (routine requests): 0.48 day
Percentage of nonroutine requests approved within 5 days: 8.0%
Average approval time for nonroutine requests (requiring additional justification): 40 days (1996)

San Jose, CA
Target: 60% of vacant position studies completed within 30 days (1997)

JOB POSTINGS

Hurst, TX
Percentage of job announcements posted within 2 days of receiving employee requisition and accurate job description: 98% (1997)

APPLICATION SCREENING/REVIEW

Lubbock. TX
Percentage of applications screened within 3 days of receipt: 90% (1995)

NEW HIRE PROCESSING

San Antonio, TX
Target: Average processing time of 3 workdays for new hires (1999)
Actual: 4.5 days (1995)

ORIENTATION

Hurst, TX
Percentage of new employee orientations conducted within 3 days of employment: 85% (1997)

San Clemente, CA
Target: Conduct at least 90% of all employee orientations within 5 days of hire
Actual: 100% (1998)

PHYSICALS

Hurst, TX
Percentage of post-offer physicals scheduled within 1 day of supervisor's request: 97% (1997)

EQUAL OPPORTUNITY

Scottsdale, AZ
Percentage of equal employment opportunity (EEO) complaints found in favor of the city: 100% (4 of 4) (1998); 100% (2 of 2) (1999)

GRIEVANCE RATE

Raleigh, NC
Formal grievances per 100 employees: 2.0 (1997), 2.0 (1998)

UP-TO-DATE PERSONNEL FILES

Palo Alto, CA
Percentage of electronic files updated within 10 days of receipt of status change or other personnel data: 85% (1997)

BENEFITS ADMINISTRATION

Palo Alto, CA
Percentage of benefits requests handled within 2 days: 100% (1997)

Norman, OK
Target: Respond to at least 90% of benefit inquiries within 2 days
Actual: 100% (1997)

Alexandria, VA
Percentage of benefits and compensation inquiries/issues resolved within 5 working days: 92% (1998)

RESPONSIVENESS

San Clemente, CA
Target: Respond within 3 days to at least 95% of all employee problems or concerns brought to HR
Actual: 100%

PRODUCTION RATIOS

San Antonio, TX
Applications processed per week per placement officer: 206 (1995)

Plano, TX
HR staff per 100 employees: 0.71 (1997)

Scottsdale, AZ
Ratio of city positions to HR staff FTE: 76.7 to 1 (1999)

COST RATIOS

Scottsdale, AZ
HR operating cost as a percentage of city payroll: 3.3% (1995), 3.1% (1997), 3.0% (1999)

15

Information Systems

In most cities, information systems—sometimes called *data processing*—is an internal service operation. Its customers are other city departments rather than citizens. As those customers become increasingly reliant on a variety of automated systems for conducting their operations, they grow to expect quick and reliable service—and become increasingly impatient when speed and reliability are absent.

Benchmarks

Many cities report precise performance statistics for their information systems operation. Among those examined for this volume, several reported system response times, response times for dealing with computing problems, statistics on system downtime, and performance records in meeting projected costs and timetables for various services.

System Response Time

Many cities tabulate system response times or declare their response targets, sometimes subdividing the statistics for local and remote transactions. Among the cities examined for this volume, most of those reporting this aspect of service expected their system to respond to users within 4 seconds (Table 15.1).

Prompt Response to Problems

Greater variation could be found in the performance targets and performance statistics for the promptness of response by information systems to computer system problems (Table 15.2). Some departments emphasized the

Table 15.1   Computer System Response Time: Selected Cities

Fort Worth, TX
Target: Average terminal response time of less than 1 second
Actual: Less than 1 second (1997)

Corpus Christi, TX
Target: Less than 2 seconds for online mainframe applications

Nashville–Davidson County, TN
Target: 2.5 seconds or less for 90% of local transactions; 3.5 seconds or less for 90% of remote
   transactions

Chesapeake, VA
Average online transaction response time: 3.22 seconds (1997)

San Jose, CA
Workstation response time: 4 seconds (1995)

Ann Arbor, MI
Target: Average response time of online mainframe applications of 5 seconds or less

speed with which they answered calls for assistance; some tracked their speed in resolving those problems; others reported both. Most expected all but the most difficult problems to be resolved within 1 or 2 days.

Among local governments participating in the ICMA's (1999b) comparative performance measurement project, most reportedly were able to resolve large portions of their help calls immediately (p. 257). The top performer among participating jurisdictions that year was Tempe, Arizona, with 78% of all help calls resolved on the spot.

Downtime

Most of the examined cities reported high computing system reliability—several claiming system downtime of less than 1% (Table 15.3). Most expected, and achieved, system availability in the vicinity of 98% or greater.

Meeting Projected Costs and Timetables

Customers of information systems want to know how long a given system upgrade, software change, or other service will take and how much it will cost. They become understandably annoyed if these projections are missed by wide margins. Some information systems units have established commendable records of meeting projected costs and timetables for various services (Table

(text continues on page 208)

Table 15.2   Prompt Response to Computer System Problems: Selected Cities

Glendale, AZ
Initial help desk response: within 15 minutes (1997)

San Jose, CA
Percentage of online user requests for assistance answered within 1 hour: 70% (1995)

Decatur, IL
Target: Respond to requests for service within 15 minutes; restore functionality with standard sys-
    tem within 2 hours
Actual: 100% (1995)

Cambridge, MA
Percentage of responses to user requests within 2 hours: 90% (1996)

Alexandria, VA
Percentage of service requests completed within 24 hours: 92% (1995)

Irving, TX
Percentage of priority 1 calls resolved within 24 hours: 75% (1998)

Corpus Christi, TX
Targets: Problem response in less than 2 hours; problem resolution within 48 hours

Cary, NC
Targets: Respond to at least 95% of trouble calls within 4 hours during business hours; resolve at
    least 90% within 24 hours of receipt (2000)

Boca Raton, FL
Percentage of calls for technical assistance responded to within 4 hours: 85% (1997)

Norman, OK
Percentage of all reported problems responded to within 4 work hours: 85% (1998)
Percentage resolved within 2 workdays: 70% (1998)

Hurst, TX
Percentage of repairs completed within 24 hours of notification: 97% (1997)

College Station, TX
1-day problem response: 91.1% (1998)

Palo Alto, CA
Target: Respond to 95% of system questions within 2 working days
Actual: 97% (1997)

Target: Address 95% of computer and software problems within 2 working days
Actual: 85% (1997)

Target: Address 98% of network connectivity problems within 2 working days
Actual: 85% (1997)

Ann Arbor, MI
Actual: 90% of high priority help desk calls resolved within 48 hours (1998)

Norfolk, VA
Percentage of problems resolved by the help desk within 7 days of request: 82% (1997)

Table 15.3   Minimizing Downtime: Selected Cities

San Antonio, TX
Percentage of mainframe processor downtime (24 hours per day/7 days per week): 0.01% (1995)

College Station, TX
Public safety dispatch system uptime (24 hours per day, 7 days per week): 99.9% (1998)
Midrange systems and wide area networks uptime (9 hours per day, 5 days per week): 98.6% (1998)

Fort Worth, TX
Target: Central computer uptime of 99.8% or greater
Actual: 99.8% uptime (1997)

Norfolk, VA
System availability (CPU): 99.8% (1997)

Winston-Salem, NC
Percentage of scheduled system availability achieved: public safety, 99.86%; general, 99.76%; computer
   aided dispatching, 99.95% (1997)

Nashville–Davidson County, TN
Target 1: At least 98% mainframe availability (24 hours per day)
Actual: 99.7% availability (1997)

Target 2: At least 98% online (CICS) availability (7 a.m. to 6 p.m., Monday through Friday)
Actual: 99.3% (1997)

Charlotte, NC
Percentage of core business hour system availability: 99.5% for nonemergency technology systems;
   99.9% for emergency systems (1999)

Bellevue, WA
Percentage of time central computers are fully functional during business hours: 99.4% (1998)

Cambridge, MA
Computer uptime: 99% (1996)

Chesapeake, VA
Percentage of time the mainframe is available: 99% (1997)
Percentage of time servers are available: 99% (1997)

Glendale, AZ
Network access availability: 99% (1997)

Little Rock, AR
Computer uptime: 99% (1995)

Oak Ridge, TN
Availability: public safety, 98%; nonpublic safety, 100% (1997)

Orlando, FL
System availability: 99% (1994)

Palo Alto, CA
Availability and accessibility of the city's servers during scheduled hours of operation (7 a.m. to 6
   p.m.): 99% (1997)

(continued)

Table 15.3   Minimizing Downtime: Selected Cities (continued)

Plano, TX
  Mainframe availability: 99% (1997)
  Network availability: 99% (1997)

Scottsdale, AZ
  Average downtime of systems: 1.0% (1999)

Albuquerque, NM
  System availability: 99% (1998, estimated)
  Percentage of uptime for citywide communications: 98% (1998)

Duncanville, TX
  Percentage of normal business hours that the system is functional: 98.42% (1998); that the network
    is functional: 100% (1998)

Philadelphia, PA
  Availability of main computer: 98.4% (1996)

Ann Arbor, MI
  Target: Mainframe computer systems downtime of 5% of operating hours or less
  Actual: 2% (1998)

Charlottesville, VA
  Target: Mainframe computer will be available between 8 a.m. and 5 p.m. 95% of the time
  Actual: 98% (1998)

  Target: Local area network (LAN) will be available between 8 a.m. and 5 p.m. 95% of the time
  Actual: 96% (1998)

Hurst, TX
  Actual: 97% operational (1997)

Irving, TX
  System availability during business hours: 97% (1998)

Reno, NV
  Target: 96% availability (1999)

Raleigh, NC
  Percentage network availability: 97.8% (1997), 95.3% (1998)

College Park, MD
  Percentage of system availability for networks (excludes scheduled downtime): 95% (1999)

15.4). Scottsdale, Arizona, for example, met its timetable 95% of the time in 1999 and remained within projected costs.

Odds and Ends

A few cities track dimensions of performance that are reported less often by others. Although less conclusive than statistics derived from multiple cases, singular performance targets or the experience of even one respected city may nevertheless be of interest to other municipalities (Table 15.5).

Table 15.4   Meeting Projected Costs and Timetables: Selected Cities

Scottsdale, AZ
  Percentage of projects completed within scheduled time: 95%
  Percentage completed within projected cost: 100% (1999)

Alexandria, VA
  Percentage of operating system and utility upgrades completed on schedule: 95% (1995)

Cambridge, MA
  Percentage of application changes completed within agreed time frame: 90% (1996)

Corpus Christi, TX
  Target: At least 90% of services completed on time

Palo Alto, CA
  Target: At least 90% of software changes completed within agreed time frame
  Actual: 66% (1997)

Table 15.5   Information Systems Odds and Ends: Selected Cities

HELP DESK

Orlando, FL
  Help desk calls handled without referral: 61% (1994), 61% (1995), 44% (1996)

Tucson, AZ
  Percentage resolved at time of call: 57% (1997)

Oakland, CA
  Percentage of help desk calls resolved during call: 50% (1996)

Savannah, GA
  Percentage of calls resolved by the help desk: 34% (1997)

COMPUTER ACCESS REQUESTS

Palo Alto, CA
  Percentage of authorized password and computer access requests processed within 2 working days:
    97% (1997)

TECHNICAL CONSULTING/
PROGRAMMING/INSTALLATION

San Jose, CA
  Percentage of requests for programming and installation completed within 20 days: 60% (1995)

Norman, OK
  Percentage of all technical consulting requests completed within 90 days: 80% (1998, estimated)

(continued)

Table 15.5   Information Systems Odds and Ends: Selected Cities (continued)

DATABASE CAPACITY MANAGEMENT

Palo Alto, CA

Target: Manage user response time by ensuring that no databases exceed 85% of capacity (1999)

SCHEDULED MAINTENANCE

Palo Alto, CA

System maintenance shutdowns planned and scheduled at least 1 week in advance: 100% (1997)

PC/PRINTER REPAIR

Savannah, GA

Average time required to repair PC/printer: 4 hours (1998, estimated)

DATA ENTRY ACCURACY

Tucson, AZ

Accuracy in keying transactions: 98% error free (1997)

Flagstaff, AZ

Data entry accuracy rate: 95% (1997)

VENDOR RESPONSIVENESS

Tucson, AZ

Percentage of desktop hardware service vendor responses within 4 hours: 100% (1997)

16

Library

In 1943, Clarence Ridley and Herbert Simon praised the advances being made in measuring the "intangibles" of library service (p. xiv). They encouraged the leaders of other municipal functions to take similar strides. Now, more than half a century later, library services continue to maintain a performance measurement edge over many other municipal services in the scope and sophistication of widely reported indicators and in the availability of detailed statistics identifying national performance norms. A citizen or municipal official wishing to assess the adequacy of the local public library has a reasonably rich array of potential benchmarks on which to draw.

History of Library Standards[1]

Throughout most of the twentieth century, the Public Library Association (PLA) and its parent organization, the American Library Association (ALA), maintained a prominent role in the development of standards for public libraries. These standards changed through the years, shifting from initial efforts to prescribe appropriate levels of financial support and staff credentials to the formulation of qualitative library goals and, eventually, quantitative standards pertaining to library collection, facilities, services, and performance.

Lists of recommended books have guided librarians through the years in the development of local collections. One buying guide published by the ALA in 1913 recommended subscriptions to periodicals such as *McClure's Magazine* and *National Geographic* and the purchase of books from an array of titles that included the now quaint-seeming *Worry, the Disease of the Age* and *Woman's Part in Government Whether She Votes or Not* at $1.35 and $1.50, respectively (Brown & Webster, 1913).

In 1921, the ALA recommended $1 per capita as the minimum reasonable support for a public library (Withers, 1974, p. 320). Through the years, the recommended minimum allocation was ratcheted upward until, in 1956, per capita expenditure prescriptions were dropped altogether, partially in acknowledgment of the uneven effects of library size and mission on resource requirements and partially in recognition of the imperfect relationship between expenditure level and service quality (Withers, 1974, pp. 322–323).

Other standards introduced by the ALA in 1933 were broad in scope but possessed only limited usefulness for evaluating a library's performance. Phrases such as "reasonably adequate library service," maintaining a collection "adequate to the needs of the community," and providing "a professional staff of high quality and adequate number" (Withers, 1974, pp. 320–321) dotted those standards and offered an imprecise yardstick against which to measure performance. Perhaps even more important to library practitioners and patrons, these qualitative "standards" offered little leverage for prying resources from the city treasury. In short, they failed to arm library directors with a persuasive means of demonstrating to budget makers local shortcomings in facilities, services, and funding. A series of quantitative standards began to emerge at midcentury, which many library directors proved more adept at applying to that purpose (Table 16.1).

The introduction of standards that prescribed acceptable travel time for library patrons, library operating hours, collection size and quality, collection maintenance practices, library staffing levels, and physical characteristics of the library facility served the cause of librarians whose operations fell below the prescribed numbers and who wished to use the standards as a peg for demonstrating and remedying local deficiencies. Such standards had several advantages—not the least of which was ease of application. It was a simple matter to compare local collection, staffing, and facility characteristics with national prescriptions and identify areas of conformance or deficiency.

Nevertheless, quantitative standards proved to be controversial. In part, they were criticized as the musings of professionals, as goals established largely without the benefit of careful research. In part, they were criticized for failing fully to recognize the ramifications of service population differences and variations in service emphasis from one community library to another. Furthermore, quantitative standards were criticized for undercutting the efforts of top-quality libraries who felt penalized for their successes when budget officials and governing bodies cited above-standard performance as the rationale for diverting resources to "needier" purposes (Baker & Lancaster, 1991, pp. 321–334; Withers, 1974, pp. 320–338).

In 1986, the International Federation of Library Associations and Institutions (IFLA) published a summary of the public library standards it had been recommending as minimal guidelines since the 1970s. In its *Guidelines for Public Libraries,* the IFLA cautioned that some of its standards were "not likely to be universally relevant" (p. 61) but proceeded nevertheless to offer

Table 16.1   Selected Midcentury Public Library Standards

- Maximum acceptable travel time to library:
  15 minutes, urban; 30 minutes, rural

- Acceptable range of operating hours:
  Community libraries serving 10,000 to 25,000 population, 45 to 66 hours per week; community libraries serving larger populations, 66 to 72 hours per week; central resource libraries, at least 66 hours per week

- Systematic "weeding" of unneeded items from collection:
  Average annual withdrawals of at least 5%

- Minimum percentage of requested adult nonfiction titles included in local collection:
  35% to 50% for libraries serving a population of less than 10,000; 50% to 65% for 10,000 to 24,999 population; 65% to 80% for 25,000 to 49,999 population; 80% to 95% for 50,000 to 99,999 population

- Minimum percentage of requested juvenile titles included in local collection:
  80%

- Maximum acceptable waiting time for reserved materials:
  6 weeks

- Minimum collection size:
  2 to 4 volumes per capita

- Minimum annual additions and replacements:
  One sixth of a volume per capita for libraries serving a population of 500,000 or less; one eighth of a volume per capita for larger libraries

- Minimum number of periodical subscriptions:
  1 per 250 population

- Minimum staff size:
  1 staff member per 2,000 population

- Library lighting intensity:
  50 foot-candles

- Minimum library floor space:
  For communities of 25,000 to 49,999 population, 5,000 square feet plus 1 additional square foot for every 10 books beyond a 50,000-volume collection; for smaller communities, 1,500 square feet plus 1 additional square foot for every 10 books beyond a 15,000-volume collection

- Minimum shelf space:
  For communities of 25,000 to 49,999 population, 6,300 linear feet plus 1 additional foot for every 8 books beyond a 50,000-volume collection; for smaller communities, 1,875 linear feet plus 1 additional foot for every 8 books beyond a 15,000-volume collection

SOURCES: American Library Association, *Minimum Standards for Public Library Systems* (Chicago: Author, 1967); American Library Association, *Interim Standards for Small Public Libraries* (Chicago: Author, 1962); F. N. Withers, *Standards for Library Service: An International Survey* (Paris: The UNESCO Press, 1974), pp. 320–338. Reproduced by permission of the American Library Association and UNESCO; copyright © American Library Association, 1967 and 1962; copyright © UNESCO, 1974.

guidance on such diverse topics as desirable minimum service population, operating hours, collection size, staff, and library distance from patrons (Table 16.2).

Even as the IFLA was moving to consolidate its quantitative standards, the PLA was moving in the opposite direction. By the late 1980s, the PLA had abandoned its quantitative standards for public libraries—in contrast not only to IFLA but also to the continued use of quantitative standards by other ALA affiliates for various types of nonpublic libraries. The PLA's departure from prescriptive numerical standards has been applauded by some librarians but has prompted concern among others who miss the evaluative usefulness of national standards and who "still want published quantitative figures to help them argue for additional resources when their services are less than optimum" (Baker & Lancaster, 1991, p. 328). Many state-level associations—at least 23 by one recent count (Moorman, 1997; see also Weech, 1988)—have filled the void by offering their own prescriptive standards for public libraries.

Performance Measures

Recommended sets of performance measures for public libraries typically include a variety of indicators pertaining to library characteristics and services, often paralleling IFLA or earlier PLA standards. Some promote the extensive use of direct feedback from library patrons and other community residents as performance indicators in addition to measures of library use, access, and general operating performance (e.g., Hatry et al., 1992, pp. 58–59). Among the leading measures of library use, material use, materials access, reference services, and library programming recommended by the PLA are the following:

Library Use

- *Annual Library Visits per Capita* is the average number of library visits during the year per person in the area served. It reflects the library's walk-in use, adjusted for the population served.

- *Registration as a Percentage of Population* is the proportion of the people in the area served who are currently registered as library users, reflecting the proportion of the people who are potential library users who have indicated an intention to use the library.

Material Use

- *Circulation per Capita* is the annual circulation outside the library of materials of all types per person in the legal service area.

- *In-Library Materials Use per Capita* is the annual number of materials of all types used within the library per person in the area served.

Table 16.2   Selected Standards for Public Libraries: International Federation of
                  Library Associations and Institutions

Minimum Population Thresholds

Preferred minimum population	150,000
Minimum population normally regarded as viable	50,000
Absolute minimum population	3,000

Minimum Service Hours

Main library (urban)	60 hours per week

Minimum Book Stocks

In smallest libraries	3 volumes per capita
In medium and larger libraries	2 volumes per inhabitant

*[When children 14 years of age or younger constitute 25% to 30% of the population, children's books
should make up one third of the total.]*

Reference books	Up to 10% of total stock (at least 100 volumes)

Annual Additions to Library Collection

In smallest libraries	300 volumes per 1,000 population
In medium and larger libraries	250 volumes per 1,000 population
Children's books	One third of additions if children (14 or younger) constitute 25% to 30% of population
Reference books in libraries serving populations of more than 50,000	10% of all additions

Periodical Holdings, Including Newspapers

In libraries serving up to 5,000 people	50
In libraries serving more than 5,000	10 per 1,000 population

[These figures include multiple copies, foreign language periodicals, and periodicals for children.]

Records and Taped Recordings of All Kinds

Stock for use within library	2,000
Annual additions	300
Records for home use	No recommendation

Foreign Language Collections

For foreign-language minorities of less than 500 persons	100 volumes
For foreign-language minorities between 500 and 2,000 persons	1 book per 5 persons
For foreign-language minorities greater than 2,000 persons	1 book per 10 persons
Foreign-language periodicals	1 per 500 foreign-language minorities

(continued)

Table 16.2   Selected Standares for Public Libraries: International Federation of Library Associations and Institutions (continued)

Staff

Total nonmanual staff (professional, clerical, administrative):

In smallest libraries	1 full-time qualified librarian with clerical assistance
In medium and larger libraries	1 per 2,000 population
In very large libraries	1 per 2,500 population

Qualified Librarians

In compact urban libraries	33% of total staff
In less compact libraries with many branches	40% of total staff

[In a large library system, one third of professional librarians should be specialized in children's work.]

Facility Location

Branch library within 1 mile of most residents

Main library within 2.5 miles of most residents

SOURCE: Adapted from *Guidelines for Public Libraries* (IFLA Publication 36) (Munich, Germany: K. G. Saur for the International Federation of Library Associations and Institutions, 1986), pp. 61–63. Reprinted with the permission of K. G. Saur Verlag (Munich, Germany).

- *Turnover Rate* measures the intensity of use of the collection. It is the average annual circulation per physical item held.

Materials Access

- *Title Fill Rate* is the proportion of specific titles sought that were found during the user's visit. It is not the proportion of users who were successful, because one user may have looked for more than one title; it is the proportion of the searches that were successful.

- *Subject and Author Fill Rate* is the proportion of searches for materials on a subject or by an author that were filled during the user's visit.

- *Browsers' Fill Rate* is the proportion of users who were browsing, rather than looking for something specific, who found something useful.

- *Document Delivery* measures the time that a user waits for materials not immediately available, including reserve and interlibrary loans. It is expressed as the percentage of requests filled within 7, 14, 30, and over 30 days.

Reference Services

- *Reference Transactions per Capita* is the annual number of reference questions asked per person in the area served.

- *Reference Completion Rate* is the proportion of reference questions answered satisfactorily.

Programming

- *Program Attendance per Capita* is the annual number of people attending programs per person in the area served. (Van House, Lynch, McClure, Zweizig, & Rodger, 1987, pp. 37–72; excerpted by permission of the American Library Association)

Benchmarks

Persons wishing to judge the adequacy of local public library facilities, collection, staff, and performance may legitimately relate current local marks to PLA standards of the past (see Table 16.1), as long as the date of those guidelines and the revised policy of PLA is acknowledged. Potential benchmarks with fewer encumbrances include the more recent IFLA standards (see Table 16.2), standards adopted by library associations at the state level, current PLA statistics on public library performance indicators, statistics compiled by the U.S. Department of Education, and standardized targets and performance indicators from individual municipalities.

The sets of standards developed by state-level library associations typically address a broad array of library characteristics and performance dimensions. The standards adopted in Ohio, for instance, prescribe desired fill rates, currency of library holdings, promptness and accuracy of responses to reference inquiries, operating hours, and adequacy of parking (Table 16.3).

The standards specified for public libraries in Iowa similarly encompass a broad array of library characteristics (Table 16.4). Although a few of Iowa's standards address service dimensions included in Ohio's set as well, a principal distinction is Iowa's practice of identifying three levels within each standard—one level regarded as "adequate," a second level representing "good," and a third level denoting "excellent."

A third example of standards developed by state-level public library associations comes from Kansas (Table 16.5). For most service dimensions, separate standards are offered for different types of libraries (e.g., gateway libraries, linking libraries, service centers, and major resource centers), serving populations of different sizes.

Individual communities wishing to find benchmarks for the local library can benefit from the descriptive statistics compiled by the PLA and the U.S. Department of Education. Both collect information on public libraries and calculate relevant statistics for libraries based on service population clusters. The remaining information in this chapter will draw extensively on those

Table 16.3   Selected Standards for Public Library Service in Ohio

- Achieve a title fill rate of 75% or greater.

- Achieve author and subject fill rate of 90% or greater.

- Provide within 7 days at least 40% of requested materials that are not immediately available; at least 80% within 30 days.

- Provide an up-to-date collection with at least 25% of holdings published within the last 5 years.

- Provide prompt and correct responses to reference inquiries, with 80% answered by the end of the business day.

- Provide patrons the means to determine what materials the library and its branches own and the location and availability of those materials.

- Provide convenient weekend and evening operating hours, with not more than 65% of operating hours coming during the daytime on weekdays.

- Provide sufficient parking to ensure the availability of at least one free parking space within one tenth of a mile during at least 80% of library operating hours.

- Provide patrons barrier-free access.

- Operate the library under the management of a professional librarian who is a graduate of an ALA-accredited library program.

SOURCE: Excerpted from *Standards for Public Library Service in Ohio: 1998 Revision* (Columbus, OH: Ohio Library Council, 1998). Used by permission.

statistics as well as standardized targets and performance indicators from individual municipalities.

Staffing Norms and Other Input "Standards": Handle With Care

As with staffing prescriptions for other municipal functions, guidelines calling for a given number of librarians per 1,000 population are input standards rather than standards by which *performance* may be assessed. A library that discovers innovative ways to achieve its objectives with less than the customary number of employees would be penalized inappropriately by the blind application of staffing standards. Nevertheless, national norms and staffing indicators from other jurisdictions do offer supplemental information of value in a more comprehensive assessment of performance by placing in context local contentions of overstaffing or understaffing.

Table 16.4   Selected Standards for Iowa's Public Libraries

	Standard		
	Adequate	Good	Excellent
Circulation per capita	n.s.	8	14
Collection turnover rate	1.0	2.0	3.0
Annual removals from collection (i.e., "weeding")	3%	4.5%	6%
New acquisitions (annual)	3%	4.5%	6%
In-library materials use per capita	n.s.	4.0	6.0
Title fill rate	n.s.	70%	80%
Author and subject fill rate	n.s.	80%	85%
Document delivery—securing materials not immediately available at time of request	n.s.	45% within 7 days; 62%, 14 days; 75%, 30 days	55% within 7 days; 72%, 14 days; 90%, 30 days
Days and hours of operation per week			
Less than 1,000 population	4 days/20 hrs	5 days/20 hrs	5 days/25 hrs
1,000–2,499	5 days/20 hrs	6 days/25 hrs	6 days/35 hrs
2,500–4,999	5 days/25 hrs	6 days/35 hrs	6 days/45 hrs
5,000–9,999	5 days/45 hrs	6 days/50 hrs	7 days/56 hrs
10,000–24,999	6 days/56 hrs	6 days/64 hrs	7 days/68 hrs
25,000 or greater	6 days/64 hrs	6 days/68 hrs	7 days/72 hrs

SOURCE: Excerpted from *In Service to Iowa: Public Library Measures of Quality* (Des Moines, IA: State Library of Iowa, 1997). Used by permission.
n.s. indicates that a specific standard is *not specified.*

The general staffing pattern among public libraries in 1996 was 11.65 full-time equivalent (FTE) staff members per 25,000 population—or slightly less than one half FTE per 1,000 population (Table 16.6). The average number of librarians on each library staff was smaller of course. Slightly fewer than four librarians were employed per 25,000 population and the rate of graduate-degreed librarians was fewer than three per 25,000 population.

Although staffing-to-population ratios are interesting and somewhat informative, more revealing are the production ratios that relate staff size to services provided (Table 16.7). These ratios provide individual communities a basis for judging whether their own staffing level is generally in line with workload. Viewed alongside measures of service quality, production ratios can be especially telling. A library with high output per staff member *and* superior service quality deserves praise for its high productivity. On the other

Table 16.5 Selected Standards for Public Libraries in Kansas

				Library Type			
	Gateway	Linking	Service Center	Major Service Center I	Major Service Center II	Major Resource Center I	Major Resource Center II
Typical Service Population	Less than 500	500–1,000	1,000–2,500	2,500–10,000	10,000–25,000	25,000–100,000	More than 100,000
Standard							
Hours of operation weekly	15	15	20	35	55	68	68
Collection size—cataloged items per capita	12	12	10	7	5	3	3
Periodical subscriptions							
total	15	20	—	—	—	—	—
per 1,000 population	—	—	35	20	15	10	10
Linear feet of shelving	1,200	1,500	2,250	5,000	11,000	n.s.	n.s.
Total floor space (square feet)	1,670	2,350	3,044	5,000	9,500	0.8 sq. ft. per capita	0.8 sq. ft. per capita
User parking spaces	5	10	18	20	30	n.s.	n.s.
Staff parking spaces	2	2	4	5	10	n.s.	n.s.

	All Libraries
Percentage of requested interlibrary loan materials delivered within 14 days	75%
Percentage of information requests answered or referred within 24 hours	90%
Minimum percentage of collection evaluated annually (i.e., "weeding")	20%
Minimum percentage removed from collection annually	1%
Percentage of periodical citations accessible (including backfiles and electronic resources)	75%

SOURCE: Excerpted from Kansas Library Association/Public Library Section, *Measurements of Quality: Public Library Standards for Kansas* (Hutchinson, KS: Author, 1995). Used by permission.
n.s. indicates that a specific standard is *not specified.*

Table 16.6   Library Staffing in 1996: National Averages

	National Average 1996
Total number of paid FTE staff per 25,000 population	11.65
Number of paid FTE librarians per 25,000 population	3.87
Number of paid FTE librarians with ALA-MLS[a] per 25,000 population	2.71

SOURCE: Adapted from Adrienne Chute and Elaine Kroe, *Public Libraries in the United States: FY 1996* NCES 1999-306 (Washington, DC: U.S. Department of Education, National Center for Education Statistics February 1999), pp. 119–120.

a. ALA-MLS: A master's degree from a program accredited by the American Library Association.

hand, a library with an overburdened staff and poor service quality may find in comparing its own production ratios to those of others confirmation of the former (overburdened staff) and a possible explanation for the latter (poor service quality).

According to PLA figures for 1997, most libraries devoted 45% to 60% of their budgets to staff salaries and 10% to 20% for library materials (Table 16.8). Remarkably little variation in the allocation pattern was evident across service population categories. As in the case of staffing ratios, such figures should be regarded as input allocation norms rather than as standards. They reflect the customary means (i.e., allocation patterns) to desired ends, rather than ends in themselves.

Adequacy of Collection

Although figures provided by the U.S. Department of Education and the PLA for average library holdings differ somewhat from one another, they nevertheless exhibit sufficient consistency to permit individual libraries to judge their collection size in a national context (Table 16.9). For example, whether the precise norm for collection size among public libraries serving populations of 25,000 to 49,999 people is the 2.9 books and serial volumes per capita reported by the Department of Education for 1996 or the 3.4 volumes per capita reported as the median in 1997 by the PLA should not be of great concern to a library in that population range wishing to assess its collection size. If its holdings per capita are less than both figures or greater than both figures, the message is clear. If its per capita holdings fall between the PLA and Department of Education figures, it may consider its collection size to be consistent with national norms.

Although the national average for collection size may be the 2.8 books and serial volumes per capita reported by the Department of Education—or a

Table 16.7   Production Ratios in Public Libraries: Selected Cities

CIRCULATION PER FTE

Denton, TX
67,048 (1991)

Corpus Christi, TX
50,162 (1991)

Fayetteville, AR
21,801 (1991)

CIRCULATIONS PER CIRCULATION DESK FTE

Ames, IA
214,453 (1997)

INTERLIBRARY LOAN (ILL)
REQUESTS PER ILL-ASSIGNED FTE

San Antonio, TX
9,975 (1995)

REFERENCE QUESTIONS PER REFERENCE FTE

Corpus Christi, TX
33,077 (1991)

Denton, TX
12,054 (adult) (1991)

Tacoma, WA
8,738 (1991)

Boise, ID
5,155 (youth) to 12,442 (adult) (1991)

TRANSACTIONS PER STAFF MEMBER

Fort Worth, TX
37,990 (1997)

slightly higher figure to reflect audio and video materials that might also be included in a count of library holdings—it is important to note the variation across population clusters. Libraries serving larger populations typically have a larger total collection but a smaller number of volumes per capita than do libraries serving smaller populations. For that reason, benchmarks for collection size should be drawn from appropriate service population clusters.

The adequacy of a library's collection may be judged not only by its sheer numbers but also by its quality. One approach to a quality-of-collection assessment tests the local holdings to see how many entries from a renowned bibli-

Table 16.8 Percentage of Library Operating Expenditures Devoted to Salaries and Materials: 1997

Library Service Population	N	Salaries as Percentage of Operating Expenditures			Materials as Percentage of Operating Expenditures		
		Lower Quartile (25%)	Median	Upper Quartile (75%)	Lower Quartile (25%)	Median	Upper Quartile (75%)
1 million or greater	21	49.0	50.9	53.3	10.3	12.3	14.4
500,000–999,999	60	46.8	51.3	54.9	12.2	14.8	18.2
250,000–499,999	75	44.3	50.0	55.4	12.4	14.6	16.8
100,000–249,999	237	46.1	50.9	55.6	12.2	14.8	17.6
50,000–99,999	248	47.0	51.1	56.7	12.0	14.4	17.3
25,000–49,999	67	47.4	54.3	58.4	13.0	15.4	17.9
10,000–24,999	30	45.9	50.7	54.3	14.5	16.5	21.7
5,000–9,999	17	45.2	46.9	52.6	11.4	13.2	15.2
Under 5,000	18	50.5	56.0	61.1	11.6	20.5	22.8

SOURCE: Adapted from Public Library Association, *Statistical Report '98: Public Library Data Service* (Chicago: American Library Association, 1998), pp. 37–38. Used by permission of the American Library Association.
N = number of libraries reporting.

ography or book list are possessed by the library. The ALA's regularly published "best books" lists may be used for that purpose, as can the *Fiction Catalog* and the *Public Library Catalog.* Additional sources may be found in the *WLN Collection Assessment Manual* (Powell & Bushing, 1992). Information is readily available on the application of this assessment technique and the identification of applicable bibliographies (e.g., Baker & Lancaster, 1991; Powell & Bushing, 1992).

Other methods of evaluating the quality of a public library's collection typically rely on indicators reflecting the popularity of the collection among the library's service population or the success of patrons in securing the materials they seek. A library with a greater than average collection turnover probably has a greater than average percentage of its collection in high-demand and well-received books and materials (Table 16.10). Circulation per capita and its variant, circulation per registered borrower, provide an even more revealing assessment of the adequacy of a library's collection as judged by community residents as a whole and by those who use or intend to use the library, respectively (Table 16.11). Not surprisingly, reported rates of circulation among

Table 16.9 Public Library Collection Size

Library Service Population		U.S. Department of Education Survey, 1996								Public Library Association Survey, 1997 Holdings per Capita		
	N	Books and Serial Vols. per Capita	N	Audio Materials per 1,000 Pop.	N	Video Materials per 1,000 Pop.	N	Serial Subscripts per 1,000 Pop.	N	Lower Quartile (25%)	Median	Upper Quartile (75%)
1 million or greater	20	2.5	20	130.7	20	33.3	20	7.1	22	1.6 vols.	2.6 vols.	2.9 vols.
500,000–999,999	52	2.6	52	122.3	52	46.1	52	6.8	59	1.7 vols.	2.5 vols.	3.2 vols.
250,000–499,999	90	2.4	90	80.2	90	38.8	90	5.4	75	1.8 vols.	2.1 vols.	3.0 vols.
100,000–249,999	313	2.2	313	86.2	313	42.6	313	5.3	234	1.7 vols.	2.3 vols.	3.1 vols.
50,000–99,999	510	2.5	510	81.2	510	49.1	510	6.3	244	1.9 vols.	2.6 vols.	3.5 vols.
25,000–49,999	863	2.9	863	94.4	863	61.2	863	8.0	67	2.6 vols.	3.4 vols.	4.8 vols.
10,000–24,9	1,649	3.5	1,679	100.9	1,679	71.8	1,679	9.6	29	3.4 vols.	3.1 vols.	4.0 vols.
5,000–9,999	1,498	4.2	1,498	101.6	1,498	89.9	1,498	12.2	16	3.6 vols.	5.5 vols.	7.0 vols.
2,500–4,999	1,327	5.3	1,327	95.5	1,327	97.9	1,327	14.9	17[a]	6.8 vols.[a]	8.3 vols.[a]	14.7 vols.[a]
1,000–2,499	1,636	7.7	1,636	109.9	1,636	140.1	1,636	21.3	—	—	—	—
Less than 1,000	958	13.8	958	191.3	958	234.6	958	33.5	—	—	—	—
National average	8,696	2.8	8,696	99.6	8,722	51.8	8,740	7.3	—	—	—	—

SOURCE: Adapted from Adrienne Chute and Elaine Kroe, *Public Libraries in the United States: FY 1996*, NCES 1999-306 (Washington, DC: U.S. Department of Education, National Center for Education Statistics, February 1999), p. 41, and Public Library Association, *Statistical Report '98: Public Library Data Service* (Chicago: American Library Association, 1998), pp. 65–66. Used by permission.

a. Includes all responding libraries having a service population of less than 5,000.

N = number of libraries.

Table 16.10 Annual Collection Turnover: 1997

Library Service Population	N	Lower Quartile (25%)	Median	Upper Quartile (75%)
I million or greater	22	1.05	1.96	2.46
500,000–999,999	59	2.25	3.49	4.40
250,000–499,999	75	1.99	2.96	3.71
100,000–249,999	231	2.02	2.80	3.52
50,000–99,999	235	1.90	2.56	3.33
25,000–49,999	64	1.66	2.14	3.27
10,000–24,999	29	2.12	2.67	3.71
5,000–9,999	16	1.72	2.10	2.30
Less than 5,000	17	0.63	1.15	1.79

SOURCE: Adapted from Public Library Association, *Statistical Report '98: Public Library Data Service* (Chicago: American Library Association, 1998), pp. 107–108. Used by permission.
N = number of libraries reporting.

card-carrying patrons are higher than among the potential service population as a whole. A word of caution is in order regarding the use of per patron figures: Assessments based on circulation per registered borrower are especially vulnerable to variations among libraries in the practice of periodically purging their patron lists. Libraries with "lean" patron lists are rewarded in this calculation.

Figures reported for circulation per capita by the U.S. Department of Education for 1996 and by the PLA for 1997 differ minimally in some population categories and substantially in others. As in the case of collection holdings per capita, jurisdictions comparing their own numbers with these figures will receive an unambiguous message if they exceed or fall short of both the PLA and Department of Education numbers in their own population cluster. They may consider their own mark to be in the average range if they fall between the reported norms.

The adequacy of a library's collection may also be gauged by fill rates—the extent of patrons' success in securing the materials of a specific or general nature that they are seeking (Table 16.12). Many of the libraries participating in the PLA study routinely approach or exceed a title fill rate of 70%, considered by Iowa standards to be a "good" level of service, even if a bit short of Ohio's 75% standard.

Table 16.11 Circulation per Capita, In-Library Use of Materials per Capita, and Circulation per Registered Borrower

Library Service Population	U.S. Department of Education Survey, 1996		Public Library Association Survey, 1997											
			Circulation per Capita				In-Library Use per Capita				Circulation per Registered Borrowed			
	N	Circulation per Capita	N	Lower Quartile (25%)	Median	Upper Quartile (75%)	N	Lower Quartile (25%)	Median	Upper Quartile (75%)	N	Lower Quartile (25%)	Median	Upper Quartile (75%)
1 million or greater	20	4.2	22	3.0	4.0	5.3	14	1.2	1.9	3.5	19	7.0	10.6	11.3
500,000–999,999	52	7.5	60	4.7	7.1	11.2	23	3.0	3.9	5.1	54	10.8	15.9	19.1
250,000–499,999	90	6.3	75	4.6	6.1	10.0	38	2.3	3.1	5.9	67	10.0	13.5	17.5
100,000–249,999	313	6.1	235	4.1	6.1	9.0	96	1.1	2.0	3.5	207	9.3	13.1	17.4
50,000–99,999	510	6.2	239	4.2	6.4	9.3	99	1.0	2.3	3.9	196	10.2	13.1	17.4
25,000–49,999	863	7.1	64	5.7	8.1	11.6	22	0.6	2.6	5.4	56	11.8	13.6	19.0
10,000–24,999	1,679	7.7	30	7.9	11.1	16.5	6	n.a.	3.1	n.a.	25	11.3	18.8	20.9
5,000–9,999	1,498	8.1	17	7.5	11.2	13.7	5	n.a.	1.5	n.a.	13	11.1	11.9	23.3
2,500–4,999	1,327	7.8	17[a]	4.3[a]	13.0[a]	20.8[a]	8[a]	n.a.	5.5[a]	n.a.	13[a]	9.4[a]	17.9[a]	24.6[a]
1,000–2,499	1,636	8.5	—	—	—	—	—	—	—	—	—	—	—	—
Less than 1,000	958	10.4	—	—	—	—	—	—	—	—	—	—	—	—
National average	8,731	6.5	—	—	—	—	—	—	—	—	—	—	—	—

SOURCE: Adapted from Adrienne Chute and Elaine Kroe, *Public Libraries in the United States: FY 1996*, NCES 1999-306 (Washington, DC: U.S. Department of Education, National Center for Education Statistics, February 1999), p. 33, and Public Library Association, *Statistical Report '98: Public Library Data Service* (Chicago: American Library Association, 1998), pp. 65–66, 107–108. Used by permission.

a. Includes all responding libraries having a service population of less than 5,000.

N = number of libraries.

n.a. = not applicable (quartiles are not recorded when the number of reporting libraries is less than 11).

Table 16.12 Fill Rates: 1997

Library Service Population	Reference Completion Rate				Title Fill Rate				Subject/Author Fill Rate				Browser Fill Rate			
	N	Lower Quartile (25%)	Median	Upper Quartile (75%)	N	Lower Quartile (25%)	Median	Upper Quartile (75%)	N	Lower Quartile (25%)	Median	Upper Quartile (75%)	N	Lower Quartile (25%)	Median	Upper Quartile (75%)
1 million or greater	1	n.a.	76.0	n.a.	1	n.a.	68.0	n.a.	1	n.a.	70.0	n.a.	1	n.a.	87.0	n.a.
500,000–999,999	9	n.a.	78.0	n.a.	7	n.a.	68.5	n.a.	6	n.a.	73.0	n.a.	4	n.a.	90.5	n.a.
250,000–499,999	15	86.8	90.0	95.2	13	63.4	73.0	78.0	11	66.5	73.0	76.5	10	n.a.	90.0	n.a.
100,000–249,999	44	83.4	90.9	96.3	31	62.5	71.0	81.5	28	72.8	78.9	83.8	26	89.8	94.0	96.0
50,000–99,999	42	88.0	92.0	94.8	35	65.0	70.0	80.5	34	73.3	76.8	85.0	33	90.1	92.0	95.0
25,000–49,999	12	85.8	92.9	95.8	8	n.a.	72.5	n.a.	8	n.a.	80.5	n.a.	8	n.a.	90.5	n.a.
10,000–24,999	5	n.a.	70.4	n.a.	2	n.a.	n.a.	n.a.	2	n.a.	n.a.	n.a.	2	n.a.	n.a.	n.a.

SOURCE: Adapted from Public Library Association, *Statistical Report '98: Public Library Data Service* (Chicago: American Library Association, 1998), pp. 107–108. Used by permission.

N = number of libraries reporting.

n.a. = not applicable (quartiles are not recorded when the number of reporting libraries is less than 11).

227

Table 16.13   Registrations as a Percentage of Library Service Population: 1997

Library Service Population	N	Lower Quartile (25%)	Median	Upper Quartile (75%)
I million or greater	19	30.7	43.7	53.1
500,000–999,999	54	45.5	55.3	66.2
250,000–499,999	67	37.1	48.7	61.6
100,000–249,999	209	37.4	51.9	64.4
50,000–99,999	202	39.3	52.2	70.2
25,000–49,999	58	43.5	61.6	73.0
10,000–24,999	25	66.1	87.0	106.9
5,000–9,999	13	42.9	65.5	94.4
Under 5,000	14	48.0	62.3	118.4

SOURCE: Adapted from Public Library Association, *Statistical Report '98: Public Library Data Service* (Chicago: American Library Association, 1998), pp. 65–66. Used by permission.
N = number of libraries reporting.

Effectiveness in Reaching
Service Population

A public library's mission typically encompasses more than just providing quality services to persons who seek out the library and avail themselves of its offerings. Customarily, libraries also encourage as many persons within their service population as they can to become registered and active patrons. Registration of one half the service population is not uncommon (Table 16.13). Patron counts in Plano, Texas; Alexandria, Virginia; and Ames, Iowa, for instance, constitute three quarters of their respective populations or more (Table 16.14). Library visits average four per capita annually, with a somewhat lower number of visits per capita among libraries serving the largest populations and higher figures among smaller libraries (Table 16.15).

Adequacy of Services

Special features such as reference services and interlibrary loan capabilities enhance a library's offerings. The number of reference transactions tends to be disproportionately greater in libraries serving large populations than in their smaller population counterparts. The averages range from fewer than one reference transaction per resident annually among the smallest libraries to more than double that rate among libraries serving the largest populations (Table

Table 16.14   Library Registrants as a Percentage of the Population:
Selected Cities

Plano, TX
  80% (1997)

Alexandria, VA
  76.0% (1998)

Ames, IA
  Borrowers as a percentage of population: 75% (1997)

Corpus Christi, TX
  64% (1998)

Avon, CT
  62% (1997)

Chesapeake, VA
  57% (1997)

Long Beach, CA
  Library cards per 1,000 residents: 568 (1996)

Tucson, AZ
  53% (1997)

Oak Park, MI
  51% (1997)

Grand Prairie, TX
  50% (1996)

San Antonio, TX
  36% (1995)

16.16). Many libraries take pride in providing prompt and thorough answers to reference inquiries, typically striving to achieve a high percentage of same-day responses (Table 16.17).

A library may be the sender or the receiver in interlibrary loan transactions. In the case of sending materials, the national average for interlibrary loans is 41.7 per 1,000 population (Table 16.18). In the case of receiving materials, the average is 43.4 loans per 1,000 population. Both types of transactions exhibit wide variation across different population clusters, particularly on the receiving end of interlibrary loans.

Libraries that promptly secure materials that are not readily available to their patrons provide a higher quality of service than those that cannot or do not respond as quickly (Table 16.19). Most libraries participating in the 1997 PLA study managed to deliver more than 80% of all requested documents within 30 days. In fact, most were able to deliver at least half of the requested documents within 7 days. Standards and targets for prompt action on requested materials adopted by selected libraries testify to the importance of that service in these communities (Table 16.20).

Table 16.15   Public Library Visits per Capita per Year

| Library Service Population | U.S. Department of Education Survey, 1996 | | Public Library Association Survey, 1997 | | | |
	N	Visits per Capita	N	Lower Quartile (25%)	Median	Upper Quartile (75%)
1 million or greater	20	3.2	22	2.4	3.0	4.2
500,000–999,999	52	3.9	60	3.6	4.3	6.0
250,000–499,999	90	3.6	75	2.5	3.8	5.2
100,000–249,999	313	3.7	238	2.8	3.9	5.4
50,000–99,999	510	4.0	248	2.8	4.3	5.8
25,000–49,999	863	4.6	67	4.1	5.4	7.9
10,000–24,999	1,679	5.0	30	4.5	9.0	10.2
5,000–9,999	1,498	5.0	17	2.0	6.2	8.7
2,500–4,999	1,327	4.7	18[a]	3.6[a]	7.1[a]	11.3[a]
1,000–2,499	1,636	4.9	—	—	—	—
Less than 1,000	958	6.2	—	—	—	—
National average	7,720	4.0	—	—	—	—

SOURCE: Adapted from Adrienne Chute and Elaine Kroe, *Public Libraries in the United States: FY 1996*, NCES 1999-306 (Washington, DC: U.S. Department of Education, National Center for Education Statistics, February 1999), p. 33, and Public Library Association, *Statistical Report '98: Public Library Data Service* (Chicago: American Library Association, 1998), pp. 107–108. Used by permission.

a. Includes all responding libraries having a service population of less than 5,000.

N = number of libraries.

Facility Standards

In addition to various facility standards regarding square footage, shelf space, adequate parking, and other physical qualities included among the sets of standards presented earlier in this chapter, standards regarding the library's lighting and its handicap accessibility are highlighted here. Standards for lighting intensity specified in Kansas call for intensity in study areas five times as great as the intensity deemed acceptable for corridors (Table 16.21).

Standards of accessibility prescribed by the federal Americans With Disabilities Act of 1990 (ADA, 42 U.S.C. § 12100 *et seq.*) allow persons with physical disabilities to enjoy the library facility. Although the full set of standards is lengthy, the Kansas Library Association has a checklist for public libraries that addresses many of the major elements (Table 16.22).

Table 16.16   Reference Transactions per Capita per Year

Library Service Population	U.S. Department of Education Survey, 1996		Public Library Association Survey, 1997			
	N	Reference Transactions per Capita	N	Lower Quartile (25%)	Median	Upper Quartile (75%)
1 million or greater	20	2.0	22	1.0	1.6	2.1
500,000–999,999	52	1.5	56	0.8	1.4	2.1
250,000–499,999	90	1.1	72	0.6	1.1	1.6
100,000–249,999	313	0.9	222	0.4	0.8	1.3
50,000–99,999	510	0.8	205	0.4	0.8	1.3
25,000–49,999	863	0.8	55	0.4	0.8	1.6
10,000–24,999	1,679	0.8	23	0.4	0.8	1.3
5,000–9,999	1,498	0.8	14	0.2	0.5	1.3
2,500–4,999	1,327	0.7	9[a]	n.a.	0.6[a]	n.a.
1,000–2,499	1,636	0.8	—	—	—	—
Less than 1,000	958	0.8	—	—	—	—
National average	7,917	1.1	—	—	—	—

SOURCE: Adapted from Adrienne Chute and Elaine Kroe, *Public Libraries in the United States: FY 1996*, NCES 1999-306 (Washington, DC: U.S. Department of Education, National Center for Education Statistics, February 1999), p. 33, and Public Library Association, *Statistical Report '98: Public Library Data Service* (Chicago: American Library Association, 1998), pp. 107–108. Used by permission.

a. Includes all responding libraries having a service population of less than 5,000.

N = number of libraries.

n.a. = not applicable (quartiles are not reported when the number of reporting libraries is less than 11).

Process Targets and Miscellaneous Benchmarks

Efficient internal processes contribute to effective library services and patron satisfaction. The prompt processing of newly received periodicals and other library acquisitions, prompt and accurate reshelving of returned materials, and prompt restoration and return to service of damaged or worn volumes enhance the likelihood that patrons will secure the materials they are seeking. The performance targets and experience of selected municipalities on these and other service dimensions can provide a useful context for planning and assessing library performance (Table 16.23).

Table 16.17 Prompt Response to Reference Inquiries: Selected Cities

Oakland, CA
Percentage answered promptly with no need for referral: 96% (1990), 92% (1991), 95% (1992)

Tacoma, WA
Percentage of telephone inquiries answered within 5 minutes: 95.2% (1991)

Flagstaff, AZ
Target: Successfully respond to at least 95% of reference questions within 1 hour (1999)

Greensboro, NC
Percentage of reference transactions completed on day of receipt: 95% (1998, estimated)

Sunnyvale, CA
Target: 95% answered on same day (1994)

Wichita, KS
Percentage of reference responses completed on day of inquiry: 92.4% (1990), 92.2% (1991)

High Point, NC
Percentage of reference questions resolved within 24 hours: 99.9% (1998)

Alexandria, VA
Percentage answered within 24 hours: 92.5% (1992), 93.0% (1993) (Note: approximately 85% answered immediately)

Table 16.18 Interlibrary Loans per 1,000 Population: 1996

Library Service Population	Interlibrary Loans per 1,000 Population		
	N	Out	In
1 million or more	20	7.0	2.8
500,000–999,999	52	20.8	13.9
250,000–499,999	90	36.3	14.2
100,000–249,999	313	31.4	27.8
50,000–99,999	510	42.9	42.1
25,000–49,999	863	64.2	72.0
10,000–24,999	1,679	93.3	110.5
5,000–9,999	1,498	81.9	115.8
2,500–4,999	1,327	61.1	126.3
1,000–2,499	1,636	42.8	137.1
Less than 1,000	958	66.0	233.6
National average	8,696	41.7	43.4

SOURCE: Adapted from Adrienne Chute and Elaine Kroe, *Public Libraries in the United States: FY 1996,* NCES 1999-306 (Washington, DC: U.S. Department of Education, National Center for Education Statistics, February 1999), p. 33. *N* = number of libraries.

Table 16.19   Document Delivery Rates: 1997

Library Service Population	Percentage Delivered Within 7 Days				Percentage Delivered Within 30 Days			
	N	Lower Quartile (25%)	Median	Upper Quartile (75%)	|N	Lower Quartile (25%)	Median	Upper Quartile (75%)
1 million or greater	1	n.a.	90.0	n.a.	1	n.a.	90.0	n.a.
500,000–999,999	1	n.a.	61.4	n.a.	2	n.a.	n.a.	n.a.
250,000–499,999	3	n.a.	55.0	n.a.	2	n.a.	n.a.	n.a.
100,000–249,999	13	33.5	60.0	64.0	12	84.4	92.6	99.3
50,000–99,999	26	38.5	52.5	72.5	25	72.0	80.0	91.5
25,000–49,999	6	n.a.	51.5	n.a.	6	n.a.	89.0	n.a.

SOURCE: Adapted from Public Library Association, *Statistical Report '98: Public Library Data Service* (Chicago: American Library Association, 1998), pp. 107–108. Used by permission.

N = number of libraries reporting.

n.a. = not applicable (quartiles are not recorded when the number of reporting libraries is less than 11).

Table 16.20   Library Performance in Securing Materials Not Immediately Available: Selected Cities

Palo Alto, CA
Percentage of interlibrary loan items secured within 21 days: 83% (1997)

Fayetteville, AR
78% of requests filled within 30 days (1991)

Greensboro, NC
Percentage of materials delivered within 7 days: 53% (1997)
Percentage of materials delivered within 30 days: 67% (1997)

Houston, TX
Targets: Fill requests from system libraries within 24 hours; achieve average within-region turn-around of 10 days and outside-region turnaround of 20 days

San Antonio, TX
Percentage of interlibrary loan requests filled: 55% (1997)
Average turnaround time: 12 days (1997)

Amarillo, TX
Targets: Average response time of 10 days for interlibrary loan items to other libraries; 20 days for items from other libraries; process requests within 24 hours

Table 16.21   Lighting Intensity Standards for Public Libraries in Kansas

	Equivalent Sphere Illumination (ESI) Levels
Study areas	100+
Casual reading/work areas	0–70
Restrooms	70
Corridors	20

SOURCE: Kansas Library Association/Public Library Section, *Measurements of Quality: Public Library Standards for Kansas* (Hutchinson, KS: Author, 1995). Used by permission.

Note

1. Much of this section on the history of library standards is drawn, with few adaptations, from David N. Ammons (1995), "Overcoming the Inadequacies of Performance Measurement in Local Government: The Case of Libraries and Leisure Services," *Public Administration Review, 55* (January/February), pp. 37–47. Used with permission.

Table 16.22　Meeting the Requirements of the Americans With Disabilities Act (ADA): A Checklist for Public Libraries

Parking

Does the library have handicapped parking spaces on a level, hard surface? Is the path between handicapped parking and the main entrance barrier free, with no curbs? If you have less than 25 parking spaces, one must be handicap accessible. If you have over 25 but less than 50, two need to be accessible. For all libraries, one parking space must be van-accessible, defined as a parking space 8 feet wide plus an aisle 8 feet wide. However, two parking spaces may share a common access aisle.

Exterior Signage

Does the library have clear, large exterior signs that can be read even by people with poor eyesight?

Main Entrance

Can patrons enter your library easily if they are in a wheelchair, on crutches, or plagued by a bad back? The accessible path to the main entrance should be a minimum of 36" wide with no slope greater than 1 in 20 (a 1" rise to a 20" run). The door should have a minimum clear opening of 32". Automatic doors are not necessary. Excessively heavy doors should be considered a barrier. Exterior steps are a serious barrier that should not be lightly dismissed. Doorway thresholds should be no more than ¾" high.

Ramps

If a ramp is necessary to access the building, it should have a maximum rise of 1 in 12 (a 1" rise to a 12" run). There can be no run greater than 30' without a level platform. A level platform at the doorway is required and it must be a minimum of 5' × 5'. Ramps should be equipped with both railings and guard curbs.

Multilevels

Can patrons access all levels of the library? If not, can nonpublic services be transferred to nonaccessible areas? If an elevator is too expensive, has a lift been investigated? Has the staff been well trained in helping patrons retrieve materials from nonaccessible collections? Are there backup services such as books-by-mail, deposit collections, service to care facilities, service to the homebound?

Stairs

Stairs should be at least 36" wide with risers no more than 11" high and with handrails on both sides. Few small libraries can afford the correction of staircases (other than the addition of handrails) but guidelines should be kept in mind for renovations. Stairs should be only a backup to elevators or lifts in all public libraries.

Restrooms

The entrance to the restroom should be a minimum of 36" wide with a clear space of 5' inside the door. The entrance to the handicapped stall, if there is one, should be at least 36" wide. The handicapped stall should be 5' × 5', though 5' × 8' is preferred. The toilet should be equipped with grab bars at the side and back; these should be 33" to 36" from the floor. The back grab bar should be 36" long and the side bar 42" long. The paper dispenser should have continuous paper delivery. All fixtures should be less than 48" off the floor, including mirror, soap dispenser, and towel rack. The sink faucet should be operated with a pushing rather than a twisting motion. The sink pipes should be either covered or wrapped to prevent burns. The restroom should be well lighted. The restroom should be equipped with both verbal and pictorial signs.

(continued)

Table 16.22   Meeting the Requirements of the Americans With Disabilities Act (ADA): A Checklist for Public Libraries (continued)

Meeting Rooms

The meeting room should have a large sign. The door should be at least 36" wide. Side and center aisles should be 40" wide when the seating is in place. There should be 1 wheelchair-accessible space for every 25 seats, and these should be located with a clear view of the front of the room.

Interior Signage

The library should have clear, large signs with maximum contrast directing people to departments, restrooms, elevators, meeting rooms, public service desks, and specific stack areas. Stack signs should be large enough and clear enough for people with poor eyesight to read.

Floors

Ripples, edges, and worn or loose places in the library carpet should be regarded as a hazard, and repaired or eliminated if possible.

Furniture

Is there a clear passage of at least 40" between one set of study tables and chairs and another? Does the library provide comfortable chairs in addition to hardback study chairs? In order for a person in a wheelchair to use reading/study seating, it must have at least a 27"-high knee clearance, a 19"-deep knee clearance, and a surface height of 28" to 34". Not all study areas need to be wheelchair accessible but part of them should be.

Public Access Catalogs

An accessible catalog should be considered a top priority since it is essential to using library services. The top of the card catalog should be no more than 48" off the floor. There should be a clear space of 36" around the catalog. If the library uses public access terminals, 1 in 5 of these should be at half height so it can be used in a seated position.

Circulation Desk

Ideally, the circulation desk should include a lower section for the use of children and disabled patrons. If a tall circulation desk cannot be restructured, a small table nearby should be used for supplementary service when needed.

Periodicals

The top row of periodicals should be no more than 50" off the floor. Most library periodical shelving is higher than this. Even if complete new shelving cannot be purchased, it is strongly recommended that the top shelf not be used since this is an essential and popular service.

Reference

It is preferable, although not legally specified, that ready reference materials be in half-height shelving. If the reference collection is in full-height shelving, the reference staff needs to be fully aware of their responsibilities to disabled citizens.

Stack Areas

Stack areas should have aisles of at least 36". The perimeter around the stacks should be 40". If there is no stack perimeter and one cannot be created, the stacks should be a minimum of 42" apart. Stack signs should be large enough to read easily. There is no legal limit on stack height.

Water Fountain

If the library has a full-height water fountain, a cup dispenser should be placed at shelf height. The addition of a water cooler is considered fuller compliance.

Telephone

The top of a public service telephone should be no more than 48" off the floor.

Copier

The copy plate of a copier should be no more than 36" off the floor. Most library copiers are in compliance.

Library Materials

Smaller libraries will have to judge very carefully what they can afford to purchase in the way of alternative format materials. Such decisions should be made after consultation with disabled citizens, since they often know best what they need in the way of library service.

Small libraries should have collections of large print materials, although in many cases they do not own these but are served through rotating collections.

Few small libraries can afford braille materials but they still have an obligation to provide these materials through interlibrary loan if requested. If the library offers telephone reference service, the purchase of a TDD (telecommunications device for the deaf) should be considered.

Sign language interpreters should be provided upon request for library programs and this service should be announced with the program.

Outreach service to those in resident institutions and to the homebound should be offered or undue hardship should be documented. Generally, the rule is that the less accessible your library is to the public, the more you should work on outreach.

SOURCE: Kansas Library Association/Public Library Section, *Measurements of Quality: Public Library Standards for Kansas* (Hutchinson, KS: Author, 1995). Used by permission.

Table 16.23   Process Targets and Miscellaneous Library Benchmarks from Selected
Cities

PROMPT ACQUISITION

High Point, NC
  Target: Acquire anticipated *New York Times* Bestseller titles within 60 days of original purchase
  request
  Actual: 71% (1998)

PROMPT PROCESSING OF ACQUISITIONS

Greensboro, NC
  Target: 80% of all books will be available for checkout within 10 days after being ordered

High Po int, NC
  Target: Add newly acquired *New York Times* Bestseller titles to database within 10 days of receipt
  Actual: 95% (1998)

San Jose, CA
  Target: 95% of new items available to the public within 30 days

Oakland, CA
  90% available for public use within 1 month of receipt (1992)

Flagstaff, AZ
  Target: Catalog all new materials within 30 days of receipt (1999)
Denver, CO
  Time from receipt of new material to availability on the shelf: 45 days (1998)

Palo Alto, CA
  Target: Make 75% of new materials available to the public within 45 calendar days of receipt
  Actual: 58% (1997)

PROMPT PROCESSING OF PERIODICALS

Lubbock, TX
  99% of periodicals processed within 24 hours (1991)

ONLINE CATALOG UPTIME

Tucson, AZ
  99.8% (1997)

Houston, TX
  97% (1992)

SPEEDY CHECK OUT

Flagstaff, AZ
  Target: Checkout desk waiting time of 5 minutes or less at least 90% of the time (1999)

ACCURATE CHECK-IN

High Point, NC
  Percentage of returned items checked in accurately: 99% (1998)

PROMPT RESHELVING OF RETURNED MATERIALS

Tacoma, WA
  Percentage within 18 hours: 60% (1991)

Burbank, CA
  Target: Within 1 working day

Glendale, AZ
  Actual: Within 24 hours (1997)

Oak Ridge, TN
  Target: Within 24 hours

Sunnyvale, CA
  Target: Within 24 hours

High Point, NC
  Target: Shelve 95% of all adult books within 24 hours of their return
  Actual: 95% of adult books on shelves or checked out again within 24 hours of being returned
    (1998)

Flagstaff, AZ
  Target: Reshelve at least 90% of materials within 24 hours of return (1999)

Palo Alto, CA
  Target: At least 75% within 2 working days
  Actual: 77% (1997)

ACCURATE RESHELVING

High Point, NC
  Percentage of adult books shelved accurately (sample): 94% (1998)

CHILDREN'S SERVICES

High Point, NC
  Number of reference questions asked per child in local population: 2.1 (1998)
  Number of reference questions per staff member: 2,426 (1998)
  Program attendance per child in local population: 1.4 (1998)
  Number of children attending programs per staff member: 1,566 (1998)

PROMPT MENDING/REBINDING

Sunnyvale, CA
  Target: 75% repaired within 30 days (1994)

Oakland, CA
  Target: 75% repaired within 2 months (1993)

OVERDUE MATERIALS

Duncanville, TX
  End-of-year overdues as percentage of annual circulation: 0.2% (1997)

COLLECTION RATE—FINES

Palo Alto, CA
  Target: 90%
  Actual: 88% (1997)

RATE OF LOSSES FROM COLLECTION

Palo Alto, CA
  Target: Of items checked out, at least 97% returned or paid for
  Actual: 100% (1997)

Tacoma, WA
  Annual loss of 0.32% of circulation (1991)

17 Management Services

Executive Offices, Budget, and Management Audit

It is no simple task to assess the performance of the three functions grouped together here under the heading *management services*. The executive, budgetary, and evaluative functions in local government call for skillful administrative leadership, attentive coordination, prudent resource allocation, conscientious evaluation, careful analysis for performance improvement, and appropriate levels of control—whether mostly centralized or mostly decentralized, whether performed formally or informally. These functions must be handled adroitly if a city wishes to attain high levels of efficiency and effectiveness.

Although it is true that many city managers, budget officers, and auditors are judged subjectively (e.g., as a "great leader," "good communicator," "shrewd," "meticulous," or "insightful"), more objective bases for assessing the performance of these functions also exist.

City Manager's Office

Some cities assign to the Office of the City Manager the responsibility for overarching objectives—that is, those objectives of the municipality that do not otherwise fit neatly within the domain of a single department. Objectives regarding intergovernmental relations, economic development, and workforce diversity are common examples. These are objectives that require leadership and coordination of efforts if they are to be achieved. Some managers also report municipal bond ratings, comparative tax rates, and statistics regarding the number of FTE municipal employees per 1,000 population as evidence of their performance.[1] Many offer statistics regarding their responsiveness to inquiries from members of the city council or to inquiries and complaints from the general citizenry. A few also provide evidence regarding the quality or completeness of the staff work performed preliminary to council action.

Most of the cities examined for this volume strive to respond quickly to inquiries and complaints, responding at least with an initial call within a day or two (Table 17.1). For example, the Office of the City Administrator in Ann Arbor, Michigan, responds to most citizen calls within 4 hours. Many cities, however, focus less on the initial reply than on completion of the response. Because providing the requested information or resolving the problem might not be possible at the time of initial contact, they set their performance targets accordingly—typically calling for completion within several working days.

Top-notch managers strive to present carefully considered, sound recommendations to their community's governing body. Blacksburg, Virginia, and Chapel Hill, North Carolina, focus on the quality and completeness of staff reports by setting ambitious standards that minimize the number deemed "insufficient" or "incomplete" by their councils (Table 17.2).

Budgeting

Just as it has established criteria for excellence in financial reporting, the GFOA (1993) similarly has designed standards for the presentation of budgetary information. "Distinguished Budget Presentation Awards" are conferred on local governments whose budgets are deemed meritorious on four dimensions:

1. As a policy document
2. As a financial plan
3. As an operations guide
4. As a communication device (pp. 1–6)

The GFOA does not attempt through its award program to prescribe a particular budget method or format, but it does regard the above functions to be key roles of the budget. Fairly specific criteria are articulated within each role, only some of these criteria being considered mandatory for recognition as a "distinguished budget" (Table 17.3).

Some of the most critical work of the budget office may involve developing recommendations that could influence community strategies and service priorities. In some cases, the budget office performs management audits or serves as an internal "consultant" to advise other departments on operational improvements. Other budgetary work of a more procedural or regulatory nature is also important for maintaining a smoothly operating organization. Such duties include handling budget adjustment requests (Table 17.4), monitoring revenue and expenditure patterns, and ensuring adherence to schedules for budget documentation and budget requests.

Table 17.1   Prompt Response to Requests and Complaints: Selected Cities

PROMPT RESPONSE TO CITY COUNCIL

Palo Alto, CA
Target: Respond to council member calls and inquiries within 8 hours (1999)

Irving, TX
Target: Establish initial contact on 95% of council and citizen inquiries within 24 hours of receipt
Actual: 95% (1995)

Overland Park, KS
Target: Initial response within 2 workdays

Lubbock, TX
Information requested by city council provided within 3 working days: 98% (1991)

Ann Arbor, MI (City Administrator)
Percentage of council requests responded to within 4 workdays: 90% (1998)

PROMPT RESPONSE TO CITIZENS

Ann Arbor, MI (City Administrator)
Percentage of calls responded to within 4 hours: 80% (1998)

Hurst, TX
Percentage of citizen requests responded to within 1 day: 99% (1997)

Overland Park, KS
Target: Initial response within 2 workdays

Lubbock, TX
Information requested by citizens provided within 3 working days: 98% (1991)

San Clemente, CA
Target 1: Respond to at least 90% of customer complaints/requests within 4 working days
Actual: 93% (1998)

Target 2: At least 85% resolved or customer informed of action plan within 14 days
Actual: 77% (1998)

Corvallis, OR
Target: Respond to citizens' requests within 5 days of receipt
Actual: 95% (1992)

Orlando, FL (Mayor's Office)
Percentage of responses to citizen requests/problems within 24 hours: 45% (1994), 73% (1996);
in 2–5 days: 37% (1994), 18% (1996); in more than 5 days: 18% (1994), 9% (1996)

Palo Alto, CA
Target: At least 50% of the complaints received by the city council will be responded to by letter
within 10 working days (1999)

Chapel Hill, NC
Target: At least 95% resolved within 30 days

Ames, IA
Percentage of citizen requests resolved in 1 day: 38% (1997); in 2–7 days: 30% (1997); in 1–4 weeks:
5% (1997); in more than 4 weeks: 27% (1997)

Table 17.2   Quality Staff Work: Selected Cities

Blacksburg, VA
  Percentage of staff reports to the town council referred back to staff due to insufficient information or analysis: 0% (1997)

Chapel Hill, NC
  Target: No more than 2 agenda item decisions per year delayed by council due to incomplete staff work

---

Management Audit

Whether performed by the City Manager's Office, by the budget office, or by a separate internal auditor, the careful review of service quality and efficiency in various municipal departments and the thoughtful consideration of alternative practices often lead to substantial performance improvements. But how can a unit specializing in internal consulting or management audits be judged? Several insights may be gleaned from a comparative study of 58 local government audit departments conducted by the National Association of Local Government Auditors (NALGA; Table 17.5).

A city wishing to maintain a relatively lean audit unit might, based on the NALGA study, pursue a staff size no greater than one auditor per 700 local government employees and a budget no greater than $1 for every $1,400 of total local government expenditures. To maintain a good turnaround on performance audits, most should be completed within about 4 months—and some much more quickly than that. To be considered one of the more effective audit units, at least 90% of its recommendations should be accepted and more than 60% should be implemented. The ratio of savings resulting from audits to audit expenditures should be at least $2.60 in savings for every audit dollar spent.

Among the cities examined for this volume, several reported the acceptance or implementation rates for management recommendations offered by their budget office or internal audit staff (Table 17.6). Many fared well in comparison with the NALGA set of local governments. Although fewer of the examined cities report benefit ratios, those that do would be among the leaders, reporting more than $5 in cost savings, revenue enhancements, or productivity gains for each dollar spent (Table 17.7).

Table 17.3   GFOA's Criteria for Exemplary Budget Documents

As a Policy Document
- The document should include a coherent statement of organization-wide financial and programmatic policies and goals that address long-term concerns and issues.
- The document should describe the organization's short-term financial and operational policies that guide the development of the budget for the upcoming year.
- The document should include a coherent statement of goals and objectives of organizational units (e.g., departments, division, offices, or programs).
- The document shall include a budget message that articulates priorities and issues for the budget for the new year. The message should describe significant changes in priorities from the current year and explain the factors that led to those changes. The message may take one of several forms (e.g., transmittal letter, budget summary section).

As a Financial Plan
- The document should include and describe all funds that are subject to appropriation.
- The document shall present a summary of major revenues and expenditures, as well as other financing sources and uses, to provide an overview of the total resources budgeted by the organization.
- The document shall include summaries of revenues and other resources, and of expenditures for prior year actual, current year budget and/or estimated current year actual, and proposed budget year.
- The document shall describe major revenue sources, explain the underlying assumptions for the revenue estimates, and discuss significant revenue trends.
- The document shall include projected changes in fund balances, as defined by the entity in the document, for governmental funds included in the budget presentation, including all balances potentially available for appropriation.
- The document should include budgeted capital expenditures and a list of major capital projects for the budget year, whether authorized in the operating budget or in a separate capital budget.
- The document should describe if and to what extent capital improvements or other major capital spending will impact the entity's current and future operating budget. The focus is on reasonably quantifiable additional costs and savings (direct or indirect) or other service impacts that result from capital spending.
- The document shall include financial data on current debt obligations, describe the relationship between current debt levels and legal debt limits, and explain the effects of existing debt levels on current and future operations.
- The document shall explain the basis of budgeting for all funds, whether GAAP, cash, modified accrual, or some other statutory basis.

As an Operations Guide
- The document shall describe activities, services, or functions carried out by organizational units.
- The document should provide objective methods (quantitative and/or qualitative) of measurement of results by unit or program. Information should be included for prior year actual, current year budget and/or estimate, and budget year.
- The document shall include an organization chart(s) for the entire organization.
- A schedule(s) or summary table(s) of personnel or position counts for prior, current, and budgeted years shall be provided, including descriptions of significant changes in levels of staffing or reorganizations planned for the budget year.

As a Communication Device
- The document should provide summary information, including an overview of significant budgetary issues, trends, and resource choices. Summary information should be presented within the budget document either in a separate section (e.g., executive summary) or integrated within the transmittal letter or other overview sections.
- The document should explain the effect, if any, of other planning processes (e.g., strategic plans, long-range financial plans, capital improvement plans) upon the budget and budget process.
- The document shall describe the process for preparing, reviewing, and adopting the budget for the coming fiscal year. It should also describe the procedures for amending the budget after adoption. If a separate capital budget process is used, a description of the process and its relationship to the operating budget should be provided.
- Charts and graphs shall be used, where appropriate, to highlight financial and statistical information. Narrative interpretation should be provided when the messages conveyed by the graphs are not self-evident.
- The document should provide narrative, tables, schedules, crosswalks, or matrices to show the relationship between different revenue and expenditure classifications (e.g., funds, programs, organization units).
- The document shall include a table of contents to make it easy to locate information in the document.
- A glossary should be included for any terminology (including abbreviations and acronyms) that is not readily understood by a reasonably informed lay reader.
- The document should include statistical and supplemental data that describe the organization and the community or population it serves, and provide other pertinent background information related to the services provided.
- The document should be printed and formatted in such a way to enhance understanding and utility of the document to the lay reader. It should be attractive, consistent, and oriented to the reader's needs.

SOURCE: Excerpted from Government Finance Officers Association, *Distinguished Budget Presentation Awards Program: Awards Criteria* (Chicago: Author, August 1993), pp. 1–6. Reprinted by permission.
NOTE: GAAP = generally accepted accounting principles.

Table 17.4   Prompt Response to Budget Adjustment Requests: Selected Cities

Greensboro, NC
Percentage of budget adjustments processed within 3 workdays: 96% (1997)

College Station, TX
Average time per budget adjustment: 5 days (1998)

White Plains, NY
Percentage of budget changes processed within 10 days of approval: 100% (1995)

Table 17.5 Selected Benchmarks for Local Government Auditors, 1998: Based on a NALGA Study of 58 Audit Departments

	Local Government Employees per Auditor (average)	Ratio of Organization Expenditure to Audit Expenditures (average)	Number of Days per Performance Audit (average and range)		Ratio of Dollar Savings to Audit Dollars Spent (average)	Percentage of Audit Recommendations Accepted (average)	Percentage of Audit Recommendations Implemented (average)
			FY97	FY98			
Small audit department (3 auditors or less)	907:1	$1,630:$1	50 (8–106)	45 (9–130)	$2.60:$1	91	73
Medium-size audit department (up to 13 auditors)	662:1	$1,219:$1	93 (8–197)	95 (20–288)	$2.85:$1	94	69
Large audit department (more than 13 auditors)	629:1	$1,257:$1	178 (21–342)	130 (24–287)	$2.65:$1	87	63
Overall	—	—			$2.74:$1	91	69

SOURCE: Peter F. Babachicos, Report on N.A.L.G.A.'s Benchmarking and Best Practices Survey for Fiscal Year 1998 (Lexington, KY: National Association of Local Government Auditors, December 1998).

Table 17.6   Acceptance/Implementation of Management Recommendations Offered by Budget or Internal Audit Staff: Selected Cities

Chesapeake, VA
Percentage of recommendations implemented: 100% (19 of 19) (1997)

Oakland, CA
Percentage of concurrence by auditees with audit recommendations: 99% (520 of 525) (1996)

Glendale, AZ
Percentage of audit recommendations accepted by management: 99% (1997)
Percentage implemented or in-process: 92% (1997)

Norfolk, VA
Percentage of report recommendations accepted/implemented by management: 96% (1997)

Orlando, FL
Percentage concurrence by auditees: 96% (1996)
Percentage of recommendations implemented: 81% (of 262) (1996)

Oklahoma City, OK
Target: At least 96% of audit recommendations accepted by management (2000)

Calgary, Alberta
Percentage of management audit recommendations accepted: 94% (1995), 95% (1996)

San Diego, CA
Percentage of audit report recommendations implemented: 92% (1998)

Overland Park, KS
Percentage of internal audit recommendations accepted by auditees: 88% (1997)

Orlando, FL
Departmental concurrence rate with management recommendations: 87% (124 of 142) (1994), 85% (56 of 66) (1996)

Portland, OR
Percentage of report recommendations implemented
Target: At least 90%
Actual: 86% (1998)

San Antonio, TX
Percentage of recommendations implemented or resolved: 71% (1995)

Shreveport, LA
Percentage of recommendations implemented: 64.9% (1994)

Table 17.7   Value of Management Recommendations From Budget or Internal
                   Audit Staff: Selected Cities

San Jose, CA
  Target: Ratio of at least 3:1 for identified cost savings or revenue enhancements to audit cost
  Actual: 5.7:1 (1995)

Greensboro, NC
  Target: Annual productivity gains equal to at least 50% of the cost of budget and evaluation services
  Actual: 535% (1997)

Orlando, FL
  Ratio of billings and annual recurring benefits to costs: 5:1 (1996)

Note

1. Although each of these measures can indeed be influenced by managerial performance, they easily can be misconstrued. Bond ratings are only partially the product of prudent financial management. Tax rate comparisons may be distorted by the timing of comprehensive property reappraisals or differing assessment-to-market value ratios. A low number of employees per 1,000 population could simply reflect a limited scope of services or a greater than average tendency to contract for services.

18

Parking Services

Cities hope to break even or perhaps even come out financially ahead on the parking spaces and facilities they operate, but public parking operations function primarily as a service to motorists and downtown businesses. Without the metered on-street spaces that many cities maintain and the parking lots and garages operated by some, the scarcity of available parking places would discourage downtown visits. Early arrivers—often the employees of downtown businesses—would capture prime unmetered spots directly outside their workplace and would have no incentive to relinquish them, thereby depriving potential customers of a chance at convenient parking. Winding endlessly through city streets in search of an elusive parking spot would persuade many prospective customers and clients to do their future business elsewhere.

Recommended Standards

Recommended standards for parking services focus on meter reliability, maintenance and collection efficiency, and achieving a favorable cost-benefit ratio (Table 18.1). Ideally, according to the prescribed cost-benefit standard, the parking services operation will return $10 in revenue for each dollar spent. In contrast, a panel assembled by the League of California Cities (LCC), considered a return of $4 for each dollar spent to be a suitably ambitious target (Table 18.2).

A 1996 survey of 37 cities confirms the reasonableness of the LCC standard for operating revenues (Kuzemka, 1997). On average, the cities collected $3 in revenue for each dollar of program expense—a figure matching the LCC standard for medium-service cities (Table 18.3). Portland, Maine, the revenue pacesetter among the 37 cities, confirmed the possibility of achieving the more ambitious 10:1 standard by collecting $16 in revenue for each dollar of program expense. Survey results provided a field test for other standards as well.

Table 18.1 Recommended Performance Standards for Parking Meter Management

	Recommended Standard
Program cost-benefit	$10 in revenue for each program dollar spent
Ratio of meters to maintenance staff	2,000 meters per maintenance employee
Average meter downtime	1%–2 %
Collection route size	200–250 meters

SOURCE: Katherine Kuzemka, "Measuring Your Parking Meter Program," *The Parking Professional* (November 1997), p.18. Used by permission.

Table 18.2 Parking Management Standards: League of California Cities

	Agency Service Level		
	High	Medium	Low
Parking Structures			
Ratio of revenues to operating costs	4:1	3:1	2:1
Ratio of paid to free parkers	15:1	10:1	5:1
Average occupancy rate	80%	60%	50%
"Full" occupants	Daily (more than 1 hour)	3 times per week (more than 1 hour)	Once per week (more than 1 hour)
Metered Street and Lot Spaces			
Cost per space to collect and maintain	5% of meter revenue	10% of meter revenue	20% of meter revenue
Actual revenue, as a percentage of potential revenue per meter	90%	75%	50%
Number of citations issued per parking enforcement officer daily	100	75	50

SOURCE: League of California Cities, *A "How To" Guide for Assessing Effective Service Levels in California Cities* (Sacramento, CA: Author, 1994), p. 27. Used by permission.

Statistics From Selected Cities

Several cities examined for this volume reported parking revenues as a percentage of operating expenses or revenues per parking space (Table 18.4). Although some reported revenues only slightly greater than expenses, others, such as Kansas City, Missouri, achieved a substantial return on investment.

Table 18.3   Parking Meter Management Norms: 1996

	1996 Survey Results (N = 37 cities)	
	Average	*Pacesetter*
Ratio of revenues to program expenses	3:1	Portland, ME (16:1)
Occupancy rate	82%	
Meter downtime	4.8%	
Meters per repair technician	667	
Meter repairs per technician day	8	Portland, OR (63)
Annual revenue per meter		
Large programs (more than 10,000 meters)	$853	New York, NY ($965)
Medium programs (5,000 to 9,999 meters)	$609	Seattle, WA ($1,042)
Small programs (less than 5,000 meters)	$549	Pasadena, CA ($1,464)
Monthly tickets per meter		
Large programs	3.9	
Medium programs	1.7	Portland, OR (4.5)
Small programs	2.3	

SOURCE: Katherine Kuzemka, "Measuring Your Parking Meter Program," *The Parking Professional* (November 1997), pp.16–23. Used by permission.

Savannah, Georgia, and Oakland, California, boast operable meters at the 98% level or greater (Table 18.5). When meters malfunction, Oklahoma City gets them repaired within 1 workday.

Some cities report rental and occupancy rates for parking facilities (Table 18.6). Others report production ratios for enforcement, collection, and maintenance personnel, as measures of efficiency (Table 18.7).

Table 18.4   Revenue Ratios for Parking Services: Selected Cities

REVENUES AS A PERCENTAGE OF OPERATING EXPENSES

Kansas City, MO
Revenues as a percentage of operating costs per parking meter:719.4% (1995)

San Antonio, TX
Garage and lot spaces: 143.4% (1997)
Parking meter spaces: 487.5% (1997)

Cincinnati, OH
196.9% (1991)

Winston-Salem, NC
Revenues for off-street parking facilities as a percentage of operating costs: 166.7% (1998)

San Jose, CA
Target: Parking fund revenues at least 10% in excess of expenses
Actual: 12% in excess (1995)

Savannah, GA
Parking revenues as a percentage of expenditures: 104.9% (1997)

REVENUES PER PARKING SPACE

San Antonio, TX
Garage and lot revenue per space per year: $690 (1995)
Parking meter revenue per space per year: $399 (1995)

Santa Ana, CA
Annual revenue per meter: $360 (1995)

Chapel Hill, NC
Revenue per space per day: $1.33 to $8.66 (range includes different lots and parking deck) (1999)

Table 18.5   Parking Meter Maintenance: Selected Cities

OPERABLE METERS

Savannah, GA
Percentage of meters functioning properly daily: 99.5% (1997)

Oakland, CA
Percentage of parking meters in working condition: 98% (1996)

New York, NY
Percentage of meters operable on-street: 91.1%; off-street: 94.0% (1998)

PARKING METER REPAIRS

Oklahoma City, OK
Percentage of faulty meters repaired within 1 workday of notification: 100% (1999)

Cambridge, MA
Percentage of responses within 48 hours to complaints of malfunctions/repairs to meters:
   80% (1995)
Percentage of meter poles and meter heads replaced within 20 days after reported knocked down:
   70% (1995)

Table 18.6   Parking Rental and Occupancy Rates: Selected Cities

SPACES RENTED

Winston-Salem, NC
Percentage of monthly lease spaces in off-street decks committed: 90% (1997)

Greensboro, NC
Percentage of parking deck spaces rented: 61% (1997)

OCCUPANCY RATE

Savannah, GA
Occupancy range among city-operated surface lots: 80% to 100% (1997)
Occupancy range among city parking garages: 91% to 148% (1997)

San Luis Obispo, CA
Percentage occupancy of garages: 57% (1994)

Table 18.7   Production Ratios in Parking Services: Selected Cities

Nashville–Davidson County, TN
Average number of tickets issued per parking patrol officer: 20,340 (1997)

Winston-Salem, NC
Number of parking tickets issued per enforcement person hour: 8.02 (1998)

Cincinnati, OH
Meters collected per week per person: 1,265 (1993)

Santa Ana, CA
Meters-to-technician ratio: 417 to 1 (1995)

19

Parks and Recreation

In 1906, the fledgling Playground Association of America (PAA) called for playground space equal to 30 square feet per child. Although the formulators of this standard acknowledged the existence of "no inherent relation between space and children," they reported the establishment of such a requirement in London and thought it reasonable (cited in Gold, 1973, p. 144; Lancaster, 1983, p. 15). Thus began a long series of standards adopted by leading associations in the parks and recreation field that have established ratios between recommended park acreage or recreational facilities and population. These ratios are the best known, but not the *only* or least controversial, standards in parks and recreation.

History of Parks and Recreation Standards[1]

Following the adoption of the 30-square-feet-per-child standard for playground space by the PAA, a successor organization subsequently modified the playground standard to 1 acre per 2,000 schoolchildren (Krohe, 1990, p. 10). In time, rule-of-thumb ratios emerged for park acreage in general, with 10 acres of recreation space per 1,000 population eventually becoming the most widely accepted norm (Bannon, 1976, p. 209; Haley, 1985, p. 176; Kraus & Curtis, 1982, p. 117; Krohe, 1990, p. 11; Lancaster, 1983, p. 11). Other normative guides also have been cited as "traditional standards"—for example, the preservation of 10% of a jurisdiction's total area as parkland and the expenditure of $8 (in 1965 dollars) per capita for public recreation and parks (Haley, 1985, p. 176)—but have been less widely accepted.

Through the years, park standards were refined by the National Recreation and Park Association (NRPA) and eventually centered on a recommendation "that a park system, at a minimum, be composed of a 'core' system of parklands, with a total of 6.25 to 10.5 acres of developed open space per 1,000 population" (Lancaster, 1983, p. 56; used with the permission of the National

Recreation and Park Association). NRPA's 1983 guidelines made further recommendations regarding an appropriate mix of park types (e.g., miniparks, neighborhood parks, and community parks), sizes, service areas, and acreage per 1,000 population (Table 19.1). In addition, the association offered standards regarding the availability of particular types of recreational facilities (Table 19.2). For example, the recommended availability of tennis courts in the 1983 guidelines was one court per 2,000 population, compared with one baseball diamond per 5,000, one football field per 20,000, and one swimming pool per 20,000 population (Lancaster, 1983, pp. 60–61).

More recent efforts of the NRPA and the American Academy for Park and Recreation Administration have focused on the development of accreditation standards for local public park and recreation agencies (National Commission on Accreditation for Local Park and Recreation Agencies [NCALPRA], 1992). In general, these accreditation standards are directed less toward outputs, outcomes, and performance results than toward planning, organizational structure, and management processes.

Performance Measures

When Ridley and Simon (1943) offered advice on measuring municipal recreation activities, they suggested three possibilities: "units of cost; units of effort expended—facilities and personnel provided; and units of performance, including records of attendance" (p. 32; reprinted by permission).

Their advice has been at least partially followed in the hundreds of municipalities whose annual budgets dutifully report the number of ball fields maintained, the number of FTE employees on the payroll, and the number of participants in municipally operated basketball leagues, softball leagues, aerobics courses, and arts and crafts classes. Fewer cities by far report unit costs of recreation services. Some, however, have incorporated measures of program quality, occasionally addressing user satisfaction directly (for suggested measures based on household or user surveys, see Hatry et al., 1992, pp. 36–37).

A point of contention in the design of performance measures for parks and recreation and, ultimately, in the identification of benchmarks, centers on the question of whether recreational facilities and parks themselves are properly considered program inputs or outputs. As noted repeatedly in this volume, measures of program inputs are poor substitutes for measures of outputs or results as gauges of performance.

Consider an example or two drawn from other municipal operations. Police cars and jails are not the products of the police function; they are tools used by police department employees as they perform that function—they are *means* to the desired ends. Similarly, a wastewater treatment plant is not the product of the sewer operation; it, too, is a tool—in other words, a resource

(text continues on page 261)

Table 19.1 Neighborhood and Community Park Acreage Standards

Component	Use	Service Area	Desirable Size	Acres per 1,000 Population	Desirable Size Characteristics
Minipark	Specialized facilities that serve a concentrated or limited population or specific group such as tots or senior citizens	Less than $\frac{1}{4}$ mile radius	1 acre or less	0.25 to 0.5 acre	Within neighborhoods and in close proximity to apartment complexes, town house developments, or housing for the elderly
Neighborhood park/playground	Area for intense recreational activities, such as field games, court games, crafts, playground apparatus area, skating, picnicking, wading pools, etc.	$\frac{1}{4}$ to $\frac{1}{2}$ mile radius to serve a population up to 5,000 (a neighborhood)	15+ acres	1.0 to 2.0 acres	Suited for intense development; easily accessible to neighborhood population—geographically centered with safe walking and bike access; may be developed as a school-park facility
Community park[a]	Area of diverse environmental quality; may include areas suited for intense recreational facilities, such as athletic complexes, large swimming pools; may be an area of natural quality for outdoor recreation, such as walking, viewing, sitting, picnicking; may be any combination of the above, depending on site suitability and community need	Several neighborhoods; 1- to 2-mile radius	25+ acres	5.0 to 8.0 acres	May include natural features, such as water bodies, and areas suited for intense development; easily accessible to neighborhood served

Recommended combined acreage: 6.25 to 10.5 acres per 1,000 population

SOURCE: Excerpted from Roger A. Lancaster (Ed.), *Recreation, Park and Open Space Standards and Guidelines* (Alexandria, VA: National Recreation and Park Association, 1983), pp. 56–57. Used with the permission of the National Recreation and Park Association.

a. Community parks are distinguished not only from miniparks and neighborhood parks but also from regional parks or open space preserved as a conservancy.

Table 19.2 Recreational Facility Development Standards

Activity/Facility	Recommended Space Requirements	Recommended Size and Dimensions	Recommended Orientation	Number of Units per Population	Service Radius	Location Notes
Basketball						
1. Youth	2400–3036 sq. ft.	46'–50' × 84'	Long axis north-south	1 per 5,000	$\frac{1}{4} - \frac{1}{2}$ mile	Usually in school, recreation center, or church facility; safe walking or bike access; outdoor courts in neighborhood and community parks, plus active recreation areas in other park settings
2. High School	5040–7280 sq. ft.	50' × 84'				
3. Collegiate	5600–7980 sq. ft.	50' × 94' with 5' unobstructed space on all sides				
Baseball						
1. Official	3.0–3.85 acre minimum	Baselines—90'; Pitching distance—60.5'; Foul lines—min. 320'; Center field—400'+	Locate home plate so pitcher throwing across sun and batter not facing it. Line from home plate through pitcher's mound runs east-northeast.	1 per 5,000; lighted— 1 per 30,000	$\frac{1}{4} - \frac{1}{2}$ mile	Part of neighborhood complex; lighted fields part of community complex

(continued)

257

Table 19.2 Recreational Facility Development Standards (continued)

Activity/Facility	Recommended Space Requirements	Recommended Size and Dimensions	Recommended Orientation	Number of Units per Population	Service Radius	Location Notes
2. Little League	1.2 acre minimum	Baselines—60'; Pitching distance—46'; Foul lines—200'; Center field—200'–250'				
Football	Minimum 1.5 acres	160' × 360' with a minimum of 6' clearance on all sides	Fall season—long axis northwest to southeast; for longer periods, north to south	1 per 20,000	15–30 minutes travel time	Usually part of baseball, football, soccer complex in community park or adjacent to high school
Golf						
1. Par 3 (18-hole)	50–60 acres	Avg. length varies— 600–2700 yards	Majority of holes on north-south axis	—	$\frac{1}{2}$–1 hour travel time	9-hole course can accommodate 350 people a day
2. 9-hole standard	Minimum 50 acres	Avg. length—2250 yards		1 per 25,000		18-hole course can accommodate 500–550 people a day
3. 18-hole standard	Minimum 110 acres	Avg. length—6500 yards		1 per 50,000		Course may be located in community or district park, but should not be over 20 miles from population center
Golf— driving range	13.5 acres for minimum of 25 tees	900' × 690' wide: add 12' width for each additional tee	Long axis southwest-northeast with golfer driving toward northeast	1 per 50,000	30 minutes travel time	Part of golf course complex; as a separate unit, may be privately operated

Activity	Size	Dimensions	Orientation	No. of units per population	Service radius	Remarks
Ice Hockey	22,000 sq. ft. including support area	Rink 85' × 200' (minimum 85' × 185'); additional 5,000 sq. ft. support area	Long axis north-south if outdoor	Indoor—1 per 100,000; outdoor—depends on climate	$\frac{1}{2}$–1 hour travel time	Climate important consideration affecting no. of units; best as part of multipurpose facility
Multiple recreation court (basketball, volleyball, tennis)	9,840 sq. ft.	120' × 80'	Long axis of courts with *primary* use is north-south	1 per 10,000	1–2 miles	
$\frac{1}{4}$ mile running track	4.3 acres	Overall width—276'; length—600.02'; track width for 8 to 4 lanes is 32'	Long axis in sector from north to south to northwest-southeast with finish line at northerly end	1 per 20,000	15–30 minutes travel time	Usually part of high school, or in community park complex in combination with football, soccer, etc.
Soccer	1.7–2.1 acres	195'–225' × 330'–360' with a 10' minimum clearance on all sides	Same as football	1 per 10,000	1–2 miles	Number of units depends on popularity; youth soccer on smaller fields adjacent to schools or neighborhood parks
Softball	1.5–2.0 acres	Baselines—60'; pitching distance—46' (men); 40' (women); fast pitch field radius from plate—225' between foul lines; slow pitch—275' (men); 250' (women)	Same as baseball	1 per 5,000 (if also used for youth baseball)	$\frac{1}{4} - \frac{1}{2}$ mile	Slight difference in dimensions for 16" slow pitch; may also be used for youth baseball

(continued)

Table 19.2 Recreational Facility Development Standards (continued)

Activity/Facility	Recommended Space Requirements	Recommended Size and Dimensions	Recommended Orientation	Number of Units per Population	Service Radius	Location Notes
Swimming pools	Varies on size of pool and amenities; usually $\frac{1}{2}$ to 2-acre site	Teaching—minimum of 25 yards x 45' even depth of 3–4 ft.; competitive—minimum of 25m x 16m; minimum of 27 sq. ft. of water surface per swimmer; ratios of 2:1 deck vs. water	None—although care must be taken in siting of lifeguard stations in relation to afternoon sun	1 per 20,000 (pools should accommodate 3% to 5% of total population at a time)	15–30 minutes travel time	Pools for general community use should be planned for teaching, competitive, and recreational purposes with enough depth (3.4m) to accommodate 1m and 3m diving boards; located in community park or school site
Tennis	Minimum of 7,200 sq. ft. single court (2 acres for complex)	36' × 78'; 12' clearance on both sides; 21' clearance on both ends	Long axis north-south	1 court per 2,000	$\frac{1}{4} - \frac{1}{2}$ mile	Best in batteries of 2–4; located in neighborhood/community park or adjacent to school site
Volleyball	Minimum of 4,000 sq. ft.	30' × 60'; minimum 6' clearance on all sides	Long axis north-south	1 court per 5,000	$\frac{1}{4} - \frac{1}{2}$ mile	Same as other court activities (e.g., badminton, basketball, etc.)

SOURCE: Excerpted from Roger A. Lancaster (Ed.), Recreation, Park and Open Space Standards and Guidelines (Alexandria, VA: National Recreation and Park Association, 1983), pp. 60–61. Used with the permission of the National Recreation and Park Association.
NOTE: Recommended standards for archery, badminton, beach areas, field hockey, handball, skeet and trap shooting, and trails are also listed in this source.

input—that is instrumental in achieving the desired product or output, clean water. The parks and recreation function, however, may constitute a special case that stands as an exception to the pattern represented by these examples.

Unlike police and sewer services, much of the parks and recreation function is designed as a self-service operation in which users simply avail themselves of municipally provided facilities. To a degree, therefore, parks and recreational facilities that are properly considered an input in the total recreation experience of community residents are nevertheless outputs as far as the municipality's role is concerned. To the degree that the public expects the city to provide facilities for "self-service" in the realm of recreation, the essential question becomes, "Are the appropriate facilities provided in sufficient quantity and quality?" Where recreation *services* are expected, the question becomes more complicated. In those cases, the facilities needed to provide the desired services are appropriately considered resource *inputs*.

Benchmarks

Benchmarks of potential usefulness to local government officials and citizens wishing to assess the adequacy of the local parks and recreation program may be found in the recommended standards of relevant associations and in the standardized targets and performance indicators of other municipalities. In this chapter, prospective benchmarks will be divided into five groups: parks and facility standards, maintenance standards for facilities and grounds, program operations standards, program offerings and administrative functions, and cost recovery.

Parks and Facility Standards

Standards that prescribe park acreages or types and numbers of recreational facilities have collected their share of critics (for example, Gold, 1973; Shivers & Hjelte, 1971). Differences from one community to another in the age profile of residents, economic conditions, and culture—not to mention climate and population density—are likely to influence the array of facilities considered practical or even desirable. Critics contend that rigid standards ignore the need for flexibility in addressing objectives that may vary from one community to the next. Furthermore, they charge that because such standards often are developed without careful research, they may be meaningless or even harmful.

Perhaps the standard with the highest profile of all in the parks and recreation field is NRPA's recommendation that combined acreage for neighborhood and community parks fall in the range of 6.25 to 10.5 acres of developed open space per 1,000 population (Lancaster, 1983, pp. 56–57)—more often expressed simply as 10 acres per 1,000 population. More than one critic has pointed out that the park acreage needed in New York City for Manhattan to comply with accepted standards would exceed the entire acreage of that island (e.g., Clawson, 1963, p. 20).

Clearly, a one-size-fits-all standard will not be suitable for cases that deviate sharply from the norm around which it was designed. But even those who formulate standards often anticipate, and even recommend, adaptation where conditions warrant. In the NRPA publication that declares most park and recreational facility standards, the author acknowledged the inappropriateness of blind application.

> Ideally, the national standards should stand the test in communities of all sizes. However, the reality often makes it difficult or inadvisable to apply national standards without question in specific locales. The uniqueness of every community, due to differing geographical, cultural, climatic, and socioeconomic characteristics, makes it imperative that every community develop its *own* standards for recreation, parks, and open space. (Lancaster, 1983, p. 37; used with the permission of the National Recreation and Park Association)

Even tailored standards may not endure the test of time, as conditions on which the standards were based begin to change. If even the most carefully customized standards assume static demographic and socioeconomic status in a community, they, too, are likely to be imperfect. Parks remain, even as the neighborhoods around them change. A reasonable beginning point for parks and recreation planning, then, is the assumption that most communities will approximate the norm but that appropriate modifications to the standards should be made for significant deviations in community characteristics. In extreme cases, a given standard may even be declared totally inapplicable.

The appropriately cautious application of park and recreational facility standards should take into consideration the nature and extent of a given community's deviation from the general characteristics of a typical U.S. community, focusing especially on those differences that are likely to affect leisure needs and tastes. Furthermore, communities attempting to apply national standards to their own parks and recreational facility ratios should be aware that *many* communities fall short of the prescribed marks. Therefore, a city that assumes that NRPA park acreage and facility standards are "minimum acceptable levels," rather than "targets of excellence," risks judging itself too harshly. The communities listed in Table 19.3 are widely respected for their municipal operations, yet a large number fall short of the popular 10 acre per 1,000 population standard or even the broader NRPA range of 6.25 to 10.5 acres per 1,000 population.

Table 19.3   Park Acres Per 1,000 Population in Selected Cities

City	Reported Park Acres Per 1,000 Population
Oak Ridge, TN	81.91 (1997)
Little Rock, AR	24.66[a] (1997)
Charlottesville, VA	24.47 (1997)
Decatur, IL	23.45 (1995)
Loveland, CO	20.7[b] (1999)
Carrollton, TX	20.43 (1993)
San Diego, CA	20.11 (2000)
Kansas City, MO	19.43 (1997)
Gainesville, GA	19.12 (1998)
Portland, OR	5.28 developed (1998) 18.08 overall (1998)
Ann Arbor, MI	18.05 (1999)
Greensboro, NC	17.99 acres of "open space" (1998)
Nashville–Davidson County, TN	17.14 (1998)
Rockville, MD	8.93 developed (1997) 17.07 overall (1997)
Orlando, FL	16.7 (1997)
Bellevue, WA	16.34 acres of parks and open space (1998)
Eugene, OR	15.95 (1997)
College Station, TX	15.38 (1996)
Ames, IA	4.85 developed (1997) 15.34 overall (1997)
Grand Prairie, TX	15.33 (1998)
Kalamazoo, MI	15.14 (1996)
Norfolk, NE	14.99 (1996)
Sterling Heights, MI	14.6 (1995)
Shreveport, LA	13.49 (1996)
Boca Raton, FL	13.24 (1998)
Iowa City, IA	12.42 (1991)
Waukesha, WI	11.5 (1992)
Cary, NC	4.8 developed (1999) 11.09 overall (1999)
Ocala, FL	10.80 (1996)
Overland Park, KS	10.75 (1996)
Corpus Christi, TX	10.49 including parks and other city sites maintained by Parks (1998)
Denton, TX	10.38 (1997)
Flagstaff, AZ	10.36 (1998)
Southfield, MI	10.33[c] (1991)

(continued)

Table 19.3   Continued

City	Reported Park Acres Per 1,000 Population
Norman, OK	8.77 developed; 10.17 overall (1999)
Irving, TX	7.69 developed; 8.70 overall (1999)
Fort Collins, CO	8.04 (1997)
Alexandria, VA	7.78 (1998)
Jacksonville, FL	7.63 (1998)
Greeley, CO	5.82 developed (1997) 7.45 overall (1997)
San Antonio, TX	6.55 (1997)
Duncanville, TX	6.14 (1999)
Tempe, AZ	5.99 (1997)
Chandler, AZ	2.33 developed 5.57 overall (1999)
Largo, FL	5.38 (1996)
Norfolk, VA	4.54 (1998)
Albuquerque, NM	3.38 developed (1996) 4.48 overall (1996)
Long Beach, CA	4.14 (1996)
Smyrna, GA	3.58 (1996)
Oak Park, MI	3.55 (1998)
Fort Smith, AR	3.45 (1997)
Reno, NV	3.12 (1997)
San Clemente, CA	3.05 including beaches (1999)

a. Including "park development."

b. Including mountain park (excluding undeveloped parkland leaves 5.8 acres of "developed" park per 1,000 population in Loveland).

c. 6.15 acres per 1,000 population, if Southfield's nature preserves are excluded.

A degree of ambiguity afflicts many parks and recreation measures, making it advisable to consider alternate interpretations when weighing these measures. In reporting park acreage, for instance, some municipalities include natural preserves, land acquired for future parks, and all other open space in the possession of the city, whereas others report only developed parkland. The implications of that ambiguity, whispered via footnotes in Table 19.3, are shouted in Table 19.4—a tabulation compiled for a 1992 study performed by Dallas, Texas, contrasting park acres and *developed* park acres per 1,000 residents in 10 cities. The ranking of Phoenix, for example, is either first or seventh, depending on whether undeveloped mountain acreage owned by that city is included in the tabulation. Four cities in the Dallas study easily exceed

Table 19.4   Two Perspectives on Local Public Parkland: "Total Acres" or
"Developed Acres"?

	Park Acres per 1,000 Residents	Rank	Developed Park Acres per 1,000 Residents	Rank
Phoenix, AZ	30.82	1	1.99	7
Austin, TX	26.50	2	13.96	1
San Diego, CA	21.42	3	6.21	2
Dallas, TX	17.47	4	5.10	3
Houston, TX	7.48	5	4.60	5
San Antonio, TX	5.75	6	5.08	4
Detroit, MI	4.98	7	4.41	6
San Jose, CA	4.09	8	1.57	9
El Paso, TX	3.10	9	1.94	8
Philadelphia, PA	1.58	10	1.51	10

SOURCE: Adapted from City of Dallas (TX) Parks and Recreation Department, *City Comparison Study* (April 1992), pp. 8–9.

the standard of 10 acres per 1,000 population if undeveloped acres are included. Among the 10 cities, only Austin, Texas, complies if undeveloped acres are excluded.

As if the ambiguity over what gets counted as park acreage were not enough, some parkland advocates argue that the ratio of park acreage to the community's total acreage is a comparably important gauge of park adequacy. A community's choice of one ratio or the other (i.e., acreage per 1,000 population or acreage as a percentage of total land) can dramatically affect its assessment of the adequacy of community parkland. As statistics compiled for the Citizens Budget Commission in New York City indicate, even a city with large parkland allocations may fare poorly in terms of acreage per 1,000 residents if the city is densely populated (Brecher & Mead, 1991). New York City ranked number one among 14 cities for park acreage as a percentage of total land at 13.5%, but slid to thirteenth when the comparison switched to park acres per 1,000 residents (Table 19.5).

Standards are sometimes presented as guidelines, sometimes as minimally acceptable levels of quality or performance, sometimes as norms, and sometimes as targets for ambitious municipalities toward which to strive. Although parks and recreation specialists may imply a "minimally acceptable level" interpretation for park acreage standards, the class roll of solid but noncomplying municipalities suggests otherwise. Although some cities manage to achieve many of the parks and facility standards and thereby demonstrate their attainability, many others fall short of the mark.

Table 19.5   Another Perspective on Local Public Parkland: "Percentage of Land Devoted to Parks" versus "Park Acreage per 1,000 Residents"

Selected Cities	Park Acreage as Percentage of Total Land	Rank	Park Acreage per 1,000 Residents	Rank
New York, NY	13.5	1	3.6	13
Baltimore, MD	12.3	2	8.5	5
Phoenix, AZ	11.9	3	34.5	1
Philadelphia, PA	11.4	4	6.0	8
San Francisco, CA	11.1	5	4.6	12
Dallas, TX	9.2	6	20.1	2
Detroit, MI	6.8	7	5.4	10
Houston, TX	5.3	8	11.1	4
Chicago, IL	5.0	9	2.4	14
Los Angeles, CA	5.0	9	4.8	11
Indianapolis, IN	4.1	11	13.1	3
San Diego, CA	3.9	12	8.4	6
San Jose, CA	3.7	13	5.8	9
San Antonio, TX	3.4	14	7.1	7
Mean	7.6		9.7	
Median	6.05		6.6	

SOURCE: Adapted from Charles Brecher and Dean M. Mead, *Managing the Department of Parks and Recreation in a Period of Fiscal Stress* (New York: Citizens Budget Commission, 1991), p. 10.

In its parks and recreation master plan, Tucson, Arizona, inventoried its parks and recreation facilities, compared those facilities with national standards, and, as recommended by NRPA standards promulgators themselves, established its own local standards for park acreage and accessibility (Table 19.6). In some cases, Tucson's holdings exceeded national standards, but in many cases they did not. The presentation of local facilities in the context of national standards affords community citizens and decision makers the opportunity to reconsider local priorities in light of professional standards and to decide which, if any, of those standards they believe applicable to their own community and worthy of pursuit.

National guidelines for types and numbers of recreational facilities provide a context for considering a community's array of facilities and the adequacy of individual components. However, consumers of these standards should not assume too much. For example, the NRPA recommendation of one tennis court per 2,000 population—one of the lowest population-per-facility ratios —attests to the popularity of that sport and the substantial demand for courts in many communities. Nevertheless, a city aspiring to reach this standard

Table 19.6 Comparison of Actual Park and Recreation Facilities and Standards in Tucson, Arizona

Facility Type	1989 Total	Facility per 1989 Population	National Standards
All Parks	3,740 acres	9.4 acres per 1,000	n.a.
Neighborhood	295 acres	0.7 acre per 1,000	2.5 acres per 1,000
District	1,493 acres	3.8 acres per 1,000	2.5 acres per 1,000
Regional	1,952 acres	4.9 acres per 1,000	20.0 acres per 1,000
Field Sports			
Baseball fields	25	1 per 15,892	1 per 5,000
Football fields	20	1 per 19,865	1 per 20,000
Little League	90	1 per 4,414	1 per 5,000
Rugby fields	4	1 per 99,325	n.a.
Soccer fields	46	1 per 8,637	1 per 10,000
Softball fields	36	1 per 11,036	1 per 5,000
Court Sports			
Handball courts	18	1 per 22,072	1 per 20,000
Horseshoe pits	24	1 per 16,554	n.a.
Multiple use courts	38	1 per 10,455	1 per 10,000
Shuffleboard courts	19	1 per 20,911	n.a.
Tennis courts	147	1 per 2,703	1 per 2,000
Volleyball courts	21	1 per 18,919	1 per 5,000
Passive Recreation			
Nature trail	1	1 per 397,300	n.a.
Picnic sites	311	1 per 1,277	n.a.
Picnic ramadas	147	1 per 2,703	n.a.
Active Recreation			
Bike paths	6	1 per 66,217	n.a.
Exercise courses	3	1 per 132,433	n.a.
Golf courses	5	1 per 79,460	1 per 50,000
Jogging paths	5	1 per 79,460	n.a.
Playgrounds	76	1 per 5,228	n.a.
Recreational centers	16	1 per 24,831	n.a.
Running tracks	2	1 per 198,650	1 per 20,000
Swimming pools	22	1 per 18,059	1 per 20,000

Tucson Definitions/Standards:
Park Size

Neighborhood park	0–15 Acres
District park	16–100 Acres
Regional park	100+ Acres

Service Area Size[a]

Neighborhood park	0.5 mile radius from park
District park	3 mile radius from park
Regional park	3 mile radius from park

General Plan Standards

Neighborhood	2.5 acres per 1,000
District	2.5 acres per 1,000
Regional	5 acres per 1,000

SOURCE: City of Tucson (AZ), *City of Tucson Parks and Recreation Master Plan 2000: Planning Our Recreational Future* (December 1989), pp. 44–45.

a. Service areas for picnic *ramadas* and recreation centers are based on 20-minute travel time to the facility.

Table 19.7   Tennis Courts in Selected Cities

City	Number of Tennis Courts per 10,000 Residents
Gainesville, GA	8.8 (1999)
Oak Ridge, TN	8.1 (1997)
Rockville, MD	7.1 (1996)
Norfolk, VA	6.5 (1998)
Oak Park, MI	5.2 (1998)
Albuquerque, NM	3.3 (1996)
Nashville–Davidson County, TN	3.2 (1998)
Norman, OK	3.1 (1999)
Tucson, AZ	3.1 (1998)
Norfolk, NE	3.0 (1997)
Little Rock, AR	2.8 (1997)
Dallas, TX	2.6 (1992)
Flagstaff, AZ	2.6 (1998)
Ames, IA	2.5 (1999)
Austin, TX	2.3 (1992)
Detroit, MI	2.3 (1992)
Fayetteville, AR	2.3 (1998)
Grand Prairie, TX	2.3 (1998)
El Paso, TX	2.1 (1992)
College Station, TX	2.0 (1996)
Orlando, FL	2.0 (1997)
Fort Smith, AR	1.9 (1997)
Jacksonville, FL	1.9 (1998)
Eugene, OR	1.8 (1999)

should realize that many of its most admirable counterparts fall short. The cities listed in Table 19.7 would need a ratio of 5 tennis courts per 10,000 residents to meet the one court per 2,000 standard. Only 5 of these 24 cities did so. If local demand for tennis facilities warrants a community's pursuit of the national standard, achievement of that ratio would be commendable and noteworthy; nevertheless, the context provided by the ratios found in many other cities offers assurances that falling somewhere short is hardly a serious blemish.

In addition to standards that use population ratios to prescribe park acreage and recreational facility needs, other standards offer guidance on the quality of facilities. One such set of standards, offered by the American Public Health Association (APHA), provides guidance for swimming pool operation, as well as for facility design and equipment capacity (Table 19.8).

Table 19.8   Selected Standards for Public Swimming Pools

Circulation system capacity:
Sufficient to clarify and disinfect entire volume of water at least 4 times per 24 hours

Filtration system capacity:
Sufficient to restore turbidity level of pool to 0.5 nephelometric turbidity units (NTUs) or less within 24 hours following peak bather load

Acceptable chemical and physical qualities of pool water:
- pH (hydrogen ion concentration) range of 7.2 to 8.0
- total alkalinity of 50 to 150 mg/L
- maximum chlorine level of 0.2 mg/L
- sufficient clarity to easily view a 6-inch black-and-white disc at the deepest point of the pool from poolside

Standard frequency of pool water testing for pH and residual disinfectant:
Within 1 hour of opening and at least 3 times per operating day from at least 2 locations (shallow and deep)

Standard ratio of attendant/lifeguards to pool area occupants:
1:75

Standard pool area load: The sum of (a) + (b) + (c), where
- (a) = one person per 15 sq. ft. of water surface in the nonswimming areas, i.e., portions of the swimming pool with a water depth of 5 ft. or less
- (b) = one person per 25 sq. ft. of water surface in the swimming areas, i.e., portions of the swimming pool with a water depth greater than 5 ft.
- (c) = 10 persons per diving board

Minimum allowances/requirements for diving boards of various heights:

	Diving Board Height		
	2 ft. or less	2ft. to 1m	1m to 3m
Water depth	8.5 ft.	10 ft.	12 ft.
Length of diving well	12 ft.	12 ft.	13 ft.
Unobstructed height above each board	13 ft.	13 ft.	13 ft.
Distance from adjacent board/side wall	10 ft./10 ft.	10 ft/10 ft.	10 ft./12ft.
Diving stand handrail	—	required	required

SOURCE: Adapted from American Public Health Association, *Public Swimming Pools: Recommended Regulations for Design and Construction, Operation and Maintenance.* Copyright © 1981 by the American Public Health Association. Adapted with permission.

Maintenance Standards

A major responsibility of most parks and recreation departments is the maintenance of park properties, recreational facilities, and sometimes the grounds surrounding other municipal facilities. As noted by a Boston analyst following a disappointing search for clear "industry standards" addressing those duties, much of "the literature . . . on the performance evaluation of

Table 19.9   Parks Maintenance Production Ratios: Selected Cities

Overland Park, KS
Acres maintained per maintenance employee: 82.6 (1998), 84.7 (1999)

Raleigh, NC
Park, greenway, and cemetery acreage per maintenance employee:
53.9 acres (1997), 53.0 acres (1998)

Glendale, AZ
Park acres maintained per park maintenance employee: 52.75 (1997)

San Antonio, TX
Park acres maintained per employee: 39.4 (1997)

Irving, TX
Acres maintained per worker year: 27.46 (1998)

Nashville–Davidson County, TN
Square feet of grounds maintained (library) per grounds maintenance full-time equivalent:
936,932 (21.5 acres) (1997)

Corpus Christi, TX
Acres maintained per full-time equivalent: 20.16 (1998)

College Station, TX
Park acres per full-time employee: 16.4 (1995), 17.3 (1998)

Tempe, AZ
Park/special facility acres maintained per employee: 16.78 (1997)

San Jose, CA
Target: 10 acres of developed park area per field maintenance employee
Actual: 10.60 (1996, estimated)

parks and recreation [is] of an extremely theoretical level (reminiscent of Veblen's *Theory of the Leisure Class*), and not pertinent to groundskeeping and Jiffy-John maintenance" (Terry, 1991, p. 14). Despite their elusiveness, however, standards, guidelines, and a few rules of thumb may be discovered in the recommendations of specialized professional associations and in the self-developed standards and performance reports of individual municipalities. For example, a report prepared by a management analysis team in Pasadena, California, concluded that a ratio of one park maintenance employee for every 7 to 10 acres should produce "A-Level" service—in other words, "a high-frequency maintenance service" (City of Pasadena [CA] Management Audit Team, 1986, p. 9.4). Standards of the maintenance-employee-per-park-acreage variety and corresponding statistics reported by individual cities, however, are complicated by the question of developed versus undeveloped park acreage noted previously and therefore should be interpreted cautiously. Among the cities examined for this volume, most of those reporting

---

Table 19.10   Sample Standards for Tree Care: Excerpts from the American National
Standard

Tools and Equipment

5.2.4.2   Hook and blade pruning tools should be used; not anvil-type pruning tools.

5.2.4.3   Climbing spurs should not be used when climbing trees, except as specified elsewhere in this standard. Climbing spur use is permissible on tree removals and in emergencies such as aerial rescue.

Mature Tree Pruning

5.3.1.2   Tree branches shall be removed in such a manner so as not to cause damage to other parts of the tree or to other plants or property. Branches too large to support with one hand shall be precut to avoid splitting or tearing of the bark. Where necessary, ropes or other equipment should be used to lower large branches or portions of branches to the ground.

5.3.1.3   When a branch is cut back to a lateral, not more than one fourth of its leaf surface should be removed. The lateral remaining should be large enough to assume the terminal role.

5.3.1.4   Not more than one fourth of the foliage on a mature tree should be removed within a growing season.

5.3.1.5   Upon completion of pruning a mature tree, one half of the foliage should remain evenly distributed in the lower two thirds of the crown and individual limbs.

Utility Pruning

5.7.2.1.4 Trees growing along the side and growing into or toward the facility/utility space should be pruned by removing entire branches. Branches that, when cut, will produce sprouts that would grow into facilities and/or utility space should be removed.

5.7.2.2.1 Climbing spurs may be used when limbs are more than throw-line distance apart, or when the bark is thick enough to prevent damage to the cambium, or there are no other practical means of climbing the tree.

SOURCE: American National Standards Institute (ANSI). This material is reproduced from ANSI 300-1995 with permission of the American National Standards Institute. No part of this material may be copied or reproduced in any form, electronic retrival system or otherwise or made available on the Internet, a public network, by satellite or otherwise without the prior written consent of the American National Standards Institute, 11 West 42nd Street, New York, NY 10036.

acreage per employee statistics stretch this resource thinner than the Pasadena model (Table 19.9).

Tree Maintenance. Many parks and recreation departments are responsible for tree maintenance, including tree trimming and removal, on park property and sometimes on other municipal property throughout the community. A variety of standards adopted by the American National Standards Institute (ANSI) relevant to this function address quality of tree care (Table 19.10). A qualitative assessment of a city's tree services could be based on the selection by local authorities of particular ANSI standards and recommended practices identified as especially relevant to municipal operations. The city would simply report the percentage of those selected standards and practices with which it complies.

Table 19.11   Tree City USA Standards

STANDARD 1: A Tree Board or Department

A tree board is a group of concerned citizens, usually volunteer, charged by ordinance to develop and administer a comprehensive community tree management program for the care of trees on public property. Tree boards usually function with the aid of professional foresters. In communities with a population of more than 10,000, city forestry departments with salaried employees are often feasible. These departments may or may not be supported by advisory boards or administrative commissions.

STANDARD 2: A Community Tree Ordinance

The community tree ordinance needs to designate the tree board or department and give them the responsibility for writing and implementing the annual community forestry work plan. The ordinance should determine public tree care policies for planting, maintenance, and removals. Ideally, the city tree ordinance will make provisions for establishing and updating a list of recommended street tree species to be planted with spacing and location requirements. A sample tree ordinance may be obtained by writing the National Arbor Day Foundation.

STANDARD 3: A Community Forestry Program With an Annual Budget of at Least $2 per Capita

Many communities begin their program by taking an inventory of the trees growing on public property. The species, location, and condition of each tree are noted (i.e., healthy, needs pruning, should be removed, etc.) and the inventory data is summarized in a written report for presentation and approval by the city council. The report should be an objective analysis of the present state of the urban forest with recommendations for future management. The essential, ongoing activity for the care of trees along streets, in parks, and in other public places is the community forestry program. The annual work plan should address planting, watering and fertilizing, dead and hazardous tree removal, safety and fine pruning, and insect and disease control. To be named as a TREE CITY USA, a town or city must annually spend at least $2 per capita for its annual community forestry program. Consider all funds spent for tree care—budget for street tree department or board, park department's tree expenditures, dead tree removal, etc.

STANDARD 4: An Arbor Day Observance and Proclamation

An Arbor Day observance can be simple and brief or an all-day or all-week observance. A proclamation issued by the mayor must accompany the observance and declare the observance of Arbor Day in your community.

SOURCE: National Arbor Day Foundation, *Tree City USA Application* (Nebraska City, NE: National Arbor Day Foundation, n.d.). Reprinted by permission.

Another organization concerned with tree care is the National Arbor Day Foundation, sponsor of the Tree City USA program. Unlike the standards and recommended practices of ANSI, which provide technical guidance to tree care specialists, the standards promoted by the National Arbor Day Foundation are directed toward local policy, governance, resource commitments, and ceremony (Table 19.11).

Reasonably precise guidance for assessing the quality of tree care is found in a checklist developed by the parks division of Sunnyvale, California (Table

Table 19.12   Park Tree Quality Standards Checklist

Safety
- There should be no hanging/cracked limbs, or other hazards obvious from the ground.
- There should be no major structural damage evidence.
- Limbs and/or foliage of semi-mature trees should not exist below the height of ten feet over pedestrian rights-of-way or below the height of fourteen feet over vehicular rights-of-way.
- Tree stumps should not present trip hazards.

Aesthetics
- Trees should display at least some semblance of the form common to the species. Some allowance shall be made for natural "character," but grossly misshapen trees will not pass muster.
- There should be no stubs, dead "flags," or other unsightly distractions.
- Stumps should not be visible. They should either be removed to at least twelve inches below ground (e.g., in open turf) or hidden by existing vegetation (e.g., planter beds).

Health
- New trees should be planted in accordance with approved specifications.
- Trees should have stakes removed as soon as the trunk is self-supporting, or one year from the date of planting, whichever comes first. Approval to either maintain stakes for periods in excess of one year, or to remove trees which are not self-supporting, shall be secured from a program supervisor prior to acting.
- Obvious and debilitating pest and/or health problems (insects, disease, man-made problems) should be nonexistent.

Utility
- Trees planted for a specific reason (e.g., to serve as visual screens, noise barriers, or to provide summer shade) should serve the intended purpose.
- Tree limbs should not interfere with buildings, utility wires, etc.
- Limbs and/or foliage should not obstruct park signs. Tree roots should not interfere with the utility of surrounding facilities or fixtures.

SOURCE: Robert A. Walker, *Parks Division Quality Standards Manual* (City of Sunnyvale, CA, 1989), pp. 9–10. Reprinted by permission.
NOTE: The City of Sunnyvale, California, has set as its target maintaining at least 80% of its park trees in compliance with the standards of safety, aesthetics, health, and utility specified in this checklist.

19.12). The local standard adopted by Sunnyvale requires that at least 80% of that community's park trees be in compliance with the standards of safety, aesthetics, health, and utility specified in the checklist. Further guidance from individual cities may be found in local performance ranging from the survival rates of newly planted trees to the promptness of tree pruning and removal. Overland Park, Kansas, for instance, tries to achieve a survival rate of at least 90%, and Durham, North Carolina, reports a survival rate of 96%. In emergency situations, Boston, Massachusetts, and Cincinnati, Ohio, expect tree pruning and removal work to be completed within 24 hours (Table 19.13). Emergency crews in Palo Alto, California, respond within 1 hour. Routine work is given much greater leeway.

Table 19.13 Tree Pruning and Removal: Standards of Promptness From Selected Cities

EMERGENCY SERVICE

Palo Alto, CA
Target: Respond to 100% of emergency requests for tree maintenance within 1 hour
Actual: 100% (1997)

Boston, MA
Percentage of emergency requests responded to within 24 hours: 100% (1992)

Cincinnati, OH
Target: By next working day following notification of tree damage from storms, accidents, or vandalism
Actual: 93% (1991)

Ann Arbor, MI
Target: Correct hazardous conditions in public trees within 48 hours of notification

ROUTINE SERVICE

Overland Park, KS
Target: Investigate within 36 hours and perform needed pruning within 24 hours to 30 days, depending on the nature of the problem

Orlando, FL
Target: Respond to citizen tree requests within 7 working days
Actual: 74% (1991)

Palo Alto, CA
Target 1: Inspect and provide tree trimming schedule to residents within 10 days of request for tree service (1999)
Target 2: Annually trim, prune, raise, and remove deadwood from 9% of street trees and 20% of park trees
Actual: 4% of street trees and 4% of park trees (1997)

Cincinnati, OH
Target: Average of 60 days or less to perform work after request for routine pruning or removal
Actual: 34-day average (1991)

Ann Arbor, MI
Target: Achieve an average 8-year pruning cycle
Actual: 9-year cycle (1998)

Grass/Weeds/Turf Mowing. The acceptable height of grass in a given community is influenced by several factors, including local taste, resource availability, and property use. Residents of an affluent community might expect their municipal soccer fields to be closely cropped, whereas slightly taller grass in other parks and grass that is taller still along roadways could be perfectly acceptable.

Some cities carefully assign different maintenance levels to parks throughout the community. Overland Park, Kansas, for example, assigns one of seven different maintenance intensity levels to each of its parks. Parks receiving "Maintenance Intensity I" are overseeded twice per year, mowed every 7 days, and have trash picked up three times each week. Parks receiving "Maintenance Intensity VII" are mowed only as needed for weed control and have trash pickup monthly (City of Overland Park [KS], *1999 Annual Budget,* pp. 6.161–6.165).

Necessary mowing frequencies are influenced, of course, by desired grass height and also by local climate and rainfall. In establishing their own local standards, communities will find the turf height standards, mowing frequencies, and rule-of-thumb labor ratios of their counterparts helpful (Table 19.14). In addition, standards used for rating the quality of turf maintenance in a study of Los Angeles–area cities are shown in Table 19.15.

Park Structures, Fixtures, and Playgrounds. The U.S. Consumer Product Safety Commission (CPSC) has developed extensive guidelines for playground design and maintenance. Checklists that summarize these guidelines are found in Table 19.16 for parks administrators and maintenance personnel and Table 19.17 for parents and community groups. Some cities use CPSC guidelines as benchmarks. Charlottesville, Virginia, for instance, reported that 68% of its playground equipment in 1998 met CPSC standards, up from 45% in 1997.

In addition, some individual communities have established their own standards for the maintenance of park structures, fixtures, and playgrounds. The parks division of Sunnyvale, California, for example, is expected to maintain at least 80% of its structures and fixtures in the conditions specified in Table 19.18. Sunnyvale's playground maintenance standard (Table 19.19) requires repair or removal of unsafe equipment within 48 hours (Walker, 1989). Several other cities have also established targets for prompt response to maintenance requests (Table 19.20).

Sports Facilities. Recreational facilities that serve as the setting for athletic contests usually require extraordinary maintenance. As in the case of other parks maintenance functions previously noted, a checklist developed by the Sunnyvale parks division may be used as a qualitative assessment tool for judging the maintenance of sports fields and other recreational facilities (Table 19.21). The assessment may be made based on the average percentage of relevant standards with which each facility complies or by the percentage of facilities in total compliance with relevant standards.

Although every municipality may wish to design its own standards to reflect local preferences and conditions, it need not start from scratch. The recreational facility standards of other communities (Table 19.22), their maintenance procedures, rating systems (Table 19.23), and rule-of-thumb labor

Table 19.14   Grass/Weeds/Turf Mowing: Selected Cities

TURF HEIGHT

Sunnyvale, CA
  Target: 1.5 to 2 inches throughout parks
College Park, MD
  Target: Maximum acceptable grass height along roadways, city property—3 inches
Cincinnati, OH
  Target: Maximum acceptable grass height along roadways, city property—4 inches
Corpus Christi, TX
  Target: At least 95% of grass in mowed areas 6 inches in height or less
  Actual: 79% (1998)
Milwaukee, WI
  Target: Ensure weed growth does not exceed 12 inches, per state statute and city ordinance

MOWING FREQUENCIES

Ann Arbor, MI
  Procedure: Mowing frequency in parks—once per week during growing season
Reno, NV
  Procedure: Mowing frequency in parks—once a week during growing season
Rock Hill, SC
  Procedure: City-owned roadsides and plant strips—every 21 days; athletic fields—every 7 days;
    parks, neighborhood centers, utility sites—every 10 days; city hall lawn and selected park areas—
    every 7 days; selected properties for class C maintenance—once per month
Fort Collins, CO
  Procedure: Mowing frequency in parks—30 times per year (1991)
  Target: Mow road shoulders at least 4 times per growing season
Corpus Christi, TX
  Target: 10-day average mowing cycle for developed parks during peak growing periods
  Actual: 20-day average cycle (1998)
Greenville, SC
  Procedure: Mowing of city street right-of-way—40-day cycle (1991)
Tallahassee, FL
  Procedure: Mow and trim street right-of-way 4 times annually

LABOR RATIOS

Little Rock, AR
  2,500 to 3,500 square yards of undeveloped property cleared of brush and mowed per day by
    2-person crew
Wichita, KS
  Park and public area acres mowed per work hour: 1.19 (1990), 1.32 (1991).
  Airfield acres mowed per work hour: 3.83 (1990), 5.04 (1991)
Winston-Salem, NC (Wastewater Collection Program)
  Work hours per mile of right-of-way mowed: 14.57 (1991), 16.80 (1992)
Study of Los Angeles–area cities (1984)[a]
  Work hours (direct labor) per acre mowed: 5.27 (sample mean); range: 2.14 to 9.87
Victoria, TX
  Work hours per acre to mow with tractor: 7.46 (1996)
  Work hours per acre to trim: 18.4 (1996)
  Work hours per acre to remove litter: 0.3 (1996)

a. Los Angeles–area statistics reported in Barbara J. Stevens (Ed.), *Delivering Municipal Services Efficiently: A Comparison of Municipal and Private Service Delivery* (Prepared by Ecodata, Inc., for the U.S. Department of Housing and Urban Development, June 1984), p. 382.

Table 19.15   Standards for Rating the Quality of Turf Maintenance

Ratings for Edging Work
   1 = *Excellent:* Grass cut smoothly at sidewalk's edge; distinct straight line
   2 = *Good:* Grass beginning to grow into edged area, although it has not grown more than $\frac{1}{2}$ inch
   3 = *Fair:* Grass has grown 1 inch into edged area
   4 = *Poor:* Grass has grown more than 1 inch into edged area

Ratings for Grass Color
   1 = Medium green
   2 = Light green
   3 = Yellow green
   4 = Brown

Ratings for Grass Height
   1 = $\frac{1}{2}$ inch to $1\frac{1}{2}$ inches
   2 = $1\frac{1}{2}$ inches to $2\frac{1}{2}$ inches
   3 = $2\frac{1}{2}$ inches to 4 inches
   4 = 4 inches to 6 inches

Ratings for Weeds
   1 = *Excellent:* No weeds found in observation area
   2 = *Good:* Weeds found in 10% or less of observation area
   3 = *Fair:* Weeds found in 10% to one third of observation area
   4 = *Poor:* Weeds found in more than one third of observation area

Ratings for Percentage of Grass Cover
   1 = 100% to 75%
   2 = 75% to 50%
   3 = 50% to 25%
   4 = 25% to 0%

SOURCE: Barbara J. Stevens (Ed.), *Delivering Municipal Services Efficiently: A Comparison of Municipal and Private Service Delivery* (Prepared by Ecodata, Inc., for the U.S. Department of Housing and Urban Development, June 1984), p. 379.
NOTE: Average quality rating in 1984 study of Los Angeles–area cities was 1.71 (range of 1.14 to 2.76).

ratios for the maintenance of various park and recreational facilities (Table 19.24), as well as labor ratio guidelines devised by the NRPA (Table 19.25), may be useful to a community deciding on its own standards, procedures, and resource requirements.

Program Operations Standards

Among the operating standards of relevance to public recreation programs are those established by APHA relating to public swimming pools. APHA standards address pool capacity, the proper ratio of attendants and lifeguards to pool area occupants, and chemical testing of the water (see Table 19.8). APHA also has established standards for lifesaving and first aid equipment at public swimming pools (Table 19.26).

(text continues on page 283)

Table 19.16   Playground Maintenance Checklist: U.S. Consumer Product Safety Commission

Surfacing
- The equipment has adequate protective surfacing under and around it and the surfacing materials have not deteriorated.
- Loose-fill surfacing materials have no foreign objects or debris.
- Loose-fill surfacing materials are not compacted and do not have reduced depth in heavy use areas such as under swings or at slide exits.

General Hazards
- There are no sharp points, corners or edges on the equipment.
- There are no missing or damaged protective caps or plugs.
- There are no hazardous protrusions and projections.
- There are no potential clothing entanglement hazards, such as open S-hooks or protruding bolts.
- There are no pinch, crush, and shearing points or exposed moving parts.
- There are no trip hazards, such as exposed footings on anchoring devices and rocks, roots, or any other environmental obstacles in the play area.

Deterioration of the Equipment
- The equipment has no rust, rot, cracks, or splinters, especially where it comes in contact with the ground.
- There are no broken or missing components on the equipment (e.g., handrails, guardrails, protective barriers, steps, or rungs on ladders), and there are no damaged fences, benches, or signs on the playground.
- All equipment is securely anchored.

Security of Hardware
- There are no loose fastening devices or worn connections, such as S-hooks.
- Moving components, such as swing hangers or merry-go-round bearings, are not worn.

Drainage
- The entire play area has satisfactory drainage, especially in heavy use areas such as under swings and at slide exits.

Leaded Paint
- The leaded paint used on the playground equipment has not deteriorated as noted by peeling, cracking, chipping, or chalking.
- There are no areas of visible leaded paint chips or accumulation of lead dust.

General Upkeep of Playgrounds
- The entire playground is free from miscellaneous debris or litter, such as tree branches, soda cans, bottles, glass, etc.
- There are no missing trash receptacles.
- Trash receptacles are not full.

SOURCE: *Handbook for Public Playground Safety,* Publication No. 325, (Washington, DC: U.S. Consumer Product Safety Commission, n.d.), p. 32.

Table 19.17 Public Playground Safety Checklist for Use by Parents and Community Groups

Here are 10 important tips for parents and community groups to keep in mind to help ensure playground safety:

1. Make sure *surfaces* around playground equipment have at least 12 inches of wood chips, mulch, sand, or pea gravel, or are mats made of safety-tested rubber or rubber-like materials.
2. Check that protective *surfacing extends* at least 6 feet in all directions from play equipment. For swings, be sure surfacing extends, in back and front, twice the height of the suspending bar.
3. Make sure play structures more than 30 inches high are *spaced* at least 9 feet apart.
4. Check for *dangerous hardware,* like open S-hooks or protruding bolt ends.
5. Make sure *spaces* that could trap children, such as openings in guardrails or between ladder rungs, measure less than 3.5 inches or more than 9 inches.
6. Check for *sharp points or edges* in equipment.
7. Look out for *tripping hazards,* like exposed concrete footings, tree stumps, and rocks.
8. Make sure elevated surfaces, like platforms and ramps, have *guardrails* to prevent falls.
9. *Check playgrounds regularly* to see that equipment and surfacing are in good condition.
10. Carefully supervise children on playgrounds to make sure they're safe.

SOURCE *Handbook for Public Playground Safety,* Publication No. 325, (Washington, DC: U.S. Consumer Product Safety Commission, n.d.), p. 43.

Table 19.18 Park Structures and Fixtures Quality Standards Checklist

Backflow Prevention Devices should be
- locked to prevent tampering
- freshly painted and graffiti-free
- camouflaged attractively where possible
- accessible to maintenance personnel

Backstops (portable) should
- be free of graffiti
- be securely anchored to the ground
- include framing, fabric, and hardware in good repair

Benches should
- offer a relatively smooth seating surface
- be secure and sturdy
- be free of unintended protrusions (e.g., nails, fasteners)
- be clean (especially seat and back supports)
- be properly sealed
- be graffiti-free

Bike Racks should
- not present a hazard to pedestrians
- be securely anchored
- be painted, where appropriate
- be clean and without graffiti
- be functional

(continued)

Table 19.18   Park Structures and Fixtures Quality Standards Checklist (continued)

Bleachers should
- offer smooth and sturdy seating surfaces
- offer railings for safety and support
- present skid-resistant foot surfaces
- be free of unintended protrusions
- be clean and free of graffiti, litter, and weeds
- be properly sealed and/or painted where applicable

Bollards should be
- visible and obvious
- tall enough to preclude pedestrian trip hazards
- securely fastened and/or locked
- free of sharp edges or unintended protrusions
- clean and free of graffiti
- properly sealed or painted where applicable
- spaced to serve the intended purpose (e.g., deter traffic, prevent soil erosion, define separate user areas, etc.)

Cigarette Butt Cans should
- be clean and free of litter and/or graffiti
- have sand replaced as needed

Drinking Fountains should
- be securely anchored
- be handicap accessible
- provide a steady flow of potable water when activated
- allow no water flow when not activated
- drain freely and completely
- be clean and free of debris
- be graffiti-free

Dumpsters and Their Enclosures
- should be properly signed
- should be free of graffiti
- dumpster should be enclosed
- dumpster should have a functional lid
- enclosure should include a functional gate
- enclosure should be clean and free of debris

Fences and Gates (chain link and wood)
- should be free of graffiti
- should be in good repair, with no boards, chain link, or hardware missing, bent, or broken
- should be sturdy and structurally sound
- should be properly sealed or painted (wood)
- gates should swing freely, with latches properly attached and functional

Flagpoles should
- be accessible and visible
- include functional cables and hooks
- be painted and free of graffiti

- flags should be in reasonable repair
- flags should be raised daily where appropriate

Light Standards and Fixtures should
- be properly anchored or secured
- be functional (bulbs and globes operational)
- include shades in good condition
- be clean and free of graffiti, cobwebs, birds' nests, etc.
- be numbered for maintenance repairs
- present maintenance personnel a secure and accessible hand hold

Pay Phones should be
- secure and operational
- clean and free of graffiti

Planter Boxes, Raised Beds, and Container Plants should be
- in good repair
- free of graffiti

Signage should be
- free of graffiti
- visible and legible
- painted or sealed where appropriate
- properly secured
- located so as not to present a hazard to park visitors

Statues, Sculptures, and Artwork should
- be clean and free of graffiti
- be properly secured
- have no loose or missing parts

Trellises, Arbors, and Gazebos should
- be free of graffiti
- be properly secured
- include appropriate hardware properly attached
- be structurally sound
- be painted/sealed

Utility Boxes should be
- clean and free of graffiti
- painted the standard color
- accessible and identifiable
- locked and securely attached to ground or structure

Waste Containers should
- be clean and free of graffiti
- be locked so as not to impede traffic or present hazards
- be free of major cracks, holes, etc.
- include functional lids and liners

SOURCE: Robert A. Walker, *Parks Division Quality Standards Manual* (City of Sunnyvale, CA, 1989), pp. 17–20. Reprinted by permission.
NOTE: The City of Sunnyvale, California, has set as its target maintaining at least 80% of its park structures and fixtures in the condition specified in this checklist.

Table 19.19   Playground Quality Standards Safety Checklist

General
- All playgrounds should contain an approved surface material [see *Play for All Guidelines* © 1987] spread over the entire playing area to a minimum depth of 10 inches (preferred 12 inches) and to within 6 inches of the curbing top.
- Surface materials should be free of hazardous materials (e.g., broken glass, animal feces).
- All playgrounds should be free of trip hazards (e.g., roots, obstacles).
- All concrete footings should be covered by surface material—none should be exposed.

All Playground Apparatus should be free of
- loose footings or loose structural supports
- sharp edges
- worn parts
- missing parts

Metal Parts should be free of
- enlarged bolt holes or loose or protruding nuts and bolts
- bent, broken, or severely worn pipe
- exposed ends of tubing intended to be plugged or capped

Wooden Parts should be free of
- splinters
- structural cracks (as opposed to cosmetic cracks or natural "checking")
- dry rot or insect (e.g., termite) damage that impacts the structural integrity of the apparatus

Synthetic Parts (e.g., **Fiberglass, Plastic, Rubber,** etc.) should be free of
- cracks, rot, or wear

Swings
- Seats, straps, and rivets should be sturdy and show no signs of deterioration.
- Chains, S-hooks, swing hangers, and clevis assemblies should be in good working order and show limited signs of wear. Chain links and S-hooks should be replaced before wearing through $\frac{1}{4}$ of the metal's diameter.
- S-hooks should be completely closed.
- Swing hangers and other moving parts should move freely.

Slides
- Apparatus should be free of structural stress or damage.
- Ladders should be sturdy and complete (no missing rungs or rails).
- Bedway exits should be parallel to the ground.
- Bedways, bedrails, and handrails should be free of foreign objects, holes, and rough edges.

Whirls
- Surface should not be worn slick or have holes.
- Bearings should move freely.
- Handrails should be sturdy and free of holes or rough edges.

Rocking and Bouncing Equipment
- All movable parts should perform their intended function.
- Spring castings should not be loose.
- Seats should be structurally sound.
- Axis of seesaws should be free of pinch or crush points.

Climbers
- Should have tight fittings; bars and pipes should be snug.
- All parts should be present and intact and serve the intended purpose.

SOURCE: Robert A. Walker, *Parks Division Quality Standards Manual* (City of Sunnyvale, CA. 1989). pp. 28–30. Reprinted by permission.

Table 19.20   Prompt Response to Maintenance Requests: Selected Cities

Ann Arbor, MI
  Target: Make all safety-related repairs on the same day as reported
  Actual: 100% (1998)

Savannah, GA
  Target: Respond to citizen reports of damage within 1 workday
  Actual percentage of repairs within 24 hours of report for athletic fields: 100%;
    for swimming pools: 100% (1997)

Duncanville, TX
  Percentage of vandalism responses within 48 hours: 100% (1998)

Palo Alto, CA
  Percentage of irrigation work orders for athletic fields completed within 72 hours: 90% (1997)

High Point, NC
  Percentage of work orders completed within 7 days: 98% (1998)

Cambridge, MA
  Average number of days to close request for unscheduled park maintenance (e.g., ball field repairs,
    playground equipment, cleanup): 17.7 (1995), 10.6 (1996, estimated)

Denver, CO
  Percentage of fountain repairs accomplished within 2 weeks: 80% (1998)
  Percentage of playground repairs accomplished within 2 weeks: 90% (1998)

Operating standards may also include proper training and credentials of operating personnel. A prime example is the special training required by most municipalities for swimming pool lifeguards. Many require American Red Cross certification or its equivalent.

Program Offerings and Administrative Functions

The success of a parks and recreation department is influenced not only by the quality of its facilities and the productivity of its maintenance operations but also by the range and adequacy of its program offerings and the proficiency with which its administrative functions are performed. Table 19.27 offers a variety of parks and recreation odds and ends reported by selected cities.

Table 19.21   Recreational Facilities Quality Standards Checklist

Athletic Fields: Ball Diamonds
- Entire area should be free of glass and litter.
- Bleachers should be clean, graffiti-free, and in safe condition.
- Backstops and chain link fences must be in safe condition—i.e., properly attached, no protrusions, snags, or holes.
- Adequate lighting should be provided.
- Permanent scoreboards should be in good condition—graffiti-free and functional.
- Playing field should be free of washboard effects, holes, and depressions.
- A minimum of 1″ of approved surface material should be maintained along base paths (all fields) and throughout infields where appropriate (additional amounts to be determined by park staff).
- Turf infield should be maintained at 1.5″ height.
- Field markings and boundaries should be accurate and clearly defined (i.e., batters box, 1st base line, foul lines—see schematic for appropriate type of field).
- All lines should be of uniform width.
- All plates should be firmly planted and level with surrounding surface material.
- Concrete dugouts should be free of debris, soil, and graffiti.
- Litter receptacle should be safe, functional, and attractive.
- Distinction between infield and outfield should be accurately measured and clearly defined.

Athletic Fields: Soccer Fields, Multipurpose Fields, Outfields
- Playing field should be reasonably level.
- There should be no holes or large dips.
- The area should be free of litter.
- During the soccer season, fields should be clearly lined according to the appropriate schematic.
- Soccer goals should be functional, secure, and in good repair. Poles should be secure and free of metal burrs.

Basketball Courts
- Fence and gates should be in good repair. Appropriate signs should be visible and securely attached.
- Nets should be in good repair.
- Hoops and backboards should be in good repair and securely fastened.
- Court surfaces should be in playable condition: smooth, without large cracks, clean, and with clearly visible and well-defined lines.
- Court lamps should be present and operational.
- Timers should be set properly and in working condition.

Bowling Green

NOTE: The City's intent is to establish quality standards that provide user groups with facilities that meet *their* needs. Because the needs of lawn bowlers *cannot* be met by normal lawn surfaces, these quality standards differ greatly from those related to general park turf. It is important that staff not judge one by standards intended for the other.

- Turf should be maintained at the lowest uniform height possible.
- Turf should be level with the height of perimeter base boards (if not level, it should be kept above rather than below board height).
- Playing surface should be firm and level.
- Playing surface should be as dry as continued maintenance of the turf will allow.
- Turf color is of little consequence: Staff should attempt to maintain a green lawn but should keep in mind that there is more than one way to accomplish this (e.g, use of dyes) and that wherever a conflict exists between standards of moisture and standards of color, highest priority shall be awarded to maintaining a surface "as dry as continued maintenance of the turf will allow."
- Turf should be free of weeds, insects, [and] disease.
- Turf should be free of litter or other debris.

- Sand gutter should be maintained at a uniform depth that allows a minimum of 3" clearance between sand and top of perimeter base boards—kept weed free; boards kept in good repair.
- Surrounding hardscape and fencing should be kept clean, graffiti-free, and functional.
- Area should be secured.

Handball

- Surface should be in playable condition: smooth, clean, with any lines clearly marked.

Horseshoes

- The pit should be of regulation size.
- Material and sand mixture around stake should be close to level with the surrounding soil.
- The stake must be secure.
- The stake should be free of sharp edges.
- The header boards should be secure and in good condition.
- Backboards should be secure and free from major damage. The support post should be secure in the ground.
- The ground should be level and free from potholes, glass, debris, litter, and weeds.

Picnic Sites

- Pedestrian surface should be level and free of trip hazards, litter, and weeds.
- Picnic areas should drain properly; there should be no standing water.
- Picnic tables and benches should be clean—free of staples, graffiti, deep carving, food residues, etc.
- Picnic tables and benches should be properly finished, dependent on material of table.
- Picnic tables and benches should be stable and secured to surface. There should be no loose hardware or sharp surfaces.
- BBQ pits should be clean and in usable condition (e.g., ashes kept to a minimum, hardware of grill in good repair, moving parts operating properly).
- All electrical outlets should be functional and covered.
- Sinks should be clean and drain properly. All plumbing fixtures should be present and functional.
- Drinking fountains should be clean and functional.
- Trash cans should be clean, functional, and safe.
- All picnic structures and fixtures should be clean and graffiti-free.

Shuffleboard

- Area should be free of litter, graffiti, and grit.
- All lines should be clearly visible.
- Playing surface should be smooth and free of large cracks.
- Area should be free of trip hazards.

Tennis

- Fence and gates should be in good repair. Appropriate signs should be visible and securely attached.
- Nets should be in good repair and at proper tension. They should measure 36" from ground to top of net at center strap, or be capable of adjustment to that height.
- Poles should be properly aligned.
- Court surfaces in playable condition: smooth, without large cracks, clean, with clearly visible and well-defined lines.
- Court reservation clock should be operating properly.
- Court lamps should be present and operating.
- Timer should be set properly and in working condition.

Volleyball

- Area should be free of glass, litter, and animal droppings.
- Support poles should be secure and free of burrs.
- Support poles' hardware should be in working order.
- Sand should be spread evenly within court; there should be no piling up of sand near the edges.

SOURCE: Robert A. Walker, *Parks Division Quality Standards Manual* (City of Sunnyvale, CA, 1989), pp. 22–26. Reprinted by permission.

Table 19.22   Golf Course Operation and Maintenance: Selected Cities

Turf height (Sunnyvale, CA)
Target: Greens, 3/16 inch; tees and fairways, 1 inch; roughs, 1.5 inch

Sand trap depth (Sunnyvale, CA)
Target: 4 inches

Putting greens (Norman, OK)
Speed of putting greens (average stimpmeter readings): 96% (1998)

Course maintenance (Reno, NV)
Procedures and performance: Mow fairway twice per week (May–October), greens three times
    per week (May–October); aerate and top-dress greens twice per month (May–October); fertilize
    greens once per month (May–October), non-greens areas once in April and once in September.
Percentage of irrigation repairs made within 1 hour of notification: 90% (1990)
Percentage of equipment repairs made within 1 day: 85% (1990)

Labor ratio (Wichita, KS)
Rounds played per labor hour of golf course maintenance: 3.32 (1990), 3.90 (1991)

Pace of play (Overland Park, KS)
Target: Achieve an overall pace-of-play program that will result in 95% of all 18-hole rounds of golf
    being completed in 4 hours and 48 minutes or less (1999)

Golf carts (Norman, OK)
Golf cart availability rate: 99.5% (1998)

Table 19.23   Standards for Rating Park Condition: Savannah

TURF

Mowing

Excellent	4	Evidence that the lawn has been recently mowed. Grass blade height is even over entire lawn. Grass height is no more than 1.75 inches overall, dependent upon grass cultivar, and 2 to 2.5 inches in areas of tractor mowing.
Good	3	Slight growth variations but grass area still visibly neat.
Fair	2	Lawn has been mowed regularly but some high seed heads or weeds need cutting. Lawn is due for regular mowing. Maximum acceptable height of grass in priority one is 2.5 inches, priority three 3 inches, and priority four 3 to 5 inches.
Poor	1	Evident that lawn is overdue for regular mowing. Grass blade height exceeds the maximum acceptable height for its priority rating.

Line Trimming

Excellent	4	Grass is neatly trimmed around trees, walls, buildings, etc., to the same height as the surrounding grass.
Good	3	Grass around trees is slightly higher than the surrounding grass. Apparent that grass has been cut one time since last trimming.
Fair	2	Grass is approximately two inches higher than surrounding grass. Apparent that grass has been cut two or more times since the last trimming.
Poor	1	Apparent that trimming has not been done for a month or more. Grass is more than three inches higher than the surrounding lawn.

Edging

Excellent	4	Grass is cut smoothly at the sidewalk edge. Very distinct straight lines.
Good	3	Grass runners beginning to grow into edged areas but appearance is still neat.
Fair	2	Grass runners obscure the edge of the walkway.
Poor	1	Grass runners obscure the entire edge of the walkway and runners over 6 inches long extend into the walkway.

Weeding

Excellent	4	Grass areas are visibly weed free.
Good	3	Weeds are visible over 5% or less of the grass area.
Fair	2	Weeds comprise 5% to 20% of the grassed area.
Poor	1	Weeds comprise more than 20% of the lawn area.

Color

Excellent	4	Even green color over entire area.
Good	3	Can have slight variations in color.
Fair	2	Grass is green but has moderate variations of color.
Poor	1	Wide variations in color. Some brown areas evident due to lack of water, fertilizer, or herbicide.

Density

Excellent	4	Turf is dense and well established.
Good	3	Turf is well established but small bare areas are evident.
Fair	2	Bare ground is evident in 10% to 20% of the turf area.
Poor	1	More than 20% of the turf area is bare ground.

LEAF REMOVAL

Excellent	4	To all appearances, there are no leaves on the ground. A few leaves are acceptable dependent upon seasonal leaf drop of tree species.
Good	3	Some leaves are apparent but do not cover over 5% of grassed area.
Fair	2	Leaves cover about 10% of grassed area where trees are located. No threat to grass.
Poor	1	Leaves cover more than 10% of the area under or near trees and threaten to smother grass.

GROUND COVER

Excellent	4	Ground cover is dense and well established. Color is even over entire area.
Good	3	Ground cover unevenly established. Some bare area, less than 5%, is evident.
Fair	2	Bare ground is evident in 5% to 20% of the area. Ground cover is healthy but has moderate variations of color.
Poor	1	Bare ground is evident in over 20% of the area. Wide variations in color. Obvious that the ground cover has not established itself.

FLOWERS

Excellent	4	Planted area is noticeably weed free and soil has been loosened by hoeing and cultivation. Dead or yellow foliage removed.
Good	3	A few weeds are apparent and a small amount of dead or yellow foliage; appears to have been recently hoed.
Fair	2	Approximately 10% weed cover and/or several pieces of dead or yellow foliage. Soil has been cultivated but requires hoeing again.
Poor	1	Heavy weed infestation and soil heavily compacted.

(continued)

Table 19.23   Standards for Rating Park Condition: Savannah (continued)

SHRUBS

Excellent	4	Hedge presents an even, well-trimmed appearance.
Good	3	Hedge has been neatly trimmed but a very small amount of new growth is apparent.
Fair	2	Hedge retains its trimmed shape but new growth is apparent over most of the hedge.
Poor	1	Does not meet the criteria for fair. Hedge definitely requires trimming.

TREES

Excellent	4	No sucker growth from the ground. No injury to trunk. No included bark in scaffold limbs. Well-balanced appearance to leaf crown. No branches interfering with traffic inside or outside park. No deadwood over 1 inch in diameter in leaf crown. No crossover limbs. No symptoms or signs of disease.
Good	3	As above except: up to 3 suckers; up to 10% circumference injury to the trunk; deadwood up to 2 inches in diameter.
Fair	2	As above except: up to 6 suckers; up to 20% circumference injury to the trunk; deadwood up to 4 inches in diameter; disease incidence less than 5% of all trees.
Poor	1	Does not meet criteria for fair.

LITTER

Excellent	4	No apparent litter on the ground.
Good	3	Up to 4 pieces of obvious litter in the field area or several nonobvious pieces of litter such as drink can tabs.
Fair	2	5 to 7 pieces of litter in the field areas or many nonobvious pieces of litter such as drink can tabs.
Poor	1	Does not meet the standard for fair.

STRUCTURES

Excellent	4	All facilities in noticeably good repair. Fountains are functioning and clean; park benches are intact; monuments are clean and graffiti-free; brick sidewalks are complete with all bricks in place. All other structures are in good order.
Good	3	One of the above categories is in need of maintenance.
Fair	2	Two of the above categories are in need of maintenance.
Poor	1	Park is neglected. Most facilities are in need of repair.

CUMULATIVE

Excellent	4	Cumulative mean of component ratings is at highest possible condition. No improvement needed.
Good	3	Cumulative mean of component ratings is 75% of highest possible condition. It is apparent the park is being maintained, but is in need of additional service.
Fair	2	Cumulative mean of component ratings is 50% of highest possible condition. Park does not appear to be receiving regular attention.
Poor	1	Cumulative mean of component ratings is 25% of highest possible condition. Park appears to be entirely neglected, not receiving any attention.

SOURCE: City of Savannah (GA) Park and Tree Department, *Condition Standards* (1999).

Table 19.24   Labor Ratios for Selected Parks and Rights-of-Way Maintenance Activities: Winston-Salem

	Labor Hours		
	Actual 1996–1997	Actual 1997–1998	Objective 1998–1999
per medium-sized tree removed (4″ to 24″ caliper)	1.47	1.68	1.75
per large tree removed (greater than 24″ caliper)	3.33	7.19	5.80
per tree pruned	0.29	0.47	0.45
per tree preventive-pruned	0.46	0.39	0.40
per shrub planted	0.07	0.13	0.15
per tree planted	1.34	0.66	1.20
per stump removed	1.13	1.11	1.50
per shrub trimmed	0.13	0.11	0.12
per cubic yard of mulch applied	0.62	0.58	0.60
per 1,000 gallons of water applied	3.37	1.96	2.80
per perennial planted	0.03	0.03	0.03
per acre of major highway rights-of-way mowed:			
tractor mowing	0.69	0.61	0.66
push mowing	1.13	0.93	1.00

SOURCE: City of Winston-Salem (NC), *1997–1998 Performance Report and 1998–1999 Business Plan: Roadway Appearance Division* (October 1998).

Recovery

Especially during periods of fiscal constraint, the question of how much of a recreation program's costs the community should recoup from fees can be a high-profile concern. Awareness of the cost recovery targets and results in other communities can provide a helpful context for local deliberations (Tables 19.28 and 19.29).

Note

1. Much of this section on the history of parks and recreation standards is drawn, with few adaptations, from David N. Ammons (1995), "Overcoming the Inadequacies of Performance Measurement in Local Government: The Case of Libraries and Leisure Services," *Public Administration Review, 55* (January/February), pp. 37–47. Used with permission.

Table 19.25   Labor Ratios for Selected Parks and Recreation Maintenance Activities

Task	Labor Hours
Mowing 1 Acre, Flat Medium Terrain at Medium Speed	
20" walking	2.8 per acre
24" walking	2.2 per acre
30" riding	2.0 per acre
72" (6-foot) riding	0.35 per acre
Bush hog	1.25 per acre
Trim	
Gas powered (weedeater)	1.0 per 1,000 lin. ft.
Planting Grass	
Cut and plant sod by hand (1.5" strips)	1.0 per 1,000 sq. ft.
Cut and plant sprigs by hand (not watered)	10.9 per 1,000 lin. ft.
Seed, by hand	0.5 per 1,000 sq. ft.
Overseeding, Reconditioning	0.8 per acre
Fertilize Turf	
24": sifter spreader	0.16 per 1,000 sq. ft.
Hand push spreader 36"	2.96 per acre
Tractor towed spreader 12"	0.43 per acre
Weed Control	
Spraying herbicide w/fence line truck, tank sprayer 2 ft. wide (1" either side of fence)	0.45 per 1,000 sq. ft.
Leaf Removal	
Hand rake leaves	0.42 per 1,000 sq. ft.
Vacuum 30"	0.08 per 1,000 sq. ft.
Planting Trees	
Plant tree 5–6 ft. ht.	0.45 per tree
Plant tree 2–2.5" dia.	1.0 per tree
Tree Removal	
Street tree removal	13.0 per tree
Street tree stump removal	3.5 per tree
Park tree removal	5.0 per tree
Park tree stump removal	2.0 per tree

Shrub Maintenance

Prune shrubs (deciduous) mature	0.50 per shrub
Prune shrubs (evergreen) mature	1.0 per shrub
Hedge, trimming by hand, includes cleanup	2.85 per 100 linear ft.
Hedge trimming, electric; includes cleanup	1.50 per 100 linear ft.

Flower Bed Preparation

Cultivating combined shrubbery and flower bed	0.9 per 100 sq. ft.
Spring bed preparation	3.3 per 1,000 sq. ft.

Planting

Annuals from a flat	0.10 per 1,000 sq ft.
Weed, no mulch	1.0 per 1,000 sq. ft.
Mulch	0.83 per 1,000 sq. ft.
Fall bed cleanup and preparation	6.6 per 1,000 sq. ft.

Ball Fields

Mowing—riding with E-10 mower	0.5 per ball field
Mowing/trimming with push-behind power mower	1.0 per ball field
Drag infield	0.75 per ball field
Clean fields, fans area, and players areas	2.0 per ball field
Drag infield, line field plus rake	2.0 per ball field
Regrade, repair, and reconstruct	8.0 per ball field
Football, soccer field, lining, general maintenance	2.5 per field

Tennis Courts

Tennis nets, check repairs; repair includes windscreen	1.0 per court
Recoat color surface	32.0 per court

Swimming Pool Preparation

Paint pool and deck (epoxy paint rolled on, including scraping, priming, painting)	5.0 per 1,000 sq. ft.
Caulking—pool and deck	120.0 each

Picnic Facilities

Picnic grill—check and clean	0.8 each
Picnic grill—repair/replace	4.0 each

Buildings

6-stool comfort station—cleaning and maintenance (2 laborers)	1.6 per building
Large double latrine/cleaning	1.0 per bilding

SOURCE: Excerpted from Ron Donahue (Ed.), *Park Maintenance Standards* (Alexandria, VA: National Recreation and Park Association, 1986), pp. 18–27. Used with the permission of the National Recreation and Park Association.

Table 19.26   Standards for Lifesaving and First Aid Equipment at Public Swimming
Pools

Standard Lifesaving Equipment
(a) One ring buoy not more than 15 in. (38 cm) in diameter or similar flotation device to which shall
     be attached a 60-ft. (18 m) length of $\frac{3}{16}$ in. (5 mm) rope or two pineapples (tightly rolled balls of
     rope) composed of $\frac{1}{4}$-in. (6 mm) rope the length of which is at least 1.5 times the maximum
     width of the swimming pool; and
(b) One life pole or shepherd's crook type pole having blunted ends and a minimum length of 12 ft.
     (3.7 m).

     At least one unit of lifesaving equipment, consisting of at least (a) and (b) above, shall be provided
     for each 2,000 sq. ft. (186 sq. m) or fraction thereof of water surface area of the swimming pool.

Standard First Aid Items

2 units—1-inch adhesive compress	1 unit—eye dressing packet
2 units—2-inch bandage compress	4 units—plain absorbent gauze, $\frac{1}{2}$ sq. yard
2 units—3-inch bandage compress	3 units—plain absorbent gauze, 24-inch × 72-inch
2 units—4-inch bandage compress	4 units—triangular bandages, 40-inch
1 unit—3-inch × 3-inch plain gauze pad	1 unit—bandage scissors, tweezers
2 units—gauze roller bandage	

SOURCE: Adapted from American Public Health Association, *Public Swimming Pools: Recommended Regulations for
Design and Construction, Operation and Maintenance.* Copyright © 1981 by the American Public Health Association.
Adapted with permission.

Table 19.27   Odds and Ends for Parks and Recreation: Selected Cities

PARTICIPATION/USAGE RATES

Portland, OR
   Target: At least 50% of youth population participating in recreation programs
   Actual: 51% (1998)

Bellevue, WA
   Percentage of all Bellevue youth served: 36% (1997)

Savannah, GA
   9.8% of local youth participating in recreation department athletic program; 32.6% in recreation program overall (1991)

Tempe, AZ
   Participation (program enrollments) per capita: 6.8 (1997)
   Percentage of citizens indicating they have used a park or participated in a city-sponsored recreation
   program in the past 12 months: 38% (1997)

Long Beach, CA
   Number of specialty recreation class (ballroom dancing, scuba diving, cooking, yoga, etc.) participants
   per 1,000 residents: 50 (1996)
   Number of day camp participants per 1,000 residents: 20 (1994), 17 (1996)
   Number of youth participating in youth sports programs per 1,000 residents: 24 (1995), 22 (1996)

PARTICIPANT SAFETY

Winston-Salem, NC
Serious injuries (requiring medical attention) per 1,000 participants at recreation centers: 0.02 (1998)
Serious injuries per 1,000 participants at swimming pools: 1 (1998)

SPACE ADEQUACY OF PUBLIC SWIMMING POOLS

San Luis Obispo, CA
283 square feet of pool surface per 1,000 population (1992)

SWIMMING POOL OPERATIONS

Duncanville, TX
Percentage of scheduled time pool is operational: 98% (1998)

SWIM LESSONS

Corpus Christi, TX
Target: At least 65% of participants in youth swim lessons Red Cross–certified at end of session
Actual: 60% (1998)

PROXIMITY TO PARKS

Eugene, OR
Percentage of residents who live within one half mile of park or open space: 52% (1998)

STAFF RELIABILITY

Duncanville, TX
Scorekeepers/gym attendants on site on time: 100% (1997), 100% (1998)

BALL FIELD PREPARATION

Duncanville, TX
Percentage of time ball fields are acceptably prepared: 100% (1997), 100% (1998)

Kansas City, MO
Percentage of ball diamonds prepared on schedule: 95% (1995)

RESTROOMS

New York, NY
Comfort stations in service: 72% (1996), 80% (1997)

IRRIGATION SYSTEM

Corpus Christi, TX
Target: All leaks repaired within 2 days of discovery
Actual: 12 days (1998)

Tucson, AZ
Percentage of turf acres irrigated with reclaimed water: 70% (1997)

PROMPT PROCESSING OF SPECIAL PERMITS

Reno, NV
99% of park, facility, field, alcohol, and special event permits processed within 48 hours (1990)

PROMPT DEPOSITING OF REVENUE

Denton, TX
90% within 24 hours of receipt (1991)

Table 19.28   Cost Recovery Goals for Recreation Activities: San Luis Obispo

	Cost Recovery Goal
High-Range Cost Recovery Activities (67% to 80%)	
Classes (Adult & Youth)	80%
Day care services	75%
Adult athletics (volleyball, basketball, softball, lap swim)	67%
Facility rentals (indoor facilities except the City/County Library)	67%
Midrange Cost Recovery Activities (30% to 50%)	
City/County Library room rentals	50%
Special events (triathlon, other city-sponsored special events)	50%
Youth track	40%
Minor league baseball	30%
Youth basketball	30%
Swim lessons	30%
Outdoor facility and equipment rentals	30%
Low-Range Cost Recovery Activities (0 to 25%)	
Public swim	25%
Special swim classes	15%
Community garden	10%
Youth STAR	0%
Teen services	0%
Senior services	0%

Table 19.29 Recreation Program "Cost Recovery" Benchmarks

Percentage of Overall Recreation Program Expenses Recovered Through Fees	Percentage of Swimming Pool Expenses Recovered Through Fees	Percentage of Athletic/Sports Program Expenses Recovered Through Fees	Percentage of Golf Course Expenditures Recovered Through Revenues	Percentage of Other Program/ Facility Expenses Recovered Through Fees
Ann Arbor, MI Target: 92% for fee-supported facilities and programs Actual: 92% (1998)	**Norman, OK** 118.4% (1997); 128.5% (1998)	**Lubbock, TX** Direct costs recovered: 125% for softball leagues; 80% for volleyball leagues; 100% for basketball leagues (1995)	**High Point, NC** 152% (1998)	**Winston-Salem, NC** 149% of coliseum operating costs (1992); 84% of stadium operating costs (1992)
Loveland, CO 84% (1998)	**High Point, NC** 109% (1998)	**Overland Park, KS** 127% for athletic programs and instructional classes (1994)	**Calgary, Alberta** 113% (1996)	**High Point, NC** 124% for tennis center (1998)
High Point, NC 56% (1997); 59% (1998)	**Duncanville, TX** 80% (1995); 52% (1998)	**Charlottesville, VA** 108% for adult softball (1998)	**Overland Park, KS** 112% (1992); 100% (1994); 100% (1997)	**Bellevue, WA** 101% of costs recovered for tennis center, aquatic center, and golf courses (1997)
Reno, NV Target: 59% (1999)	**Lubbock, TX** 70% (1995)	**Ames, IA** 84% (1997)	**Corpus Christi, TX** 108% (1991)	**Waukesha, WI** Targets: 100% of operating costs (excluding administration) for youth programs; 100% (including administration) for adult programs
San Clemente, CA Target: 50% Actual: 52% (1998)	**Avon, CT** 66.5% (1997)	**Tallahassee, FL** 65.1% for major adult sports; 36.0% for major youth sports (1991)	**Macon, GA** 103% (1995)	
Tempe, AZ 49% (1997)	**Ames, IA** 66% (1997)	**College Station, TX** 60.4% (1995)	**Ames, IA** 100% (1997)	**Lubbock, TX** 95% of direct costs of tennis center recovered (1995)
Cincinnati, OH 45% (1991)	**Overland Park, KS** 66% (1994)	**Eugene, OR** 60% (1996); 49% (1997)	**San Clemente, CA** 100% (1998)	**Reno, NV** Target: 90% of aquatics instructional classes at least breaking even (1999)
Denton, TX 42% (1997)	**Wichita, KS** 52.4% (1998)	**Savannah, GA** 49% for adult sports (1991)	**Orlando, FL** 87.6% (1991); targeted profit margin of 25% or greater on pro shop merchandise sales	**Vancouver, WA** Target: At least 90% of operating and debt service expenses of tennis and racquetball center
Cary, NC 40% (1999)	**College Station, TX** 41% (1998)	**High Point, NC** 33% (1998)	**Winston-Salem, NC** 85% (1991)	
San Luis Obispo, CA 40% (1990)	**Oak Ridge, TN** Indoor pool: 32% (1997) Outdoor pool: 41% (1997)		**Tallahassee, FL** 67.1% (1991)	**Calgary, Alberta** 81% for ice arenas; 71% for leisure centers (1996)
	Eugene, OR 37% (1997)		**Greensboro, NC** 44% (1997)	
	Reno, NV Target: 35% (1999)			

(Continued)

Table 19.29 Recreation Program "Cost Recovery" Benchmarks (Continued)

Percentage of Overall Recreation Program Expenses Recovered Through Fees	Percentage of Swimming Pool Expenses Recovered Through Fees	Percentage of Athletic/Sports Program Expenses Recovered Through Fees	Percentage of Golf Course Expenditures Recovered Through Revenues	Percentage of Other Program/Facility Expenses Recovered Through Fees
				Alexandria, VA 72% of recreation center costs (1998) **Overland Park, KS** 70% for youth programs (1994) **College Station, TX** 55.3% for instructional programs; 111% for concessions; 12.8% for special events (1995) **Tallahassee, FL** 40.7% for park center room rental; 35.6% for tennis center; 23.8% for gymnastics program; 17.7% for arts and crafts; and 8.4% for multi-purpose centers (1991) **Victoria, TX** 33% of expenditures recovered for community center (1996) **Duncanville, TX** 30% of utility costs recovered via light fees (1997) **Greensboro, NC** 30% cost recovery at recreation centers (1997) **Savannah, GA** Target: Recover at least 30% of tennis program costs Actual: 27% (1991) **Wichita, KS** 26.0% for recreation centers (1998)

20

Police

Despite an abundance of potential benchmarks for police services, their use as measuring rods for judging or comparing municipal police departments has been controversial. When headlines in the local newspaper, for instance, proclaim that the community's crime rate is below the state or national average, the police chief and other local officials are understandably proud. A low rate of crime is touted as another example of the community's high quality of life and a testimony to its competent police department. On the other hand, when local news media bemoan a community's "crime problem," charge that police efforts are ineffective, and cite higher-than-average crime rates to substantiate their contentions, several predictable arguments against the use of crime rates invariably are raised by police officials and supporters of the local police force.

Nationally collected crime statistics, formally known as Uniform Crime Reports (UCR), are not above challenge. Critics point out that crime statistics can only be as accurate as the data provided to the Federal Bureau of Investigation (FBI) by individual police departments. If individual departments are sloppy in their data collection or reporting, or if they intentionally misrepresent their performance statistics, the quality of the UCR is thereby diminished. Furthermore, the UCR includes statistics only on *reported* crimes. Victimization studies indicate that many crimes go unreported and, therefore, are excluded from the UCR. Finally, critics of the use of UCR crime rates for evaluating police departments correctly contend that a host of community factors other than police performance also contribute to a community's rate of crime.

So, where does this leave a local government official searching for appropriate police benchmarks? Should the FBI's nationally collected UCR crime rates be used or not? The simple answer is, "Yes, they should"—but users should be clear about the limitations of UCR data, and in most cases, the national average might not be the best figure to use. Often, more precise community benchmarks can be found.

FBI Crime Statistics

Approximately 17,000 law enforcement agencies serving 96% of the U.S. population participate in the FBI's uniform crime reporting program (U.S. Department of Justice, 1999, p. 1). Their collective input generates an extensive set of data providing insights into four categories of violent crime (murder and nonnegligent manslaughter, forcible rape, robbery, and aggravated assault) and four property crimes (burglary, larceny-theft, motor vehicle theft, and arson). These crimes (sometimes excluding arson) constitute what are known as "Part I" crimes, with more than 12 million incidents reported annually.

The FBI itself (U.S. Department of Justice, 1999) warns against "simplistic and incomplete analyses" (p. iv) that merely compare crime index figures for one community with those of another. Unless cities are carefully matched or unless composite statistics take region and community size into consideration, too many other factors could explain the crime rate difference—even assuming equal diligence in reporting.

Police officials who contend that their department's efforts are only one component in the battle against crime are correct. Nevertheless, those officials must remember that their department's reason for existence is to reduce the incidence of crime. It is reasonable for the public and public officials to desire a local crime rate near, or perhaps below, the norm for similar communities. Most police officials probably expect that themselves.

Recommended Performance Measures and Standards

Appropriately, most sets of recommended performance measures for police departments have not focused exclusively on crime rates and the incidence of particular types of crime but have also incorporated indicators of other dimensions of police performance (e.g., Drebin & Brannon, 1992; Hatry et al., 1992). Measures that directly address the efficiency and quality or effectiveness of police services can be of considerable value to benchmarkers as they search for top performers and standards of performance.

Standards of performance differ from *process standards,* such as those recommended by the Commission on Accreditation for Law Enforcement Agencies (CALEA). Although commendable for their value in encouraging comprehensive policies and progressive practices, the focus of most CALEA standards is on the various methods *presumed to lead to favorable results* rather than on the desired levels of performance outcomes themselves. An individual police department wishing to assess its performance based on the results it is achieving will have to find most of its *performance standards* elsewhere.

Staffing Norms—If You Insist!

Like the adoption of recommended procedures or processes, the decision to employ police officers and support personnel is a decision reflecting strategy—a choice made by local decision makers who anticipate that the employment of these personnel will be an effective *means* of achieving important program objectives. It is not an end in itself. Accordingly, the employment of law enforcement personnel should not be granted status as an output objective or even as a measure of service quality. Police department staffing is a form of input. It reflects police performance only indirectly, if at all, and should not be regarded as a primary benchmark. The major benchmarking value of staffing information is as a diagnostic device in exploring possible reasons for performance differentials.

Despite potential hazards in relying too much on staffing norms, staffing information is reported here for three reasons. First, staffing information does have diagnostic value. If a police department consistently ranks below its counterparts on various performance indicators, the possibility of understaffing is one potential factor worth exploring. Second, public officials simply are interested in staffing levels and frequently are buffeted at budget time by claims of inadequate staffing. And third, officials who use national staffing data—or who have it used *on them*—should be acquainted with existing information and be aware of how an appropriate norm for their community may differ from the national average.

The FBI reports that in 1998, municipal police departments had an average of 2.4 sworn officers per 1,000 residents and 3.1 law enforcement employees overall (sworn and civilian) per 1,000 population (U.S. Department of Justice, 1999, p. 291). Individual cities, however, vary from the mean, influenced by local economics, perceived crime problem, and community values. The Washington, D.C., Metropolitan Police, for example, had 6.7 sworn officers per 1,000 in 1998—more than two-and-one-half times the national average. Many other cities, of course, had fewer than the national average. Stockton, California, for example, had only 1.5 sworn officers per 1,000 population.

Higher average staffing numbers are found among the largest cities (i.e., cities of 250,000 or greater) and the smallest cities (i.e., those with less than 10,000 population; Table 20.1). Lower average staffing levels are found among cities of intermediate size. Furthermore, higher staffing levels were reported for 1998 by northeastern and southern cities, and lower staffing levels were reported by midwestern and western cities.

Individual cities often report police staffing level not only as a sign of their commitment to public safety but also as an indicator of officer availability in

Table 20.1 Law Enforcement Staffing Levels in U.S. Cities, 1998

| Region | All Cities | | Full-Time Law Enforcement Employees and Law Enforcement Officers per 1,000 Inhabitants, by Population Cluster | | | | | | | | | | | |
| | | | 250,000 or greater | | 100,000–249,999 | | 50,000–99,999 | | 25,000–49,999 | | 10,000–24,999 | | Less than 10,000 | |
	Employees	Officers	Employees	Officers	Employees	Officers	Employees	Officers	Employees	Officers	Employees	Officers	Employees	Officers
All cities	3.1	2.4	4.1	3.2	2.6	2.0	2.4	1.8	2.3	1.8	2.4	1.9	4.0	3.1
Northeast	3.5	2.8	6.3	5.0	3.6	3.0	2.5	2.1	2.4	2.0	2.1	1.8	2.7	2.3
New England	2.7	2.2	5.3	3.8	3.7	3.1	2.5	2.2	2.3	2.0	2.2	1.9	2.9	2.3
Middle Atlantic	3.8	3.1	6.4	5.1	3.4	2.9	2.5	2.1	2.5	2.1	2.1	1.8	2.6	2.3
Midwest	2.8	2.3	4.4	3.5	2.4	2.0	2.1	1.7	2.0	1.6	2.3	1.8	3.1	2.5
East North Central	2.9	2.4	4.6	3.8	2.5	2.0	2.2	1.7	2.1	1.7	2.3	1.8	3.1	2.5
West North Central	2.5	2.0	3.7	2.7	2.3	1.8	1.8	1.4	1.9	1.4	2.2	1.7	3.0	2.5
South	3.6	2.7	3.6	2.7	2.9	2.2	2.9	2.3	2.9	2.2	3.1	2.4	5.6	4.3
South Atlantic	4.1	3.2	4.3	3.3	3.2	2.4	3.4	2.5	3.3	2.5	3.5	2.8	6.8	5.3
East South Central	3.8	2.8	4.8	2.8	3.0	2.1	3.1	2.4	3.0	2.4	3.0	2.4	4.8	3.8
West South Central	3.0	2.3	3.0	2.3	2.5	1.9	2.4	1.8	2.4	1.8	2.6	2.0	4.7	3.4
West	2.5	1.8	2.9	2.1	2.0	1.4	2.0	1.4	2.1	1.5	2.3	1.7	4.7	3.5
Mountain	2.7	1.9	2.9	1.9	2.2	1.6	2.1	1.5	2.3	1.7	2.6	1.9	4.3	3.2
Pacific	2.5	1.8	2.9	2.1	1.9	1.3	1.9	1.3	2.0	1.4	2.1	1.6	5.1	3.7

SOURCE: U.S. Department of Justice, Federal Bureau of Investigation, *Crime in the United States 1998: Uniform Crime Reports* (Washington, DC: Government Printing Office, 1999), pp. 292–293.

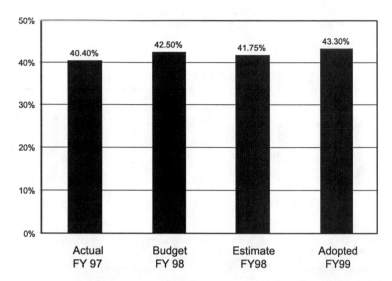

The Patrol Availability Factor is a measure of the actual time a patrol officer is available to patrol. This does not include the time needed to respond to calls for service or administrative duties.

Figure 20.1. Patrol Availability Factor: San Antonio
SOURCE: City of San Antonio (TX), *Annual Budget, Fiscal Year 1998–99,* p. 448.

the community. Higher levels of staffing are presumed to mean more officers *on the street.* In reality, however, the number of officers employed is one of two important factors in that equation. A second and often overlooked ingredient is police management proficiency in actually getting available officers out of the station and onto the street, where they can respond to calls or engage in undirected patrol—the latter of which constitutes a community's *patrol availability factor* (Figures 20.1 and 20.2).

How much of a patrol officer's time should be spent on patrol rather than on assigned responses or various administrative duties? A panel assembled by the LCC (1994) suggests that officers in high-service-level departments are able to devote at least 45% of their time to patrolling the field uncommitted; officers in medium-service departments have 30% to 45% for uncommitted patrol; and those in low-service departments have less than 30% (p. 34). The evidence from reporting cities, however, indicates that departments able to commit one third of the typical patrol officer's time to actual patrol are doing rather well (Table 20.2).

The use of officer-to-population ratios for staffing decisions would imply that resident head counts can serve as good indicators of demand for law enforcement services. Although a relationship between population size and the need for police officers clearly exists, in reality population provides only a general clue to likely demand for services. More precise indicators are available through direct measures of demand. Accordingly, ratios of calls for service per officer or arrests per officer provide demand and workload informa-

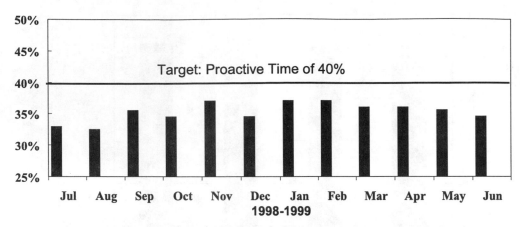

Goal: To achieve a 40% proactive time rate and to spend that time working with community members to solve identified neighborhood crime problems.

Percentage of time goal was met: For FY 1999, the department goal was not met.

Note: The "Proactive Time Rate" is the percentage of total officer time available to be used for field-initiated activities. This time is estimated by subtracting the amount of time officers spend on committed/out-of-service duties from the total time. Committed/out-of-service time has three elements: (1) calls for service and related activities; (2) administrative activities (meetings, court appearances, etc); (3) line-up/end-of-shift activities.

Figure 20.2. Proactive Time Rate for Police Officers in San Diego
SOURCE: City of San Diego (CA), *Fiscal Year 1999 Semi-Annual Performance Report*, p. 7.

Table 20.2   Patrol Availability: Selected Cities

City	Patrol Availability Factor[a]
San Antonio, TX	32.7% (1991), 40.6% (1995), 40.4% (1997)
San Clemente, CA	Target percentage for preventive patrol: 40%
	Actual: 40% (1998)
Savannah, GA	Target: 40%
College Station, TX	Percentage of patrol officers' time that is proactive: 39%; reactive: 61% (1998)
Peoria, AZ	37.7% (1991)
Portland, OR	37% (1998)[b]
Sacramento, CA	34% (1990)

a. Percentage of shift that an officer is available for proactive patrol (excludes assigned responses, administrative time, miscellaneous duties, etc.). For example, Savannah tries to devote 40% of patrol officers' time to preventive patrol, directed patrol, and community-oriented policing (City of Savannah [GA], *1992 Program of Work*, p. 69).
b. "Average time available for self-initiated activity and problem solving" (City of Portland [OR], *Adopted Budget Fiscal Year 1999–2000* [Vol. I], p. 121).

tion that in many ways may be more revealing than staffing ratios based simply on population (Table 20.3). Comparative studies of 1998 data in North and South Carolina revealed averages of 767 calls per officer among selected cities in North Carolina and 897 calls per officer among selected counterparts in South Carolina (Table 20.4).

Crime Rates

Variations in crime rates are influenced not only by police efforts and economic factors but also by whether a community is rural or urban, its population, and the state or region in which it is located. Rates of violent crime tend to be lower in cities lying outside metropolitan statistical areas, and crime rates for violent crimes and property crimes are lower still in rural counties (Table 20.5). Crime rates tend to be higher in big cities than in small towns (Table 20.6), and tend to be greater in the South and West than in other regions (Table 20.7).

With variables such as population and region possessing such obvious significance for crime rate tendencies, should not those factors be taken into consideration as a city chooses—or establishes—an appropriate benchmark? One way to do so is by calculating a *projected norm* using the crime rate for a given community's population cluster and adjusting the rate on the basis of that community's state or region (Table 20.8). By doing so, a city can gauge how its actual crime rate compares with the rate that could be expected on the basis of its size and locale.

Consider, for example, the plight of city officials in Independence, Missouri, as they might have attempted to assess their crime statistics for 1998. Independence, a city of 111,505 inhabitants, reported 676 violent crimes that year for a rate of 606.3 per 100,000 population (U.S. Department of Justice, 1999, p. 129). Officials could have compared their local mark with the national average of 566.4 or with the state average of 555.7 violent crimes per 100,000 population and been discouraged; or they could have stacked the local figure up against the national average of 758.2 violent crimes per 100,000 population among cities of 100,000 to 249,999 population and been rather pleased. Which average should they have used? Recognizing that cities of Independence's size face a more formidable crime challenge than smaller communities, they might have started with the average crime rate for their population cluster but then could have adjusted that rate because Independence is located in a state with a lower-than-average rate of reported crimes. Through such an adjustment, a more appropriate projected norm of 743.9 violent crimes per 100,000 population could have been calculated (see Table 20.8). Because Independence's actual rate of 606.3 fell comfortably below the projected norm of 743.9, community leaders might have congratulated police

Table 20.3   Officer Demand: Indicators From Selected Cities

CALLS FOR SERVICE

Tallahassee, FL
1,712 calls per uniformed officer (1991)

Raleigh, NC
Calls for service per beat officer: 1,489 (1997), 1,579 (1998)

Savannah, GA
1,017 calls per sworn officer in Patrol Division (1997)

Rochester, NY
929 calls per uniformed officer (1995)

Corpus Christi, TX
826.9 calls per sworn officer (1997)

Ames, IA
759.2 calls per patrol division FTE (1997)

San Antonio, TX
716 per patrol officer (1997)

Calgary, Alberta
Calls for service per sworn staff member: 691 (1996)

College Station, TX
Average number of calls for service per officer: 681 (1995)

Tempe, AZ
Emergency calls per patrol officer: 671 (1997)

Norfolk, NE
605 calls per officer (1995)

Alexandria, VA
Calls dispatched to patrol per patrol officer: 456 (1998)

Portland, OR
Dispatched incidents per precinct officer: 451 (1997)
Officer-initiated incidents per precinct officer: 245 (1997)

Kingsport, TN
431.5 calls per officer (1997)

Kansas City, MO
260 calls per police officer (1996)

Chandler, AZ
Calls for service per sworn officer: 197 (1999)

ARREST RATIO

Rochester, NY
50.0 arrests per uniformed officer (1995)

Kingsport, TN
49.9 arrests per officer (1997)

Bloomington, MN[a]
18.2 arrests per FTE sworn staff member (1998)

San Mateo, CA[a]
15.2 arrests per FTE sworn staff member (1998)

Mankato, MN[a]
14.8 arrests per FTE sworn staff member (1998)

Merced, CA[a]
11.9 arrests per FTE sworn staff member (1998)

Colorado Springs, CO[a]
10.4 arrests per FTE sworn staff member (1998)

Modesto, CA[a]
10.0 arrests per FTE sworn staff member (1998)

Oak Ridge, TN
9.9 Part I arrests per staff year (1997)

a. Selected arrest ratio statistics were drawn from International City/County Management Association, *Comparative Performance Measurement: FY 1998 Data Report* (Washington, DC: Author, 1999), pp. 29, 51.

Table 20.4 Officer Demand: Statistics From Comparative Performance Projects, 1998

	Averages	
	Calls for Service per Patrol Officer	Dispatched Calls per Patrol Officer
7 North Carolina cities of 63,000–260,000 population	767	674
14 North Carolina cities of 2,000–62,000 population		663
10 South Carolina cities of 12,000–75,000 population	896.5	

SOURCES: *Interim Report on Services for Seven Cities* (Chapel Hill, NC: Institute of Government/University of North Carolina, 1998); *Performance and Cost Data: Phase III City Services* (Chapel Hill, NC: Institute of Government/University of North Carolina, 1999); *South Carolina Municipal Benchmarking Project: 1998* (Columbia, SC: Institute of Public Affairs/University of South Carolina, 1999). Reprinted with permission of the Institute of Government, The University of North Carolina at Chapel Hill.

officials on their success and then challenged them to achieve a benchmark set at, say, 80% of the projected norm in the future.

In contrast, Tallahassee, Florida's rate of 1,279.4 violent crimes per 100,000 population in 1998 (U.S. Department of Justice, 1999, p. 119) was far greater than the national average of 566.4, its population cluster average of 758.2, or the Florida average of 938.7 violent crimes per 100,000 population. Applying the prescribed calculations, however, yields a projected norm

(text continued on page 311)

Table 20.5 Crime Rates for Metropolitan, Suburban, and Rural Areas, 1998

Rates per 100,000 Inhabitants

		Violent Crimes					Property Crimes			
	Crime Index Total	Subtotal	Murder and Nonnegligent Manslaughter	Forcible Rape	Robbery	Aggravated Assault	Subtotal	Burglary	Larceny Theft	Motor Vehicle Theft
United States, total	4,615.5	566.4	6.3	34.4	165.2	360.5	4,049.1	862.0	2,728.1	459.0
Metropolitan statistical areas	4,975.0	630.5	6.7	36.1	197.9	389.7	4,344.6	899.9	2,911.4	533.2
Cities outside metropolitan areas	4,987.3	444.3	4.0	35.6	66.0	338.7	4,543.0	884.9	3,435.1	223.0
Rural counties	1,998.0	226.6	4.6	22.9	16.5	182.6	1,771.4	596.9	1,045.4	129.1

SOURCE: U.S. Department of Justice, Federal Bureau of Investigation, *Crime in the United States 1998: Uniform Crime Reports* (Washington, DC: Government Printing Office, 1999), p. 65.

Table 20.6 Crime Rates in U.S. Cities by Population Cluster, 1998

Rates per 100,000 Inhabitants

	Crime Index Total	Violent Crimes					Property Crimes			
		Subtotal	Murder and Nonnegligent Manslaughter	Forcible Rape	Robbery	Aggravated Assault	Subtotal	Burglary	Larceny Theft	Motor Vehicle Theft
All cities	5,704.0	738.2	7.8	39.9	242.8	447.6	4,965.9	993.7	3,361.8	610.5
1 million or greater	6,152.6	1,286.8	13.5	37.3	513.9	722.1	4,865.8	942.6	3,016.2	907.1
500,000–999,999	7,591.6	1,153.9	16.3	60.9	432.5	644.2	6,437.7	1,347.7	4,016.4	1,073.6
250,000–499,999	8,017.0	1,168.1	13.6	62.6	419.5	672.5	6,848.8	1,425.2	4,344.4	1,079.2
100,000–249,999	6,406.7	758.2	8.6	45.2	251.8	452.6	5,648.5	1,190.7	3,769.8	688.0
50,000–99,999	5,301.7	589.7	5.3	39.6	176.6	368.1	4,712.0	958.2	3,235.2	518.6
25,000–49,999	4,701.5	454.1	3.7	32.1	118.1	300.2	4,247.4	842.6	3,014.6	390.2
10,000–24,999	4,286.2	373.1	3.3	28.6	82.7	258.5	3,913.1	732.9	2,893.9	286.4
Less than 10,000	4,646.0	397.1	3.0	28.4	59.8	305.8	4,249.0	779.7	3,226.3	243.0

SOURCE: U.S. Department of Justice, Federal Bureau of Investigation, *Crime in the United States 1998: Uniform Crime Reports* (Washington, DC: Government Printing Office, 1999), pp. 188–189.

Table 20.7 Crime Rates by Region and State, 1998

Rates per 100,000 Inhabitants

Area	Crime Index Total	Violent Crimes					Property Crimes			
		Subtotal	Murder and Nonnegligent Manslaughter	Forcible Rape	Robbery	Aggravated Assault	Subtotal	Burglary	Larceny Theft	Motor Vehicle Theft
United States	**4,615.5**	**566.4**	**6.3**	**34.4**	**165.2**	**360.5**	**4,049.1**	**862.0**	**2,728.1**	**459.0**
Northeast	**3,473.5**	**500.5**	**4.3**	**23.7**	**184.8**	**287.7**	**2,973.1**	**589.2**	**2,011.0**	**372.8**
New England	3,388.1	422.5	2.5	26.5	86.0	307.5	2,965.6	608.4	2,008.8	348.5
Connecticut	3,786.5	366.3	4.1	22.2	133.8	206.2	3,420.2	665.9	2,366.3	388.1
Maine	3,040.7	125.8	2.0	18.1	21.1	84.6	2,914.9	666.8	2,126.8	121.3
Massachusetts	3,453.9	621.3	2.0	27.4	96.6	495.2	2,814.6	607.3	1,777.7	429.5
New Hampshire	2,419.8	107.2	1.5	33.8	21.5	50.4	2,312.7	325.1	1,863.2	124.4
Rhode Island	3,517.8	312.1	2.4	35.5	66.7	207.5	3,205.7	653.0	2,165.1	387.6
Vermont	3,139.1	106.3	2.2	27.6	9.5	67.0	3,032.8	671.1	2,213.9	147.9
Middle Atlantic	3,503.5	527.8	4.9	22.7	219.4	280.8	2,975.7	582.5	2,011.8	381.4
New Jersey	3,654.1	440.1	4.0	20.0	186.2	230.0	3,213.9	671.1	2,109.3	433.6
New York	3,588.5	637.8	5.1	21.1	270.3	341.3	2,950.7	576.7	1,998.9	375.1
Pennsylvania	3,273.0	420.5	5.3	26.9	164.9	223.5	2,852.4	531.4	1,965.4	355.5
Midwest	**4,379.4**	**494.1**	**5.7**	**36.8**	**141.4**	**310.2**	**3,885.3**	**775.7**	**2,713.4**	**396.2**
East North Central	4,441.2	536.9	6.4	37.5	161.1	331.9	3,904.3	789.4	2,685.7	429.2
Illinois	4,872.8	807.7	8.4	34.0	248.5	516.9	4,065.0	826.1	2,799.4	439.5
Indiana	4,169.4	431.0	7.7	33.1	111.2	279.0	3,738.4	789.2	2,590.1	359.2
Michigan	4,682.9	620.8	7.3	50.4	155.8	407.3	4,062.1	837.8	2,630.0	594.3
Ohio	4,327.5	362.5	4.0	40.5	133.5	184.5	3,965.0	810.1	2,771.1	383.8
Wisconsin	3,543.1	249.0	3.6	19.9	85.6	139.9	3,294.1	569.3	2,452.8	272.0

West North Central	4,233.2	392.9	4.3	35.1	94.8	258.7	3,840.3	743.4	2,778.8	318.1
Iowa	3,500.6	311.5	1.9	25.4	50.9	233.3	3,189.1	673.7	2,306.6	208.7
Kansas	4,858.8	397.0	5.9	42.6	86.8	261.8	4,461.7	892.6	3,341.4	227.8
Minnesota	4,046.5	310.2	2.6	49.9	92.5	165.2	3,736.3	687.5	2,723.6	325.2
Missouri	4,826.4	555.7	7.3	26.9	149.2	372.2	4,270.7	872.5	2,948.4	449.8
Nebraska	4,405.2	451.4	3.1	25.1	77.6	345.7	3,953.8	634.0	2,971.7	348.0
North Dakota	2,681.0	89.3	1.1	33.2	10.2	44.8	2,591.7	356.4	2,058.6	176.6
South Dakota	2,624.1	154.3	1.4	35.0	20.2	97.8	2,469.8	468.6	1,897.8	103.4
South	**5,223.5**	**1,633.0**	**7.8**	**38.1**	**169.1**	**418.0**	**4,590.5**	**1,045.4**	**3,074.7**	**470.4**
South Atlantic	5,565.9	703.5	7.8	37.5	197.8	460.5	4,862.4	1,088.4	3,269.3	504.7
Delaware	5,363.2	762.4	2.8	67.1	194.2	498.3	4,600.8	859.5	3,313.0	428.2
District of Columbia	8,835.6	1,718.5	49.7	36.3	689.5	943.0	7,117.0	1,216.3	4,657.7	1,243.0
Florida	6,886.0	938.7	6.5	49.6	242.7	639.9	5,947.4	1,361.7	3,886.8	698.9
Georgia	5,463.0	572.7	8.1	30.4	187.2	347.0	4,890.3	990.8	3,342.8	556.6
Maryland	5,365.7	796.6	10.0	33.4	298.7	454.5	4,569.1	922.9	3,096.8	549.4
North Carolina	5,322.2	579.4	8.1	30.6	160.8	379.9	4,742.8	1,324.6	3,092.0	326.2
South Carolina	5,777.0	903.2	8.0	45.7	154.9	694.6	4,873.8	1,162.7	3,295.4	415.7
Virginia	3,660.4	325.7	6.2	26.7	105.6	187.2	3,334.7	560.9	2,503.5	270.3
West Virginia	2,547.2	248.6	4.3	18.7	37.3	188.3	2,298.6	613.5	1,497.9	187.2
East South Central	4,297.6	507.6	8.0	37.1	131.9	330.6	3,790.0	953.1	2,467.4	369.5
Alabama	4,597.1	512.1	8.1	33.2	130.9	339.9	4,085.0	964.3	2,779.0	341.7
Kentucky	2,889.4	284.0	4.6	29.3	75.4	174.7	2,605.3	637.4	1,750.1	217.8
Mississippi	4,384.0	410.7	11.4	37.3	123.3	238.6	3,973.3	1,144.5	2,490.0	338.7
Tennessee	5,034.4	715.0	8.5	45.8	178.0	482.8	4,319.4	1,075.9	2,726.1	517.4
West South Central	5,173.2	586.8	7.7	39.7	142.8	396.6	4,586.4	1,025.9	3,090.5	470.0
Arkansas	4,283.4	490.2	7.9	35.2	96.2	350.9	3,793.2	928.3	2,581.8	283.2
Louisiana	6,098.3	779.5	12.8	36.8	198.0	531.9	5,318.8	1,172.1	3,605.1	541.6
Oklahoma	5,003.9	539.4	6.1	45.2	92.0	396.1	4,464.5	1,143.4	2,915.8	405.3
Texas	5,111.6	564.6	6.8	40.0	145.1	372.6	4,547.0	986.3	3,071.7	489.1

(continued)

309

Table 20.7 Crime Rates by Region and State, 1998 (Continued)

Rates per 100,000 Inhabitants

Area	Crime Index Total	Violent Crimes					Property Crimes			
		Subtotal	Murder and Nonnegligent Manslaughter	Forcible Rape	Robbery	Aggravated Assault	Subtotal	Burglary	Larceny Theft	Motor Vehicle Theft
West	4,879.4	593.1	6.1	35.4	167.1	384.5	4,286.3	895.7	2,810.0	580.6
Mountain	5,409.5	490.2	6.4	40.2	119.8	323.8	4,919.4	978.8	3,383.4	557.1
Arizona	6,575.0	577.9	8.1	31.1	165.2	373.6	5,997.0	1,209.5	3,922.4	865.1
Colorado	4,487.5	377.9	4.6	47.4	81.5	244.4	4,109.5	786.5	2,917.9	405.1
Idaho	3,714.6	282.2	2.9	31.4	21.5	226.4	3,432.5	693.1	2,553.7	185.7
Montana	4,070.7	138.8	4.1	17.8	19.9	96.9	3,931.9	511.5	3,191.6	228.9
Nevada	5,280.5	643.6	9.7	52.1	254.9	326.8	4,636.9	1,137.6	2,711.3	788.0
New Mexico	6,719.1	961.4	10.9	55.1	163.4	732.0	5,757.7	1,394.0	3,743.9	619.9
Utah	5,505.9	314.2	3.1	41.7	66.0	203.5	5,191.7	812.9	4,012.1	366.7
Wyoming	3,807.7	247.6	4.8	27.7	16.2	199.0	3,560.1	560.5	2,860.5	139.1
Pacific	4,674.3	633.0	6.0	33.6	185.4	408.0	4,041.3	863.5	2,588.1	589.7
Alaska	4,777.0	653.9	6.7	68.6	86.6	492.0	4,123.1	667.4	3,031.1	424.6
California	4,342.8	703.7	6.6	29.9	210.6	456.6	3,639.1	823.5	2,217.1	598.5
Hawaii	5,333.0	246.9	2.0	29.5	102.7	112.7	5,086.1	936.2	3,681.0	468.9
Oregon	5,646.6	419.8	3.8	39.8	105.2	271.0	5,226.8	927.5	3,773.3	526.0
Washington	5,867.4	428.5	3.9	48.2	115.6	260.8	5,438.9	1,062.5	3,757.7	618.7

SOURCE: U.S. Department of Justice, Federal Bureau of Investigation, *Crime in the United States 1998: Uniform Crime Reports* (Washington, DC: Government Printing Office, 1999), pp. 66–73.

Table 20.8   Establishing Community Benchmarks From UCR Information

---

Uniform Crime Report (UCR) statistics often reveal broad differences in key variables (e.g., staffing levels, crime rates, and clearance rates) among regions, states, and population clusters. Using the "national average" as a community benchmark may be a poor choice.

A better baseline for crime and clearance rates, for example, may be calculated as a "projected norm," as follows:

$$\text{Projected Norm} = \text{Statistic for population cluster} \times \frac{\text{Statistic for state or region}}{\text{national average}}$$

Then, a community benchmark may be set in relation to that baseline.

For example, a California community of 40,000 that wishes to set a suitable benchmark for reported robberies would first calculate the projected norm as a baseline. On the basis of 1998 statistics, the following calculation would be made:

$$\text{Projected Norm} = (118.1 \text{ robberies per } 100,000 \text{ population}) \times \frac{210.6}{165.2} = 150.6 \text{ robberies per } 100,000$$

If the community's rate of reported robberies exceeds 150.6 per 100,000 inhabitants, then the projected norm might serve as a suitable benchmark. If the community's rate is less than the projected norm, a more useful benchmark might be set at 90% or 80% of the projected norm.

A similar technique may be used for establishing community benchmarks for local clearance rates or other performance statistics using UCR information.

---

of 1,256.6 violent crimes per 100,000 population, a figure that recognizes that Tallahassee police officials are situated in a state and region that have higher rates of reported crimes than the national average. The projected norm appears to be a much more reasonable—and clearly more attainable—benchmark for Tallahassee than any of the three more usual choices (i.e., national average, population cluster average, and statewide average).

Response Time

Municipalities often judge their police department in part on the promptness with which they respond to emergencies. High-service-level departments, according to the LCC panel (1994), should respond to emergencies within 5 minutes (p. 34). Coincidentally, the median response time among the 11 cities examined for this volume that reported comprehensive response time (i.e., the time from initial call until arrival) was 5 minutes and 37 seconds—just slightly higher than the LCC prescription (Table 20.9). The quickest average

Table 20.9   Response Times to Police Emergencies: Selected Cities

Time From Call Receipt at 911 Until Arrival at Scene

Fort Collins, CO 3.0-minute average (1999)	**Duncanville, TX** 4-minute 38-second average (1999)	**Tucson, AZ** 5-minute 53-second average (1999)	**Dayton, OH** 7.0-minute average (1999)
Oak Ridge, TN 3-minute 18-second average (1999)	**Corpus Christi, TX** 4-minute 59-second average (1997)	**New York, NY** 6-minute 18-second average to "critical" crimes in progress; 10-minute 18-second average to crimes in progress overall (1999)	**Charlotte, NC** 11.0-minute average (1999)
Vancouver, WA 4-minute 37-second average (1998)	**Tempe, AZ** 5-minute 37-second average (1999, estimated)		**Orlando, FL** 15-minute 12-second average (1997)

Time From Dispatch Until Arrival at Scene (Excludes Dispatch Time)

Germantown, TN 2-minute 40-second average[a] (1996)	**Fayetteville, AR** 3-minute 51-second average[a] (1997)	**Tucson, AZ** 4-minute 42-second average[a] (1998)	**College Station, TX** 5-minute 36-second average[a] (1998)
Blacksburg, VA 2-minute 59-second average[a] (1997)	**Kansas City, MO** 4.0-minute average (1998, estimated)	**Scottsdale, AZ** 4-minute 48-second average[a]; 76% within 6 minutes (1999)	**Newnan, GA** 6.0-minute average (1999)
Palo Alto, CA 92% within 3 minutes[a] (1997)	**Gresham, OR** 4-minute 8-second average (1999)	**Wichita, KS** 4-minute 48-second average[a] (1998)	**Durham, NC** 6-minute 9-second average[a]; 49.8% within 5 minutes (1998)
Corvallis, OR 3-minute 5-second average (1999)	**Long Beach, CA** 4-minute 12-second average[a] (1996)	**Reno, NV** Target: At least 75% within 5 minutes (1999)	**Plano, TX** 6-minute 12-second average[a] (1997)
Lubbock, TX 3-minute 6-second average (1997)	**San Clemente, CA** 4-minute 15-second average[a] (1998)	**Chandler, AZ** 58% within 5 minutes[a] (1998)	**Greensboro, NC** 6-minute 18-second average[a] (1997)
Bellevue, WA 3-minute 8-second average to life-threatening emergencies (1998)	**Cary, NC** 4-minute 25-second average[a] (1999)	**San Antonio, TX** 5-minute 1-second average[a] (1997)	**Nashville–Davidson County, TN** 6-minute 29-second average (1998)
Cincinnati, OH 3-minute 24-second average (1999)	**Chapel Hill, NC** 4-minute 29-second average[a] (1999)	**Portland, OR** 5-minute 6-second average (1999)	**Philadelphia, PA** 6-minute 46-second average[a] (1998)
Norman, OK 3-minute 24-second average[a] (1998)	**Glendale, AZ** 70% within 4 minutes 30 seconds[a] (1997)	**Raleigh, NC** 5-minute 21-second average (1999)	**San Diego, CA** 6-minute 54-second average[a] (1998)
Duncanville, TX 3-minute 25-second average[a] (1998)	**San Jose, CA** 4-minute 31-second average (1999)	**Flagstaff, AZ** Target: 5-minute average Actual: 5-minute 30-second average[a] (1997)	**Boston, MA** 56% of priority 1 calls responded to within 7 minutes[a] (1996)
Danville, KY 3-minute 30-second average[a] (1997)			**Chesapeake, VA** 8-minute 31-second average (1999)

a. "Response time" is undefined in the reporting document for this city and is presumed conservatively to indicate the time from dispatch until arrival at the scene.

response times were reported by Fort Collins, Colorado, with an average of 3 minutes, and Oak Ridge, Tennessee, with an average of 3 minutes and 18 seconds from the initial call to 911 until arrival of an officer at the scene.

Some cities—in fact, most cities—exclude dispatch time from the response time statistics they commonly report. Rather than declaring comprehensive response times, they instead report the average time from the dispatch of a unit until arrival. Although taking this more limited view of response time permits persons in charge of police patrol to focus more sharply on that aspect of response most fully within their control, the omission of dispatch time distorts the meaning of response time from the perspective of most citizens. Among cities examined for this volume, the median time from dispatch until arrival was an average just slightly greater than 4.5 minutes and the quickest averages were less than 3 minutes. Assuming an average dispatch time of 1 minute, these response time statistics would remain impressive even if reported in a more comprehensive fashion.

Various studies of comparative performance confirm the reasonableness of a 5-minute standard for excellent police emergency response—especially if response time is perceived to include only the time from dispatch until arrival (Table 20.10). Average cities in three recent studies have achieved sub-5-minute response time averages and some have reported averages of 3 minutes or less.

Cities often differentiate between calls requiring emergency response and those of a less urgent nature. Responses to lower-priority and nonemergency calls typically range from 10- to 30-minute averages.

Clearance Rates

Cases are said to be "cleared by arrest" when primary suspects have been arrested for that offense and turned over to the court or "cleared by exception" when the victim refuses to cooperate, extradition is denied, the offender is deceased, or other extraordinary circumstances preclude the placing of charges. The police discontinue their pursuit of suspects in such cases. Obviously, a high clearance rate is desired. Clearance rates tend to be higher in small cities (Table 20.11) and to vary slightly by region (Table 20.12). Projected norms for clearance rates may be calculated in the same fashion as norms for crime rates.

Potential Benchmarks for Selected Police Activities

Among the performance targets and performance records reported by various municipalities are potential performance benchmarks for a variety of police functions and support services.

Table 20.10 Emergency Response Time From Dispatch to Arrival at the Scene: Statistics From Comparative Performance Projects

	Average Time From Dispatch Until Arrival at the Scene	Pacesetter
28 U.S. and Canadian cities and counties of 100,000 population or greater	4.5 minutes	**Houston, TX** (3.0-minute average)
15 U.S. cities and counties of 100,000 population	4.2 minutes	**Santa Monica, CA Merced, CA** (2.9-minute averages)
7 North Carolina cities of 63,000–260,000 population	4.3 minutes	**Winston-Salem, NC** (2.6-minute average)
14 North Carolina cities of 2,000–62,000 population	3.9 minutes	**Roanoke Rapids, NC** (1.5-minute average)
10 South Carolina cities of 12,000–75,000 population	4.7 minutes	**Aiken, SC** (3.5-minute average)

SOURCES: *Comparative Performance Measurement: FY 1997 Data Report* (Washington, DC: International City/County Management Association, 1999), pp. 29, 79; *Interim Report on Services for Seven Cities* (Chapel Hill, NC: Institute of Government/University of North Carolina, 1998); *Performance and Cost Data: Phase III City Services* (Chapel Hill, NC: Institute of Government/University of North Carolina, 1999); *South Carolina Municipal Benchmarking Project: 1998* (Columbia, SC: Institute of Public Affairs/University of South Carolina, 1999). Reprinted with permission of the Institute of Government, The University of North Carolina at Chapel Hill.

Traffic Enforcement. Some cities report traffic enforcement statistics relating the number of citations issued to other variables, such as population, traffic volume, the number of accidents, and the number of injury accidents. The police department in Wichita, Kansas, for example, issued 251 traffic citations per 1,000 population during one recent year. San Clemente, California, strives to maintain an "enforcement index" of at least 15, where the index consists of citations for hazardous moving violations divided by the number of injury accidents. College Station, Texas, reported an enforcement index of 46 at high accident locations, where its index consists of the number of citations issued divided by the number of accident reports from those locations. The idea in each case is to ensure that the deployment of traffic enforcement resources is, and remains, commensurate with traffic enforcement needs.

DUI/DWI Enforcement. Driving under the influence of intoxicants—or driving while intoxicated—is a serious problem that has prompted special enforcement efforts in many communities. Several cities report performance statistics that can serve as potential benchmarks for others (Table 20.13).

Parking Enforcement. Effectiveness in the enforcement of parking regulations is reported in the form of compliance rates, error rates for parking citations, apprehension rates for scofflaws, and other relevant statistics (Table 20.14).

Table 20.11 Clearance Rates in U.S. Cities by Population Cluster, 1998

		Percentage of Offenses Cleared by Arrest or by Exception									
		Violent Crime						Property Crimes			
City Population	Crime Index[a]	Subtotal	Murder and Nonnegligent Manslaughter	Forcible Rape	Robbery	Aggravated Assault	Subtotal[a]	Burglary	Larceny Theft	Motor Vehicle Theft	Arson
All cities	21.3	47.7	67.6	50.0	28.0	57.6	17.5	13.1	19.5	13.3	16.1
1 million or greater	20.4	46.8	70.1	52.6	29.2	59.0	13.7	12.3	15.7	8.8	10.6
500,000–999,999	18.2	40.3	59.5	57.6	22.5	50.5	14.3	11.7	15.8	12.2	10.5
250,000–499,999	18.8	43.9	65.6	53.5	25.5	54.2	14.5	11.2	16.5	10.8	15.4
100,000–249,999	20.9	48.8	68.7	49.3	28.6	59.6	17.2	12.9	19.2	13.2	16.2
50,000–99,999	21.5	46.8	64.6	43.6	28.6	55.7	18.3	12.4	20.9	12.9	16.3
25,000–49,999	22.0	50.0	70.9	45.0	30.4	57.7	19.1	13.3	21.2	15.1	19.1
10,000–24,999	25.2	55.0	76.0	49.2	34.7	61.6	22.4	15.5	24.1	22.6	25.5
Less than 10,000	23.5	59.7	76.3	50.6	37.1	64.6	20.1	16.5	20.4	27.6	24.2

SOURCE: U.S. Department of Justice, Federal Bureau of Investigation, *Crime in the United States 1998: Uniform Crime Reports* (Washington, DC: Government Printing Office, 1999), pp. 201–202.

a. Excludes arson figures (reported by fewer agencies than other categories of crime).

Table 20.12 Clearance Rates by Region, 1998

| | | Percentage of Offenses Cleared by Arrest or by Exception | | | | | | | | | |
| | | Violent Crimes | | | | | Property Crimes | | | | |
Region	Crime Index[a]	Subtotal	Murder and Non–Negligent Manslaughter	Forcible Rape	Robbery	Aggravated Assault	Subtotal[a]	Burglary	Larceny Theft	Motor Vehicle Theft	Arson
Total, all agencies	21.3	49.1	68.7	49.9	28.4	58.5	17.4	13.6	19.2	14.2	16.3
Northeast	23.8	50.7	78.4	54.1	31.4	62.7	19.1	16.0	21.4	12.0	17.1
New England	22.0	51.2	62.8	46.7	28.0	58.3	17.7	14.0	19.7	12.9	16.0
Middle Atlantic	24.4	50.5	81.2	56.8	31.9	64.3	19.6	16.6	22.0	11.7	17.4
Midwest	18.9	44.1	64.2	45.9	22.3	53.2	15.9	11.3	17.4	14.7	15.0
East North Central	16.8	39.8	61.3	42.8	20.4	48.6	13.8	10.4	15.1	12.4	13.1
West North Central	22.7	53.6	72.6	52.8	28.0	62.2	19.6	13.1	21.1	20.5	19.6
South	22.1	50.3	70.9	54.4	29.5	58.1	18.4	14.3	19.8	18.0	20.2
South Atlantic	22.0	49.9	70.1	55.5	29.0	58.6	18.3	15.2	19.2	18.9	19.7
East South Central	22.8	49.0	68.0	49.9	30.2	56.3	19.0	14.3	21.0	18.0	24.1
West South Central	22.0	51.2	72.8	55.1	30.0	58.1	18.3	13.5	20.1	17.1	19.8
West	20.1	49.0	61.6	44.5	27.1	58.9	16.0	12.4	18.1	11.5	12.7
Mountain	18.8	42.0	56.2	36.4	24.1	49.3	16.4	11.1	18.5	13.7	17.1
Pacific	20.6	50.9	63.6	47.8	27.8	61.5	15.8	12.9	17.9	10.8	11.7

SOURCE: U.S. Department of Justice, Federal Bureau of Investigation, *Crime in the United States 1998: Uniform Crime Reports* (Washington, DC: Government Printing Office, 1999), pp. 203–204.

a. Excludes arson figures (reported by fewer agencies than other categories of crime).

Table 20.13   DUI/DWI Enforcement: Selected Cities

Bellevue, WA
DUI arrests versus DUI-related accidents: 8.6 to 1 (1997)

College Station, TX
Ratio of DWI arrests to alcohol-related traffic accidents: 8.2 to 1 (1995)
Percentage of accidents that are alcohol related: 1.87% (1998)

Scottsdale, AZ
DUI arrests versus accidents in which the driver had been drinking: 5.7 to 1 (1999)

Wichita, KS
DUI arrests per 1,000 population: 7.94 (1998)

NOTE: DUI = driving under the influence of intoxicants; DWI = driving while intoxicated.

Recovery of Stolen Property. Police effectiveness in pursuing thefts may be judged in part by the recovery of stolen items. Recovery rates of 25% to 40% are not uncommon (Table 20.15).

Community-Oriented Policing. Many cities have adopted community-oriented, neighborhood policing strategies. Some have begun to report performance statistics focusing on outputs and results, such as targeted crime reduction and evidence of community cooperation and assistance (Table 20.16).

Call Diversion. To maximize the productive use of sworn officers and to reduce the need for additional officers, many communities have adopted practices that divert work, when appropriate, from patrol units to non–sworn personnel who take reports by telephone or otherwise handle nonemergency police calls without requiring officers. Some departments have quantified the success of these efforts (Table 20.17).

Crime Lab. How quickly should crime evidence be processed and analyzed? The expectations and experience of Santa Ana, California, and Scottsdale, Arizona, are shown in Table 20.18.

Police Records. Potential benchmarks for the police records function are provided in Table 20.19.

Internal Affairs

Police officers deal with many situations involving discretion and requiring good judgment. Occasional complaints are inevitable. Still, the effectiveness of some cities in holding complaints to a minimum and dealing promptly with those that do occur are points of pride that are reported in their performance statistics. Durham, North Carolina, for example, reported receiving only 1.6 citizen complaints per 1,000 police contacts during one recent year. Similarly, Scottsdale, Arizona, reported receiving only 0.06 complaints per 1,000 popu-

Table 20.14   Parking Regulations Enforcement: Performance Targets and Actual Experience of Selected Cities

COMPLIANCE WITH PARKING REGULATIONS

Palo Alto, CA
Target: Achieve a 95% compliance rate for parking time limits in the downtown district
Actual: 97% (1997)

Savannah, GA
Target: No more than 10% of vehicles parked in violation of regulations in controlled parking district
Actual: 7% in violation (1990); 8% (1991)

PARKING ATTENDANT/OFFICER OUTPUT

Corpus Christi, TX
751 parking citations monthly per attendant (1991)

PARKING CITATION ERROR RATE

Corpus Christi, TX
0.29% of all citations (1991)

College Park, MD
Target: Void fewer than 2% of parking tickets due to error

RATE OF PARKING CITATIONS CONTESTED

Fort Collins, CO
0.4% (1991)

CAPTURE RATE

Boston, MA
Percentage of illegally parked vehicles that are ticketed at any given time: 30.8% (1996)

PARKING TICKET CLEARANCE RATE

Fayetteville, AR
68% (1997)

APPREHENSION OF SCOFFLAWS

Corpus Christi, TX
9.2% of vehicles on scofflaw list immobilized (1991)

ENFORCEMENT AND COLLECTION: "RETURN ON INVESTMENT"

San Jose, CA
Target: Costs for citation issuance and collection should not exceed 50% of the value of fines collected
Projected: 42% (1996)

lation. But when they do happen, how promptly are serious matters handled? And what are the results? Internal affairs investigations against officers often were completed within 30 days among the cities reporting such statistics (Table 20.20) and typically involved disciplinary actions in at least one case out of five.

(text continued on page 322)

Table 20.15   Recovery of Stolen Property: Selected Cities

City	Recovery Rate of Stolen Property
League City, TX	20.7% (1990), 83.8% (1991)
Eugene, OR	37.1% (1997)
Chesapeake, VA	35% (1997)
Chandler, AZ	34.9% of value (1991)
Shreveport, LA	30% (1991)
Overland Park, KS	29.2% (1994)
Raleigh, NC	27% (1998)

Table 20.16   Signs of Success in Community-Oriented/Neighborhood Policing

Boston, MA
Reported Part I crimes vs. previous 3-year average in districts with neighborhood policing:
90% (1996)

Cincinnati, OH
Target: At least 97% conviction rate for arrests from CrimeStopper tips
Actual: 97.4% (1993)

Denver, CO
Target: At least 90% voluntary compliance in dealing with property or vehicles identified as public nuisances
Actual: 98.9% (1998)

Winston-Salem, NC
CrimeStopper investigative referrals per 1,000 population: 9.03 (1997)

Table 20.17   Diverting Calls From Patrol Response: Selected Cities

Alexandria, VA
Reduction in patrol workload due to reports written by Telephone Reporting Unit: 18% (1998, estimated)

Norman, OK
Percentage of reports taken by telephone to reduce patrol workload: 12.2% (1997), 12.5% (1998)

Philadelphia, PA
Percentage of calls handled by Differential Response Unit:[a] 6.45% (1998)

Scottsdale, AZ
Percentage of calls for service handled by teleserve: 5.75% (1999)

a. Police reports taken by telephone.

Table 20.18   Prompt Work by Crime Lab: Santa Ana and Scottsdale

Santa Ana, CA
Target: Photographic work completed within 5 days
Actual: 100% (1995)
Target: Latent fingerprint work completed within 14 days
Actual: 85% (1995)
Target: Firearm examinations completed within 5 days
Actual: 5% (1995)

Scottsdale, AZ
Percentage of photo requests processed within 10 working days: 100% (1999)
Percentage of latent print evidence processed and compared within 5 days: 98% (1999)
Percentage of priority drug seizure analyses completed within 10 working days: 100% (1999)
Percentage of other drug seizure analyses completed within 30 working days: 100% (1999)
Percentage of blood analyses completed within 14 days: 100% (1999)
Percentage of serology cases analyzed within 30 days: 96% (1999)
Percentage of trace cases analyzed within 60 days: 97% (1999)
Percentage of crime scene calls responded to within 3 hours: 91% (1999)

Table 20.19   Police Records: Performance Targets and Actual Experience of Selected
Cities

PROMPT PROCESSING OF POLICE REPORTS INTO RECORDS SYSTEM

Oklahoma City, OK
Percentage of reports entered within 4 hours of receipt: 50% (1992)

Lubbock, TX
Percentage of crime, supplement, and accident reports entered within 12 hours of log-in: 95% (1997)

College Station, TX
Percentage of citations processed by 3 p.m. of the day following issuance: 95% (1995), 91% (1998)

Overland Park, KS
Target: Within 24 hours of receipt

Ann Arbor, MI
Target: Input general incident/accident reports within 2 days
Actual: 2 days (1998)

Palo Alto, CA
Target: Enter at least 80% of reports into system within 3 working days (1999)

Cary, NC
Average turnaround time for data entry of police reports: 72 hours (1999)

Savannah, GA
Target: Process all police reports, including data entry, by the 7th day of the following month
Actual: 99% (1997)

Table 20.19   Police Records: Performance Targets and Actual Experience of Selected Cities (Continued)

PROMPT PROCESSING OF WARRANTS

Eugene, OR

Average time to process district and circuit court (criminal) warrants, from receipt through data entry: 20 min. (1998, estimated)

PROMPT RESPONSE TO REQUESTS FOR POLICE REPORTS

Overland Park, KS

Target: Fulfill in-person requests within an average of 10 minutes

Tucson, AZ

Targets: Process 90% of phone and counter requests within 15 minutes; 90% of written requests within 10 workdays

Largo, FL

Targets: Fulfill internal requests within 24 hours; external requests within 36 hours

Duncanville, TX

Average time to comply with Open Records Act requests: 3 days (1996), 2 days (1998)

Raleigh, NC

Target: Process mail requests within 3 workdays

Palo Alto, CA

Target 1: Scan, index, and make available at least 80% of all police reports to the public and the department for further processing within 2 working days
Target 2: Process at least 90% of mail-in requests for police reports within 4 working days
Actual: 97% (1997)

Reno, NV

Target: Process at least 92% within 5 workdays (1999)

Chandler, AZ

97% within 5 business days (1998)

PROMPT RECORD CHECKS

Chandler, AZ

Target: Respond to criminal history requests by the end of the following shift
Actual: 98.8% (1995), 100% (1998)

Savannah, GA

Targets: Complete 95% of record checks within 24 hours; 100% within 48 hours
Completed within 24 hours: 90% (1993), 96% (1997)
Completed within 48 hours: 100% (1997)

Reno, NV

Target: Respond to at least 45% of requests for information on criminal history within 5 workdays (1999)

Overall Feelings of Security

Top-notch police departments strive to achieve low rates of crime, high degrees of satisfaction with police service, and strong feelings of safety throughout the city. A variety of community conditions in addition to good police work influence success on these measures, but some cities maintain a general barometer by periodically asking a sample of residents how safe they feel (Table 20.21). Many are able to report favorable results.

Table 20.20   Internal Affairs Investigations: Performance Targets and Actual
Experience in Selected Cities

PROMPT INVESTIGATION OF
COMPLAINTS AGAINST OFFICERS

Long Beach, CA
90% of complaints investigated within 28 days (1995)

Ann Arbor, MI
95% of citizen complaint investigations completed within 30 days (1998)

Oakland, CA
95% of investigations by the citizens' complaints board completed within 30 days (1992)

Tucson, AZ
Targets: 50% of investigations completed within 5 workdays; 80% within 10; 90% within 15;
    95% of all formal citizen complaints within 30 days

College Station, TX
Percentage of complaints processed within 30 days: 78% (1995)

Alexandria, VA
Internal affairs cases completed within 30 days: 73% (1991)

Fort Collins, CO
33% completed within 30 days (1996)

Scottsdale, AZ
Percentage of external complaints completed within 30 business days: 27% (1999)

Shreveport, LA
Percentage of internal affairs complaints resolved within 30 days: 20% (1994)

New York, NY
Civilian Complaint Review Board case completion time averages: excessive force complaints,
    252 days; abuse of authority complaints, 178 days; discourtesy complaints, 134 days (1997)

Table 20.20   Internal Affairs Investigations: Performance Targets and Actual
Experience in Selected Cities (Continued)

INVESTIGATIONS RESULTING
IN DISCIPLINARY/CORRECTIVE ACTION

Fayetteville, AR
82% (9 of 11 investigations) (1991)

Scottsdale, AZ
Percentage of external complaints sustained: 45% (5 of 11) (1999)

Fort Collins, CO
28% (7 of 25 investigations) (1991)

Cincinnati, OH
26% (144 of 554 investigations) (1991)

Winston-Salem, NC
Sustained citizen complaints: 24.1% (14 of 58 complaints) (1997); 26.6% (17 of 64 complaints) (1998)

Wichita, KS
18% (23 of 128 investigations) (1990); 23% (31 of 133 investigations) (1991)

Greenville, SC
Sustained complaints against officers: 20% (12 of 60 investigations) (1990)

Table 20.21   Sense of Security: Selected Cities

Bellevue, WA
Percentage of citizens who feel safe or moderately safe: 91.8% (1998)

Plano, TX
Percentage of residents who feel safe in Plano: 89% (1997 survey)

Raleigh, NC
Percentage of respondents feeling safe in Raleigh: 89% (1997), 84% (1998)

Portland, OR
Percentage of residents feeling "safe" or "very safe" walking alone in their neighborhood during the
day: 88% (1998); at night: 48% (1998)

Tucson, AZ
Percentage of respondents who feel safe in their neighborhood during the day: 86% (1997)
Percentage of respondents who feel safe being out alone in their neighborhood at night: 32% (1997)

Eugene, OR
Percentage of residents who feel their neighborhood is very safe or somewhat safe: 79.8% (1996)

21

Property Appraisal

Property taxes constitute a major source of revenue for most cities. The rationale for that tax, although imperfect, is simple: Those having the greatest stake in the community and those having the greatest ability to pay should bear a commensurate share of the cost of municipal government and its services. The value of property is used as the gauge for both criteria.

Although critics have challenged the fairness of the property tax and the validity of property value as a proxy for ability to pay, state restrictions have left municipal governments with few viable options for replacing such a major source of revenue, even if local officials wished to do so. The objective in most communities is to administer the property tax fairly and efficiently, diverting as little of the revenue for tax administration purposes as possible.

Property Appraisal and Tax Administration Benchmarks

Fairness of the property tax depends on fairness of property appraisals. Persons owning properties of greater value should pay higher taxes than those owning properties of lesser value; persons owning properties of equal value should pay equal property taxes.

How, then, may the performance of a municipal property tax office be judged? In several ways, according to the International Association of Assessing Officers (IAAO; Almy, Gloudemans, & Thimgan, 1991). The IAAO has established a set of professional standards for property tax administration, including, for example, the maintaining of municipal appraisals at 100% of current market value (Table 21.1). The extent to which appraisers are able to achieve that high level of accuracy can be tested by the periodic calculation of "sales ratios" that compare the prices from actual sales transactions with the appraised values of those properties. No municipality should expect absolute

324

Table 21.1 Selected Professional Standards for Property Tax Administration

Accurate Appraisal of Market Value
- 100% of current market value

Frequent Ratio Studies
- Monitor the level and uniformity of appraised values through ratio studies at least annually

Frequent Reinspection
- All properties should be physically inspected at least every 6 years

Prompt Updating of Records
- (1) Update ownership and legal description information in tax office records within 30 days of recorded sale; (2) Parcel splits and combinations should be noted on cadastral maps within one month of a new deed's recording

Accessible Property Records
- Accessible (i.e., indexed) by parcel identifier, address, and owner

Sufficient Resources and Staffing
- Rules of thumb: Approximately 1.5% of property tax collections or at least $10 (1991 dollars) per parcel
- Staffing level that provides an appraiser-to-parcel ratio as follows:
  1 appraiser per 1,000 to 1,500 parcels in small jurisdictions (fewer than 10,000 parcels)
  1:2,500—cities with 10,000 to 20,000 parcels
  1:3,000 to 1:3,500—cities with more than 20,000 parcels
  Higher in the largest jurisdictions

SOURCE: Richard R. Almy, Robert J. Gloudemans, and Garth E. Thimgan, *Assessment Practices: Self-Evaluation Guide* (Chicago: International Association of Assessing Officers, 1991). Reprinted by permission.

accuracy, but some have demonstrated an ability to come reasonably close overall (Table 21.2).

Outstanding quality in appraisal work occurs when a community's overall assessment-to-sales ratio approaches 100% and when there is high uniformity among individual property ratios. In other words, accuracy should be high and dispersion or variability in the level of accuracy should be low. A city would prefer to achieve an overall 98% assessment-to-sales ratio on the strength of appraisals that consistently fall in the 90% to 106% range relative to market value rather than as the mathematical product of appraisals ranging widely from, say, 68% to 128% of market value.

A measure recommended by the IAAO for gauging variation in appraisal accuracy is called the *coefficient of dispersion* (COD) (*Standard on Ratio Studies*, 1990, p. 18). Communities using this measure typically suggest that CODs of less than 10 or 15 reflect acceptable levels of dispersion (Table 21.3).

The IAAO recommends that ratio studies be conducted frequently, that property reinspection occur at least every 6 years, that property records be updated and accessible, and that sufficient resources—including personnel—be

Table 21.2   Accurate Appraisal of Market Value

HOW CLOSE TO MARKET VALUE?

International Association of Assessing Officers[a]
Recommended target: 100% of current market value

Boston, MA
Accuracy rate of data quality and assessment: 97% (1992)[b]

Milwaukee, WI
Assessment levels as a percentage of market value: 96.4% (1997)

Charlottesville, VA
Assessment-to-sales ratio: 96% (1998)

Alexandria, VA
Assessment-to-sales ratio: 95.4% (1996), 95.8% (1997), 95.0% (1998)

Calgary, Alberta
Residential assessment-to-sales ratio: 90.94% (1996)
Nonresidential: 93.58% (1996)

a. The IAAO recommended target is reported in Richard R. Almy, Robert J. Gloudemans, and Garth E. Thimgan, *Assessment Practices: Self-Evaluation Guide* (Chicago: International Association of Assessing Officers, 1991), p. 3.
b. Boston statistics are based on 5% random sample of reviewed abatement filings.

Table 21.3   Accuracy of Appraisals: Coefficients of Dispersion in Selected Cities

Charlottesville, VA
7 (1998)

Kalamazoo, MI
Target: 15 or less
Actual: 10.6 (1997)

Milwaukee, WI
Target: less than 10
Actual: 11.3 (1997)

committed to the tax administration function. Offered as rules of thumb for adequate budget and staffing are resource commitments equal to 1.5% of property tax collection and staffing levels ranging from one appraiser per 1,000 parcels in small jurisdictions to one appraiser per 3,500 (or more) parcels in the largest cities. The reasonableness of these rules of thumb, however, may be drawn into question by municipalities that have managed to perform

Table 21.4   Typical Production Rates for Property Assessor (Parcels per Day)

Property Type	Type of Appraisal Activity			
	New Work	Field Canvass	Field Review	Model Review
Homes	8–10	16–20	50–100	200–250
Apartments	2–4	5–10	15–20	100
Commercial	2–4	5–10	15–20	100
Agricultural	2–4	4–6	n.a.	50
Vacant land	30–50	50–75	100–150	400

SOURCE: Richard R. Almy, Robert J. Gloudemans, and Garth E. Thimgan, *Assessment Practices: Self-Evaluation Guide* (Chicago: International Association of Assessing Officers, 1991), p. 20. Reprinted by permission.

NOTES: *New work* is fieldwork necessary to describe and appraise new construction. *Field canvass* is an orderly parcel-by-parcel detailed inspection program. *Field review* is a drive-by inspection. *Model review* is an in-office review of computer-assisted mass appraisal (CAMA) model value estimates, including the flagging of selected parcels for field review or canvass.

n.a. = not applicable.

well despite more modest resource commitments. For example, Alexandria, Virginia, has reported an appraiser-to-assessable-parcels ratio ranging from 1:4,362 in 1993 to 1:6,615 in 1989.

The work of a property appraiser varies in difficulty, depending on the type of property appraised and the type of appraisal activity conducted. The IAAO has offered guidance to municipal officials attempting to determine a fair day's work for appraisal operations by declaring typical production rates according to appraisal activity and property type (Table 21.4). According to this guideline, for example, it is reasonable to expect an appraiser doing "new work" on newly constructed homes to appraise 8 to 10 homes during one full day. Similarly, a "field canvass" of 5 to 10 commercial properties should be expected to consume a full day's work for a typical appraiser.

In addition to the market test of appraisal accuracy provided by periodic ratio studies, yet another gauge of quality is the extent to which the public deems property appraisals to be acceptable. Some communities boast extremely low rates of assessment objections and formal appeals (Table 21.5). Perhaps even more significant is a favorable record for having municipal appraisals upheld when appeals are made.

Office Space Standards

Although much of an appraiser's work is conducted in the field, suitable office space is also needed. The IAAO recommends minimally 50 to 100 square feet of office space per appraiser, with greater allocations for supervisors, other

Table 21.5   Appeals of Property Appraisal

MINIMAL APPEALS AS AN INDICATION OF ACCEPTABILITY

Milwaukee, WI

Assessment objections to the board of appeals as percentage of real and personal property assessments: 1.76% (1993), 1.1% (1997)

Formal assessment appeals to the board of review as percentage of real and personal property assessments: 0.20% (1993), 0.19% (1997)

Alexandria, VA

Appeals to the appraisal department, as percentage of total parcels: 1.32% (1994), 1.03% (1998)

Appeals to the board of equalization, as percentage of total parcels: 1.27% (1994), 0.52% (1998)

DEFENDABILITY OF APPRAISALS ON APPEAL

Kalamazoo, MI

Percentage of appeals lowered by the board: 21% (1997)

Alexandria, VA

Percentage upheld/adjusted at departmental appeal: 70%/30% (1990), 60%/40% (1993), 61%/39% (1994), 65%/35% (1995)

Percentage upheld/adjusted at the board of equalization appeal: 49%/51% (1990), 45%/55% (1993), 45%/55% (1994), 47%/53% (1995)

Table 21.6   Space Standards for Property Assessors' Offices

Position	Space Requirement
Assessor	200 square feet of private space (approximately 14 ft. by 14 ft.)
Chief deputy or manager	170 square feet (approximately 12 ft. by 14 ft.)
Supervisor	150 square feet (approximately 12 ft. by 12 ft.)
Appraiser	50 to 100 square feet of space

SOURCE: Richard R. Almy, Robert J. Gloudemans, and Garth E. Thimgan, *Assessment Practices: Self-Evaluation Guide* (Chicago: International Association of Assessing Officers, 1991), p. 16. Reprinted by permission.

administrators, and the chief assessor (Table 21.6). The question of how much space to provide is, of course, a resource input issue that may influence performance or results, but office space standards should not be confused with *performance* standards.

22

Public Health

Although public health services are a function of county governments in many parts of the country, they are provided by city governments in some regions. Services typically span a broad spectrum but often include health clinics, disease outbreak investigations, community programs promoting hygiene and healthy practices, vermin and mosquito control, and the regular inspection of food establishments.

Recommended Performance Measures

Public health was among the local government services examined in the SEA reporting project of the GASB. On the basis of that study, assorted measures of input, output, outcome, and efficiency have been recommended for several public health activities, including programs addressing maternal and child health care, sexually transmitted diseases, AIDS, chronic diseases, and the control of stress and violent behavior (Carpenter, Ruchala, & Waller, 1991).

Compiling a good set of performance measures is a fundamental step toward performance management, but relevant standards or points of comparison are needed in order to assess local accomplishments and judge performance proficiency. Cities attempting to establish local standards for public health and wishing to have the benefit of external benchmarks as they proceed should anticipate encountering some difficulty in assembling information on the performance targets and actual experience of municipalities in public health operations. Many municipalities simply do not engage in public health services, leaving them instead to county governments, where more abundant performance information may reside.

A readily available external peg for local conditions, however, has been provided through the development of national public health targets, many of which are output or outcome related. A sample of national public health targets is provided in Table 22.1. A community wishing to establish its own

Table 22.1   National Public Health Targets

Public Health Focus	National Target
Reduction of death rate from coronary heart disease	No more than 100 per 100,000 people (baseline: 135 per 100,000 in 1987)
Reduced incidence of overweight	No more than 20% among people 20 years of age and older; 15% among adolescents (baseline: 26% for 20- to 74-year-olds in 1976–1980)
Reduced cancer death rate	No more than 130 per 100,000 people (baseline: 133 per 100,000 in 1987)
Food choices in restaurants	At least 90% offering identifiable low-fat, low-calorie food choices (baseline: 70% in 1989)
School and child care breakfasts and lunches	At least 90% with menus consistent with the *Dietary Guidelines for Americans*
Reduced lung cancer death rate	No more than 42 per 100,000 people (baseline: 37.9 per 100,000 in 1987)
Reduction in cigarette smoking	No more than 15% of people aged 20 or older (baseline: 29% in 1987)
Reduction in adolescent pregnancy	No more than 50 per 1,000 girls aged 17 or younger (baseline: 71.1 per 1,000 girls 15 through 17 years old in 1985)
Reduction of suicides	No more than 10.5 per 100,000 people (baseline: 11.7 per 100,000 in 1987)
Reduction in maltreatment of children	Less than 25.2 incidents of maltreatment per 1,000 children younger than age 18 (baseline: 25.2 per 1,000 in 1986)
Reduced incidence of foodborne infections	Incidence rates from key foodborne pathogens:

	1987	
	Target	Baseline
Salmonella species	16 per 100,000	18
Campylobacter jejuni	25 per 100,000	50
Escherichia coli 0157:H7	4 per 100,000	8
Listeria monocytogenes	0.5 per 100,000	0.7

Dental care for children	Untreated caries in permanent or primary teeth among no more than 20% of children aged 6 through 8; no more than 15% among 15-year-olds (baseline: 27% of children, 6 through 8, and 23% of 15-year-olds in 1986–1987)
Optimal water fluoride levels	Between 0.7 and 1.2 parts fluoride per 1 million parts water, depending on daily air temperature (baseline: 62% in 1989)
Use of topical or systemic fluorides	At least 85% of persons who do not receive optimally fluoridated public water (baseline: 50% in 1989)
Reduced infant mortality	No more than 7 per 1,000 live births (baseline: 10.1 per 1,000 in 1987)
Prenatal care	At least 90% of all pregnant women receiving prenatal care in first trimester (baseline: 76% of live births in 1987)

SOURCE: Adapted from American Public Health Association, *Healthy Communities 2000: Model Standards*, 3rd edition. Copyright © 1991 by the American Public Health Association. Adapted with permission.

local standards may simply adopt the national target or, if the target seems overly ambitious given local conditions and current public health record, a point between the community's current mark and the national target.

Among the national health statistics, as well as occasional performance targets and performance records reported by selected municipalities, are potential performance benchmarks for a variety of public health programs and functions.

Disease Rates. National statistics on the incidence rates of various diseases provide a set of benchmarks against which local rates may be compared (Table 22.2).

AIDS Cases. National statistics on the incidence of AIDS and its prevalence by gender, race, and age allow health officials to view the scale of the local problem in the context of its magnitude elsewhere (Table 22.3).

Substance Use. In their effort to promote healthy practices and curtail unhealthy ones, many public health departments develop programs designed to reduce the use of cigarettes, alcohol, marijuana, and cocaine by adolescents and young adults. National statistics on the use of these substances provide a benchmark for local progress (Table 22.4).

Dental Health. Programs focusing on dental health may use national statistics regarding the incidence of dental problems as a means of gauging local conditions and tracking local progress (Table 22.5).

Healthy Weight. The problems of overweight and obesity among Americans are serious concerns for public health officials. Programs designed to encourage proper nutrition and physical fitness may be judged in part on their success with specific individuals and in part on their success with the general citizenry in comparison to national statistics (Table 22.6).

Immunization. Public health programs often focus on the immunization of children to prevent various diseases. The success of these efforts is ultimately gauged by the absence of these diseases. On an intermediate level, success is measured by vaccination rates—compared to previous years, compared to national averages (Table 22.7), or compared to programs of other communities (Table 22.8).

Food Service Inspections. Periodic restaurant inspections are conducted to ensure cleanliness, proper food handling procedures, and proper conditions for food preparation in an effort to protect patrons from foodborne illnesses. Standards of inspection frequency (Table 22.9) and inspector efficiency (Table 22.10) developed in individual communities may be helpful to others attempting to establish their own performance expectations.

Miscellaneous Facets of Public Health Performance

The array of public health performance targets established in Corpus Christi, Texas, offers guidelines for key services in that community (Table 22.11). The

Table 22.2   Selected Notifiable Disease Rates: National Statistics for 1996 and 1997

	Cases per 100,000 Population	
Disease	1996	1997
Diphtheria	0.01	0.01
Haemophilus influenzae, invasive	0.45	0.44
Hepatitis A	11.70	11.22
Hepatitis B	4.01	3.90
Lyme disease	6.21	4.79
Meningococcal disease	1.30	1.24
Mumps	0.29	0.27
Pertussis (whooping cough)	2.94	2.46
Poliomyelitis, total	0.01	0.01
Paralytic	0.01	0.01
Rocky Mountain spotted fever	0.32	0.16
Rubella (German measles)	0.10	0.07
Rubeola (measles)	0.20	0.06
Salmonellosis, excluding typhoid fever	17.15	15.66
Shigellosis	9.80	8.64
Tuberculosis	8.04	7.42
Sexually transmitted diseases:		
Syphilis	20.10	17.50
Primary and secondary	4.30	3.20
Early latent	7.60	6.30
Late and late latent	7.70	7.70
Congenital	0.50	0.40
Chlamydia	192.60	207.00
Gonorrhea	123.10	122.50
Chancroid	0.10	0.10

SOURCE: U.S. Department of Health and Human Services, *Health, United States, 1999* (Washington, DC: Author, 1999), p. 203.

even broader assortment of performance targets and results reported in Table 22.12 offers additional points of comparison from other communities.

Table 22.3   AIDS Cases, Gender, Race, and Age at the Time of
                  Diagnosis: National Statistics for 1998

	Cases per 100,000 Population
Overall	19.6
Males, 13 years of age or older	38.5
White, non-Hispanic	20.0
Black, non-Hispanic	145.3
Hispanic	67.2
American Indian or Alaska Native	18.1
Asian or Pacific Islander	10.0
Females, 13 years of age or older	10.2
White, non-Hispanic	2.6
Black, non-Hispanic	54.1
Hispanic	17.6
American Indian or Alaska Native	4.5
Asian or Pacific Islander	1.6
All children, under 13 years	0.7
White, non-Hispanic	0.2
Black, non-Hispanic	3.1
Hispanic	0.9
American Indian or Alaska Native	0.2
Asian or Pacific Islander	0.2

SOURCE: U.S. Department of Health and Human Services, *Health, United States, 1999* (Washington, DC: Author, 1999), p. 204.

Table 22.4   Use of Selected Substances: National Statistics for 1997

	Use of Selected Substances in the Past Month				
	Cigarettes	Alcohol	Heavy Alcohol	Marijuana	Cocaine
12- to 17-year-olds					
Male	19%	21%	10%	10%	0.9%
Female	21%	20%	7%	8%	1.1%
18- to 25-year-olds					
Male	47%	66%	39%	17%	1.9%
Female	35%	51%	17%	8%	0.5%

SOURCE: U.S. Department of Health and Human Services, *Health, United States, 1999* (Washington, DC: Author, 1999), pp. 215–216.

Table 22.5   Untreated Dental Caries Among Children: National Statistics for
1988–1994

	Untreated Dental Caries	
	Children 2 to 5 Years of Age	Children 6 to 17 Years of Age
Overall	18.7%	23.1%
Male	19.2%	22.6%
Female	18.1%	23.7%

SOURCE: U.S. Department of Health and Human Services, *Health, United States, 1999* (Washington, DC: Author, 1999), p. 225.

Table 22.6   Weight Among Persons 20 Years of Age and Over: National Statistics
for 1988–1994

	Healthy Weight	Overweight	Obesity
Overall	41.7%	54.6%	22.6%
Male	39.1%	59.4%	19.9%
Female	44.3%	49.9%	25.1%

SOURCE: U.S. Department of Health and Human Services, *Health, United States, 1999* (Washington, DC: Author, 1999), p. 223.
NOTES: Healthy weight = a body mass index (BMI) of 19 to less than 25 kilograms/meter^2. Overweight = a BMI greater than or equal to 25. Obesity = a BMI greater than or equal to 30.

Table 22.7   National Vaccination Rates Among Children 19–35 Months of Age:
1997

	Percentage Receiving Combined Series Vaccination
Overall	76
Economic Status	
Below Poverty	71
At or Above Poverty	79
Location of Residence	
Central City	74
Other—Inside MSA[a]	78
Outside MSA	77

SOURCE: U.S. Department of Health and Human Services, *Health, United States, 1999* (Washington, DC: Author, 1999), p. 200.
NOTE: The 4:3:1:3 combined series consists of 4 doses of diphtheria-tetanus-pertussis (DTP) vaccine, 3 doses of polio vaccine, 1 dose of a measles-containing vaccine, and 3 doses of the *Haemophilus influenzae* type b (Hib) vaccine.
a. MSA stands for *metropolitan statistical area*.

Table 22.8   Immunization of Children in the Community: Selected Cities

Boston, MA
Percentage of children immunized by the time they start school: More than 97% (1992)

Alexandria, VA
Percentage of children with immunizations that are current (based on sample of records): 95% (1993)

Corpus Christi, TX
Target: At least 90% of enrolled children under 2 years of age to be fully immunized; 95% of those 2 or older (1999)

New York, NY
Percentage of entering students completely immunized: 88% (1992), 92% (1998)

Houston, TX
Target: At least 90% of children under age 2 having had complete immunization series

San Antonio, TX
Immunization rate for infant/child patients at public health clinics: 85% (1995)

Nashville–Davidson County, TN
24-month-old children completely immunized, as a percentage of the sample surveyed: 83.3% (1998)

Philadelphia, PA
Percentage of 2-year-olds with up-to-date immunizations: 68% (1996), 81% (1998)

Cincinnati, OH
Percentage of Health Department pediatric patients who are up-to-date with their immunizations: 80% (1993)

Table 22.9   Frequency of Food Service Inspections: Selected Cities

Alexandria, VA
Inspection visits per food service establishment (annual): 12 (1995)

Corpus Christi, TX
Target: Average of 4 regular inspections per year

Overland Park, KS
Target: At least 3 routine inspections of all food service establishments annually

Boston, MA
Target: To inspect high-risk food establishments 3 times annually
Actual: 99.65% (1996)

Kansas City, MO
Target: Inspect all food service establishments and retail food stores at least twice a year

Wichita, KS
Inspections per establishment: 1.8 (1990), 1.5 (1991)

Philadelphia, PA
Interval between routine food inspections: 17.6 months (1996), 15.6 months (1998)

Table 22.10   Food Service Inspections Program Statistics: Garland, Texas, 1990–1991

Permitted establishments per full-time equivalent employee	278.10
Inspections per citation	11.85
Inspections per closure	64.44
Inspections per permitted establishment	1.77

SOURCE: Pat Fowler, "The Establishment of Health Indicators in Municipal Health Departments," *Texas Town & City*, 79 (February 1992), pp. 20–24. Reprinted by permission.

Table 22.11   Selected Public Health Performance Targets in Corpus Christi

Target 1: 40% of all food-related complaints will be investigated within 1 working day; 100% within 6 working days

Target 2: Communicable diseases classified as immediately reportable will be investigated within 1 working day

Target 3: At least 95% of all tuberculosis cases will be evaluated monthly for compliance, correct drug regimen, bacteriology, and appropriate laboratory studies

Target 4: 100% of all high-risk pregnant women will be referred for genetic screening and education

Target 5: 100% of all adolescents who require family-planning services will be accommodated on a walk-in basis

Target 6: Patients having abnormal tests will be followed within 72 hours

Target 7: Larvicide within 5 days following a rain all accessible bodies of water having mosquito larvae

SOURCE: City of Corpus Christi (TX), *Annual Quality Plans: FY 1995–96*.

Table 22.12 Odds and Ends in Public Health: Selected Cities

WAITING TIME FOR APPOINTMENT

Philadelphia, PA

Percentage of appointments for non–acute care made within 2 weeks: 68% (1996)

Percentage of appointments within 3 weeks: 75% (1998)

Cincinnati, OH

Average waiting time for primary care service appointment: routine appointment, 4 weeks; urgent appointment, 1 day (1993)

ADULT IMMUNIZATIONS

Nashville–Davidson County, TN

Percentage of audited charts reflecting adequate immunization (including flu/pneumonia/hepatitis B in adult population): 90% (1997), 89% (1998)

LAB SERVICE ACCURACY

Kansas City, MO

Target: Achieve a 98% or better score on all laboratory proficiency testing

Actual proficiency testing scores: Dairy, 95% (1995); tuberculosis, 90% (1995)

Error rate: Venereal disease, 8% (1995)

TUBERCULOSIS TREATMENT

San Antonio, TX

Percentage of TB contacts completing preventive therapy: 90% (1997)

HIV TESTING

San Antonio, TX

Percentage of high-risk STD clients tested for HIV: 92% (1995)

New York, NY

Average turnaround time for HIV-1 tests: negative reports, 1.5 days (1997); positive reports, 2.4 days (1997)

VITAL RECORDS

Cincinnati, OH

Target: Issue copies of birth or death certificates within 15 minutes of request at counter; within 15 days of request by mail

San Antonio, TX

Percentage of customers for birth or death certificates served within 30 minutes: 90% (1996, estimated)

Lubbock, TX

Percentage of birth records filed within 5 days of receipt: 95% (1997)

Percentage of death records filed within 5 days of receipt: 95% (1997)

New York, NY

Average response time for mailed requests: Birth certificates, 6 days (1997), 4 days (1998); death certificates, 9 days (1997), 7 days (1998)

PRENATAL SERVICE

San Antonio, TX

Percentage of prenatal patients receiving first trimester care: 48% (1995), 73.8% (1997)

(continued)

Table 22.12   Odds and Ends in Public Health: Selected Cities (continued)

Corpus Christi, TX
   Targets: (1) At least 80% of maternity clients begin prenatal services in the first 3 months of
   pregnancy, with at least 10 subsequent prenatal visits; (2) At least 90% of clients who are referred
   for abnormal conditions complete the referral within 2 weeks; (3) At least 90% of clients in need
   of follow-up care who miss appointments and are unable to be contacted by phone receive home
   visits (1999)

LOW BIRTHWEIGHT

San Antonio, TX
   Low-birthweight infants as a percentage of all births to public health clinic patients: 5.72% (1997)

INFANT MORTALITY

Milwaukee, WI
   Infant deaths per 1,000 live births (5-year average): 11.9 (1996)

FOOD SERVICE INVESTIGATION

Denton, TX
   Percentage of general complaints investigated within 2 workdays: 93% (1997)

FOLLOW-UP ON FOOD PROTECTION VIOLATIONS

Kansas City, MO
   Target: Reinspect establishments and stores in violation of critical items within 15 days

RAT INVESTIGATIONS

Kansas City, MO
   Percentage of rat complaints investigated within 3 days: 60% (1995)

AIR QUALITY

Alexandria, VA
   Percentage of air quality complaints responded to within 1 day: 100% (1995), 100% (1998)
   Percentage of air quality complaints closed within 10 days: 48% (1995), 30% (1998)

VOLUNTARY COMPLIANCE WITH ENFORCEMENT NOTICES

Norfolk, VA
   Percentage of environmental abatements after first notice: 95% (1997, estimated)

23

Public Transit

New York City has been the scene of many "firsts" in public transit. The first horse-drawn urban stagecoach line appeared there in 1827, followed in 1832 by the first horse-drawn street railway line and in 1905 by the first motor bus line, the Fifth Avenue Coach Company (American Public Transit Association [APTA], 1996, pp. 46–49). By the 1920s most large cities had transit systems featuring streetcars and buses; but in 1923 the smaller cities of Bay City, Michigan; Everett, Washington; and Newburgh, New York, became the first to replace all their streetcars with motor buses. In 1933, San Antonio, Texas, became the first large city to do so.

Rapid escalation in the ownership of personal automobiles, accompanied by the suburbanization of America, caused transit ridership to plummet from its peak of 23.4 billion riders in 1946 to a low of 5.3 billion in 1972 (APTA, 1996, pp. 48–49). Nevertheless, public transit has long been a necessity to sizable segments of the urban working class. Today, its appeal is broadening as public transit is increasingly viewed as an important element in the strategy to combat traffic congestion and air pollution.

Many cities operate some form of public transit service. Medium-sized and large cities often provide bus services with regular routes on major thoroughfares and principal arteries. Many of the largest cities also offer heavy rail, commuter rail, and light rail transit service. Relatively small cities sometimes offer transportation services on demand—occasionally on a contractual basis with local taxi companies or other private entities—rather than by means of regular transit routes. The focus of this chapter is primarily on bus services, a staple among cities that offer regular public transit services.

Performance Measures and Standards

Research undertaken on behalf of the GASB revealed a fairly wide assortment of productivity indicators espoused by transit authorities or currently in use

Table 23.1   Productivity Indicators for Mass Transit

Passengers per vehicle
Passengers per vehicle hour
Passengers per revenue hour
Revenue passengers per vehicle hour
Passengers per trip
Passengers per revenue mile
Passengers per mile
Passengers per service mile
Revenue passengers per mile
Operating revenue per operating expense
Revenue vehicle hours per operating expense
Revenue: cost by route
Revenue per vehicle hour
Passenger miles (and trips) per directional miles
Passenger miles (and trips) per vehicle operated in maximum service
Passenger miles (and trips) per revenue vehicle hour
Passenger trips per employee
Passenger trips per nonrevenue mile
Vehicle hours per employee
Vehicle miles per peak vehicles required
Vehicle miles per maintenance employee
Annual vehicle miles per annual maintenance expense
Annual vehicle miles per gallon of fuel (mpg) or per kilowatt hour of power
Maintenance work done a second time
Miles between work orders (bus repairs through inspection)
Miles per mechanic equivalent

SOURCE: Wanda A. Wallace, *Mass Transit: Service Efforts and Accomplishments Reporting—Its Time Has Come* (Norwalk, CT: Governmental Accounting Standards Board, 1991), p. 11. Reprinted by permission.

(Wallace, 1991; Table 23.1). If collected widely and tabulated centrally, many of these indicators could have considerable value as performance benchmarks.

Relevant standards for bus systems recommended in a U.S. Department of Transportation report (Attanucci, Jaeger, & Becker, 1979) address major facets of transit operations (Table 23.2). The authors of that report contended, for example, that passenger fares should equal at least 20% of operating expenditures, that buses should neither depart from a stop ahead of schedule nor be more than 5 minutes late, that buses should average at least 1.5 passengers per vehicle mile, and that buses should arrive at each stop at least hourly.

Table 23.2   Recommended Standards for Bus Systems

Fares as a percentage of operating expenditures	20% to 50%
Adherence to route schedule	Zero minutes early to 5 minutes late
Bus stop spacing	660 to 2,000 feet apart
Passengers per vehicle mile	At least 1.5
Service frequency	At least every 30 minutes during peak periods; at least hourly during off-peak periods

SOURCE: John P. Attanucci, Leora Jaeger, and Jeff Becker, *Bus Service Evaluation Procedures: A Review—Short Range Transit Planning*, Special Studies in Transportation Planning (Washington, DC: U.S. Department of Transportation, April 1979), pp. 10–13.

National Norms and Benchmarks

National data compiled by the U.S. Department of Transportation, Federal Transit Administration (FTA), provide operating statistics aggregated for large and small agencies. Those statistics, for example, indicate that transit agencies serving populations of less than 200,000 in 1996 derived an average of 22% of their operating funds from passenger fares, whereas transit agencies serving larger populations averaged 40% (Table 23.3). Overall, the average transit bus nationwide was between 8 and 9 years old. Furthermore, transit agencies averaged 1.71 to 2.96 unlinked passenger trips per vehicle revenue mile and 23.82 to 37.73 unlinked passenger trips per vehicle revenue hour for bus services, depending on the population of the urban area served.[1]

Summary figures for transit agencies serving populations in excess of 200,000 showed an average cost recovery of 40% from fares. Traditionally, the portion of this group serving the very largest populations recovers a slightly higher percentage. Relatively few agencies, however, report recovering more than half of all operating expenses through fares.

The set of 11 cities featured in Table 23.4 offers statistics generally consistent with the Department of Transportation report's prescription that 20% to 50% of expenditures be recovered through fares. Among these cities, only High Point, North Carolina, and Alexandria, Virginia, approach the 50% cost recovery mark.

Selected municipal transit operations report 13.8 to 43 passengers per service hour and 1.44 to 3.45 passengers per revenue mile (Table 23.5). At least one factor likely to influence such numbers is the degree to which a transit operation maintains its routes on schedule. Almost all of the municipalities examined for this volume that reported on-time statistics indicated an on-

(text continues on page 344)

Table 23.3 Selected Operating Statistics of Public Transit Agencies, 1996

Service Population	N	Revenue From Fares as a Percentage of Operating Funds Expended	Average Fleet Age in Years				Unlinked Passenger Trips per Vehicle Revenue Mile				Unlinked Passenger Trips per Vehicle Revenue Hour			
			Motor Bus	Heavy Rail	Commuter Rail	Light Rail	Motor Bus	Heavy Rail	Commuter Rail	Light Rail	Motor Bus	Heavy Rail	Commuter Rail	Light Rail
Transit agencies serving populations greater than 200,000	272	40%	8.3	20.1	20.5	16.0	2.96	4.09	1.59	7.06	37.73	84.50	52.76	103.00
Transit agencies serving populations of less than 200,000	190	22%	8.8	n.a.	n.a.	8.0	1.71	n.a.	n.a.	5.55	23.82	n.a.	n.a.	25.90

SOURCES: U.S. Department of Transportation, Federal Transit Administration (1996a), *Transit Profiles: Agencies in Urbanized Areas Exceeding 200,000 Population for the 1996 National Transit Database Report Year* (Washington, DC: Author, 1996), pp. A.2–A.4; U.S. Department of Transportation, Federal Transit Administration (1996b), *Transit Profiles: Agencies in Urbanized Areas With a Population of Less Than 200,000 for the 1996 National Transit Database Report Year* (Washington, DC: Author, 1996), pp. A.2–A.4.

Table 23.4   Cost Recovery Targets and Experience Among Selected Municipal
Transit Systems

City	Cost Recovery via Operating Revenue
High Point, NC	46% (1998)
Alexandria, VA	44% (1998)
Wichita, KS	Cost recovery from fares: 31.6% (1998)
Lubbock, TX	31.2% (1995)
Chapel Hill, NC	30% (1995), 28% (1999)
Iowa City, IA	30% (1991)
Phoenix, AZ	Target: Fare box revenues equal to 30% of costs
Ames, IA	29.7% for fixed route service (1997)
Winston-Salem, NC	28% (1998)
Raleigh, NC	Fare box revenues as a percentage of expenditures: 23% (1997), 25% (1998)
Greensboro, NC	Fare box revenues as a percentage of direct costs for fixed route service: 21% (1997)

Table 23.5   Passengers per Service Hour and Revenue Mile: Selected Municipalities

Passengers per Service Hour	*Passenger Trips per Revenue Mile*
Calgary, Alberta Passengers per operating hour: 43 (1996)	**Tempe, AZ** Boardings per revenue miles: 3.45 (1997)
San Luis Obispo, CA Passengers per service hour: 41 (1994)	**Iowa City, IA** Passengers per revenue mile: 2.95 (1991)
Charlotte, NC Passenger boardings per revenue hour: 39.8 (1994, 5-month record)	**Winston-Salem, NC** Passenger trips per revenue mile: 2.50 (1991), 2.43 (1992)
Ames, IA Passengers per revenue hour: 39.2 for fixed route service (1997)	**Chapel Hill, NC** Passenger trips per service mile: 2.46 (1995), 2.47 (1999)
Tucson, AZ Passenger rides (includes transfer trips) per vehicle hour of service: 29.3 (1997)	**Tucson, AZ** Passenger rides (includes transfer trips) per vehicle mile: 2.37 (1997)
Wichita, KS Passenger trips per revenue hour: 18.9 (1990), 18.2 (1991)	**Kansas City, MO** Passengers per mile operated: 1.96 (1995)
Lubbock, TX Average city route passengers per hour: 18 (1995)	**Fort Collins, CO** Passengers per revenue mile: 1.63 (1991)
Waukesha, WI Passengers per vehicle hour: 13.8 (1992)	**Decatur, IL** Passengers per mile: 1.52 (1991)
	Waukesha, WI Passengers per vehicle mile 1.44 (1992)

Table 23.6   Routes on Schedule: Selected Cities

Blacksburg, VA
100% on time (1997)

Oklahoma City, OK
99.89% on schedule (1999)

Charlotte, NC
99.5% on time (1998)

Chapel Hill, NC
Target: 98%

Charlottesville, VA
Target: 92%
Actual: 97% (1998)

High Point, NC
95% on time (1998)

Wichita, KS
95% on schedule (1990); 94.1 % on schedule (1991)

Phoenix, AZ
Target: 95% on schedule

Lubbock, TX
93% on schedule (1995)

Boise, ID
90.4% within 5 minutes of schedule (1991)

Tempe, AZ
90% on time (bus); 76% (dial-a-ride) (1997)

Raleigh, NC
87% on schedule (1997); 85% on schedule (1998)

schedule rate in excess of 90%, with several reporting performance targets or actual on-time results of 95% to 100% (Table 23.6). Oklahoma City, Oklahoma, reported a "missed trips" rate of only 0.11% during a recent year. Greensboro, North Carolina, reported an even lower rate of 0.01%.

Safety rate—that is, an agency's ability to avoid bus accidents—is yet another potential performance benchmark for transit operations. Among examined performance documents that provided transit system accident rates, several cities reported rates of less than two accidents per 100,000 bus miles (Table 23.7). Raleigh, North Carolina, and Charlottesville, Virginia, reported "preventable" accident rates of less than one per 100,000 miles.

Table 23.7 Avoiding Bus Accidents: Selected Cities

City	Accidents per 100,000 Bus Miles
Raleigh, NC	0.30 preventable accidents (1997); 0.40 preventable accidents (1998)
Charlottesville, VA	0.5 preventable accidents (1998)
Blacksburg, VA	1.3 (1994), 1.9 (1997)
Greensboro, NC	1.61 (1997)
Knoxville, TN	1.65 (1991)
Oklahoma City, OK	1.7 preventable accidents (1999)
Tucson, AZ	1.9 (1997)
Chapel Hill, NC	1.98 preventable accidents (1996); 2.97 (1999)
Phoenix, AZ	2.34 (1991)
Wichita, KS	2.5 (1991)
Lubbock, TX	2.8 (1991)
Decatur, IL	2.99 (1995)

Note

1. Each time a passenger climbs aboard a bus, it is considered an "unlinked trip." If traveling across the city requires a passenger to ride a total of three buses, the journey counts as three unlinked trips.

24

Public Works
Engineering and Miscellaneous Services

T ypically, public works is one of the largest municipal departments. The department is usually an amalgam of functions, often including engineering and a variety of services directed toward the repair and maintenance of municipal equipment, properties, and infrastructure. Among the repair and maintenance functions often found in public works departments are five that will be reviewed in this chapter: street cleaning, street lighting, custodial services, building maintenance, and radio maintenance. Other functions sometimes included in public works departments but treated in separate chapters of this volume are streets, storm drainage, solid waste collection and disposal, traffic engineering and control, public utilities, parks maintenance, and vehicle and equipment maintenance.

Sources of potential benchmarks for public works operations are varied. One logical source is the American Public Works Association, which, in fact, does endorse a set of "standards." Those standards, however, primarily are directed toward the adoption of "best practices" and the establishment of written policies and procedures covering various public works duties and issues. Although the adoption of modern methods and the establishment of written policies are undoubtedly worthwhile, process standards have limited value as performance benchmarks.

Fortunately, other options are available. In some cases, "engineered standards" have been developed in the private sector for operations that correspond closely to public works functions. In a few other cases, performance standards have been engineered specifically for public works operations. In still other instances, performance targets and performance records of respected municipalities can be used as the foundation for public works benchmarks.

346

Public Works Administration

The specific duties of public works administrators differ from one city to another. Many differences are attributable to variations in assigned functions—for instance, some public works directors are responsible for public utilities, refuse collection, and parks maintenance, whereas others are not. In all cases, however, public works administrators are responsible for coordination and control of assigned functions. Those in charge are expected to see that departmental objectives of efficiency and effectiveness are met and that operating units are responsive.

How quickly should public works departments respond to inquiries, complaints, or requests for services? Hurst, Texas, reported responding in one recent year to all "action line" calls within 5 working days. The Rockville, Maryland, public works department had an average turnaround time of 3 days on citizen calls or requests. Blacksburg, Virginia, responded to 98% of all citizen contacts within 48 hours. Even quicker response was the norm in Duncanville, Texas, where the public works department responded to 90% of all citizen complaints within 24 hours, and in Orlando, Florida, where all citizen complaints were investigated within 24 hours.

Engineering

The skillful, conscientious, and prompt performance of engineering duties can produce major short-term and long-term benefits for a municipality. On the other hand, poor engineering performance can have major adverse ramifications for public relations, developer relations, and capital budgeting, as well as serious long-term consequences for community development. Unfortunately, quality engineering often goes unnoticed.

Several municipalities have established performance targets or reported experience on selected dimensions of engineering (Table 24.1). Some report impressive local standards of performance. Raleigh, North Carolina, for example, reported reviewing preliminary site and subdivision plans in an average of less than 1 day in 1997 and reviewing construction plans in less than 5 days on average. Orlando, Florida, attempts to review and process building plans in 5 working days and subdivision plans in 3 working days. Standards published by the LCC (1994) call for the checking of plans within 2 weeks by high-service engineering departments and within 4 weeks by medium-service departments (p. 39).

Some individual cities have established their own responsiveness targets for a variety of engineering services. The targets of Orlando, Florida, are listed in Table 24.2. The targets and performance records of San Jose, California, and Savannah, Georgia, are listed in Tables 24.3 and 24.4, respectively.

Table 24.1   Engineering: Prompt Review of Plans

Raleigh, NC
  Average working days to review preliminary site/subdivision plans: 2.0 (1996), 0.6 (1997)
  Average working days to review construction plans: 3.1 (1996), 4.8 (1997)

Savannah, GA
  Target: Review all subdivision plats and site plans within 1 week
  Actual: 100% (1990), 98% (1991)

Orlando, FL
  Percentage of sewer capacity applications and building plans reviewed within 5 working days:
    98% (1991)
  Percentage of final plats and development construction plans processed within 3 working days:
    98% (1991)

Irving, TX
  Percentage of final plat construction plans reviewed within 5 working days: 95% (1991)

Peoria, AZ
  Percentage of routine plats reviewed within 1 week: 100% (1991)

Ann Arbor, MI
  Target: Review site plans within 1 week
  Actual: 60% within 1 week (1998)
  Target: Review construction plans within 4 weeks
  Actual: 70% within 4 weeks (1998)

Hurst, TX
  Percentage of plans reviewed within 10 days: 100% (1997)

San Clemente, CA
  Percentage of development project applications reviewed within 10 working days: 92% (1998)
  Percentage of proposed development improvement plans reviewed within 10 working days:
    93% (1998)

Palo Alto, CA
  Target 1: Review plans and submit written comments to the planning department within 10 working
    days of receipt of private development application
  Actual: 88% (1997)
  Target 2: Process certificates of compliance within 25 working days from the initial application date
  Actual: 88% (1997)

Norman, OK
  Percentage of construction plans reviewed within 10 workdays: 75% (1998)
  Percentage of final plats reviewed within 10 workdays: 100% (1998)

Durham, NC
  Percentage of construction plans reviewed within 14 days: 100% (1998)

Overland Park, KS
  Targets: Review of plans for developer-constructed streets, storm sewers, and sidewalks within
    2 weeks (inspection within 3 work hours of request); review of right-of-way work permit
    applications within 3 workdays; review of building site plans within 2 weeks

New York, NY
Sewer design reviews within 14 days: 100% (1997)
Water and sewer site connection proposal reviews within 30 days: 87% (1997)
Private drainage proposal reviews within 90 days: 100% (1997)
Private sewer plan reviews within 30 days: 100% (1997)

Vancouver, WA
Target: Review developer requests for water/sewer connections or plan checks within 15 days

Reno, NV
Target: Review site plans within 30 days of filing (1999)

Winston-Salem, NC
Percentage of subdivision and site plan reviews completed within 38 days: 91% (1997)

Table 24.2   Responsiveness Goals for Engineering and Survey Services:
City of Orlando

	Responsiveness Goal Within ...
Engineering Development and Review	
Building plans reviewed and processed	5 workdays
Subdivision plans reviewed and processed	3 workdays
Sanitary sewer capacity allotment applications reviewed and processed	7 workdays
Citizen complaints investigated	24 hours
Survey Services	
Land and boundary, topographic, and "as-built" surveys	15 days
Lake elevations	24 days
Construction layouts	15 days
Plat checks	5 hours

SOURCE: City of Orlando (FL), *Annual Budget 1995–1996*, pp. XIII.20–21.

The timely updating of maps is a function that has drawn special attention in several cities (Table 24.5). "Timely," however, can mean within 4 weeks in some communities and within a year in others.

Several municipalities focus on holding engineering costs at reasonable levels. Some report engineering or design costs not as a raw figure, but instead as a percentage of total construction costs for municipal projects (Table 24.6). Cincinnati, Ohio, for example, tries to hold engineering costs for contract administration and quality control within 7% of construction costs for city projects.

Engineers often are asked to provide cost estimates for capital projects. Accordingly, several cities track the accuracy of those projections. Although LCC

Table 24.3   Engineering Responsiveness Targets and Actual Performance:
City of San Jose

	Target	Compliance With Target in 1995
Review geologic hazard application	2 weeks	95%
Process grading permit	3 weeks	100%
Review improvement plans—residential projects	45 workdays	100%
Review improvement plans—nonresidential projects	38 workdays	100%
Review improvement plans—private street projects	40 workdays	100%
Conduct/review traffic analysis for development project	14 days	95%
Testing of construction materials	within 24 hours of request	100%
Mapping of city-owned infrastructure	30 days	50%
Issuance of utility permits	10 days	90%
Marking of underground facilities	within 48 hours of request	100%

SOURCE: City of San Jose (CA), *Proposed 1996–97 Operating Budget*, pp. 237, 243.

Table 24.4   Engineering Effectiveness: City of Savannah

	FY 1997 Actual
Projects completed in compliance with specifications	100%
Projects initiated on schedule	85%
Projects completed on schedule	79%
Projects completed within budget	81%
Surveying requests completed within 10 workdays	92%
Project cost estimates provided within 10 workdays	95%
Geo-Data maps provided within 5 workdays	100%
Design sketches provided within 10 workdays	95%
General mapping information provided within 5 workdays	100%
Field and office consultations provided within 7.5 workdays	99%

SOURCE: City of Savannah (GA), *1999 Service Program and Budget*, p. 306.

(1994) standards suggest that the estimates of top engineering departments
should be within 5% of the mark and those of medium-grade departments
should be within 10% (p. 37), the average variance among cities reporting
those statistics in the documents examined for this volume ranged from 7%
to 25% (Table 24.7). Engineering involvement in major capital projects often extends beyond initial cost estimates to also include supervision of those

---

Table 24.5   Prompt Updating of Maps

Hurst, TX
  Target: Update all relevant maps within 4 weeks after receiving "as-builts"
  Actual: 80% of relevant maps updated within 4 weeks (1997)

Overland Park, KS
  Target: Updated monthly to be at least 95% current with revisions to easements, rights-of-way, new
    plats, new construction

Charlotte, NC
  Percentage of maps produced within 30 days of request: 93% (1997), 95% (1998)

Tucson, AZ
  Target: Update water system map within 90 days of system upgrade/modification

Irving, TX
  Percentage upgrading of water maps and sanitary sewer maps within 6 months of installation: 10%
    (1991)

Sunnyvale, CA
  Target: Enter at least 95% of all physical facility modifications on 50-scale engineering maps and
    records within 1 year of modification (1994)

---

projects. Project completion on time and within budget is an important objective of that supervision—an objective met high percentages of the time by several of the cities listed in Table 24.8.

Street Cleaning

The scope of street cleaning activities and the size of workforce needed to perform these activities are influenced by a variety of factors, including the frequency of street cleanings, miles of street requiring cleaning, terrain, climatic conditions, vehicle traffic, pedestrian traffic, parking conditions, building density, and the use of sand during the winter (Rubin, 1991, p. 55).[1]

Street sweeping cycles varied widely among the cities whose documents were examined for this volume. Boston sweeps downtown streets "almost daily" and posted neighborhood streets twice monthly, whereas some cities maintain weekly, monthly, or 45-day cycles (Table 24.9).

The factors such as terrain, traffic, and on-street parking that affect street sweeping activities generally, also influence street-sweeping production ratios reported by various communities. One authority has suggested as a rule of thumb that street sweepers should average 20 to 25 curb miles per shift (Foster, 1978, p. 6.5). The reasonableness of this guideline was corroborated

Table 24.6   Engineering Cost Ratios: Selected Cities

CONTRACT ADMINISTRATION/QUALITY CONTROL/ENGINEERING COSTS AS A PERCENTAGE OF CONSTRUCTION COSTS

Cincinnati, OH
Percentage for contract administration/quality control: 6.3% (1991)

Rock Island, IL
Target: In-house engineering costs not to exceed 15% of project construction costs
Actual: 6.3% (1990), 5.1% (1991)

Norman, OK
8.8% (1998, estimated)

Winston-Salem, NC
Engineering costs including consultant, architect, testing, and in-house staff as percentage of
   project costs for building construction: 11% (1997); for street construction: 13% (1997);
   for utility construction: 13% (1997)

DESIGN AND CONSTRUCTION ADMINISTRATION COSTS AS A PERCENTAGE OF CONSTRUCTION COSTS

Reno, NV
Target: 16% or less (1999)

DESIGN/PLAN COSTS AS A PERCENTAGE OF CONSTRUCTION COSTS

Corpus Christi, TX
Design cost as percentage of bid: 5.8% (1991)

Cincinnati, OH
Target: 6% or less for water main installation projects
Actual: 6% (1991)

Rock Island, IL
Target: 7% or less
Actual: 5.3% (1990), 3.9% (1991)

Lubbock, TX
Target: 13% or less

by a federally sponsored study of Los Angeles–area cities that discovered an average of 2.71 curb miles cleaned per hour by city employee operators, 3.46 curb miles per hour by contractors, and 3.09 curb miles per operator hour overall (Stevens, 1984, p. 52). A higher standard was suggested in a study conducted by the City of Pasadena, California, Management Audit Team (1986) that showed the operations in neighboring communities averaging 31.5 lineal street miles per sweeper day and 3.9 street miles per sweeper hour (p. 5.2). Worker day averages among the examined documents ranged from 17 to 35 curb miles.

Table 24.7   Accurate Cost Projections for Capital Projects: Selected Cities

Winston-Salem, NC
Bid amount as percentage of estimated costs for major projects: 93% (1997)

Lubbock, TX
Percentage of projects constructed within 10% of estimated cost: 90% (1998, estimated)

Peoria, AZ
Average variance between estimated and actual costs of city projects: 10% (1991)

San Clemente, CA
Target: At least 80% of projects within 10% of engineering estimates (1999)

Norman, OK
Percentage of projects constructed within 15% of original budget: 100% (1999, estimated)

Greenville, SC
Percentage of construction project estimates within 15% of actual bid: 71% (1989), 80% (1990)

Rock Island, IL
Target: Preliminary cost estimates within 20% of bid
Actual: +3% (1990), +25.5% (1991)

Sunnyvale, CA
Target: At least 85% of projects within 25% of cost estimate (1994)

Table 24.8   Effective Supervision of Capital Projects: Selected Cities

San Clemente, CA
Percentage of capital projects completed within budget: 100% (1998)

Danville, KY
Percentage of capital projects completed on schedule: 95% (1996), 80% (1997)

Chandler, AZ
Percentage of projects completed within 5% of original contract cost: 81% (1995)
Percentage of projects completed on schedule: 90% (1995), 87.5% (1998)

Rockville, MD
Percentage of capital projects completed on time: 88% (1997)

Norman, OK
Percentage of projects completed on schedule: 86.6% (1999, estimated)

Fayetteville, AR
Percentage of projects completed within budget and on time: 85% (1991)

Corvallis, OR
Percentage of capital project design and construction schedules met: 85% (1991), 94% (1992)

Savannah, GA
Percentage of capital projects completed on schedule: 79% (1997)
Percentage of capital projects completed within budget: 81% (1997)

Table 24.9   Street Sweeping Cycles and Production Ratios: Selected Cities

Street Sweeping Cycles	*Street Sweeping Production Ratios*
Boston, MA Targets: Posted neighborhood streets twice monthly; arterial streets weekly; downtown streets almost daily	**Long Beach, CA** 5.0 miles of street swept per employee work hour (1996)
Reno, NV Target: Sweep streets every 4 weeks during nonwinter months; sweep at least 50% of all streets within 7 days following a snow event (1999)	**Irving, TX** Lane miles swept per crew hour: 5.0 (1991), 4.9 (1998)
Richmond, VA Target: Commercial streets are swept 5 nights weekly; residential streets, every 4 to 6 weeks	**Oak Ridge, TN** Curb miles cleaned mechanically per staff hour: 4.47 (1991), 2.82 (1992)
Palo Alto, CA Percentage of residential and arterial routes swept weekly: 94% (1997) Percentage of business district streets swept 3 times per week: 100% (1997)	**Santa Ana, CA** Curb miles machine swept per direct worker hour: 3.97 (1987)
Orlando, FL Target: Sweep downtown area at least 3 times per week; industrial/commercial areas, every 9 workdays; residential areas, every 16 workdays	**San Clemente, CA** Target: 28 to 32 miles per operator day (1999) **Peoria, AZ** Curb miles swept daily per sweeper operator: 29 (1991)
Peoria, AZ Target: 3 times per week downtown and roadside parking areas; 17-day cycle for arterial and residential streets	**Little Rock, AR** Curb miles per operator per 8-hour shift: 15 to 35
Kansas City, MO Downtown streets swept 3 times weekly: 10% (1995) Arterials swept once a month for 9 months: 33% (1995)Residential streets swept twice a year: 95% (1995)	**Decatur, IL** Central business district curb miles per staff hour: 2.93 Priority curb miles per staff hour: 2.30 Neighborhood curb miles per staff hour: 0.63 (1995)
Fort Collins, CO Targets: Sweep arterials, collectors, and bike lanes twice per month, downtown 8 times per month, and residential streets twice per year; additional sweeping after snowstorms	**Iowa City, IA** Curb miles per unit per day (7 hours per day): 20 (1990), 20 (1991)
Corpus Christi, TX Targets: Downtown streets swept weekly; arterial streets monthly (1999)	**Fort Collins, CO** 2.5 lane miles per worker hour (1996)
Lee's Summit, MO Target: Sweep downtown streets weekly; other curb-and-guttered streets once per year	**Charlotte, NC** Target: At least 17 miles per worker day Actual: 18.6 miles per worker day (1994, 5-month record)
St. Charles, IL Target: Average of once per week during nonwinter months	**Chesapeake, VA** 3,200 curb miles annually per FTE (1997)

Street Sweeping Cycles	Street Sweeping Production Ratios

Ames, IA
  Targets: Business district streets, 32 times per
    year; arterial and collector streets, 16 times
    per year; residential streets, 8 times per year
    (1999)
Sunnyvale, CA
  Target: Average of 26 sweepings per street per
    year
Overland Park, KS
  Targets: All streets having curb and gutter will
    be swept at least 3 times per year; downtown
    and high-priority streets, every 2 weeks; sand
    and debris swept from first-, second-, and
    third-priority streets following snow/ice storm
Lubbock, TX
  Percentage of downtown swept monthly:
    88% (1997)
  Percentage of thoroughfares swept monthly:
    64% (1996)
Hurst, TX
  Target: Maintain a 45-day street cleaning cycle
  Actual: 91% of streets cleaned within 45-day
    cycle (1997)

Street Lighting

Not only can street lighting affect a community's traffic safety, it can also in-fluence the incidence of crime, the likelihood that offenders will be appre-hended, and the general sense of security among residents. Proper illumina-tion for traffic safety, however, differs from one roadway to another. Key factors are the luminance of a given road surface (Table 24.10), the type of road (e.g., expressway, collector street, etc.), and the commercial or residen-tial nature of the area (Table 24.11). A community's choice among options for achieving desired illumination may have budgetary and aesthetic ramifica-tions—different lighting sources yield different lighting efficiencies, intensi-ties (Table 24.21), and even different colors. Low-pressure sodium lamps, for example, emit a yellowish glow, compared to the more orange-hued color of high-pressure sodium or the white-blue of metal halide.

Most cities distinguish between the lighting needs of thoroughfares and residential streets. In Cleveland, Ohio, light fixtures on thoroughfares carry 400 watts, whereas those in residential areas carry 150 to 175 watts (Couret, 1999, p. 56). Some cities adopt streetlight policies with crime control in mind.

Table 24.10   Luminance of Various Roadway Surfaces

Road Surface Class	Luminance Coefficient (Q_o)	Description	Mode of Reflectance
R1	0.10	Portland cement concrete road surface. Asphalt road surface with a minimum of 15% of the aggregates composed of artificial brightener (e.g., Synopal) aggregates (e.g., labradorite, quartzite)	Mostly diffuse
R2	0.07	Asphalt road surface with an aggregate composed of a minimum 60% gravel (size greater than 10 mm)	Mixed (diffuse and specular)
R3	0.07	Asphalt road surface (regular and carpet seal) with dark aggregates (e.g., trap rock, blast furnace slag); rough texture after some months of use (typical highways)	Slightly specular
R4	0.08	Asphalt road surface with very smooth texture	Mostly specular

SOURCE: *American National Standard Practice for Roadway Lighting.* ANSI/IESNA RP-8-83 (1983, reaffirmed 1993). Published by the Illuminating Engineering Society of North America, 120 Wall Street, 17th Floor, New York, NY 10005. Reprinted by permission.

Savannah, Georgia, for example, prescribes a minimum lighting level of 0.25 foot-candles for high-crime areas, 0.04 foot-candles for high-density areas, and spacing of no greater than 500 feet between streetlights in low-density areas (City of Savannah [GA], *1992 Program of Work,* p. 39).

Streetlight ratios among the examined documents ranged from 17 streetlights per 1,000 population in Sterling Heights, Michigan, to 167 streetlights per 1,000 residents in Oak Ridge, Tennessee (Table 24.13). White Plains, New York, reports 36.5 streetlights per mile. Several cities report prompt investigation of requests for new streetlights and prompt restoration of malfunctioning streetlights (Table 24.14).

Custodial Services

Fairly detailed production standards have been developed for custodial services (Table 24.15). The application of such standards to a community's total custodial workload offers local officials a basis for judging the adequacy of current staffing.

A simpler but less precise gauge of custodial staffing adequacy may be obtained by comparing local staffing with other communities' ratios of custodians to square footage maintained (Table 24.16). A city with a ratio of 13,000

(text continues on page 360)

Table 24.11 Recommended Roadway and Walkway Illumination[a]

Roadway and Walkway Classification		Average Illuminance Pavement Classification[b]						Uniformity Avg./Min.
		R1		R2 & R3		R4		
		Foot-Candles	Lux	Foot-Candles	Lux	Foot-Candles	Lux	
Expressway[c,d]	Commercial	0.9	10	1.3	14	1.2	13	3:1
	Intermediate	0.7	8	1.1	12	0.9	10	
	Residential	0.6	6	0.8	9	0.7	8	
Major[d]	Commercial	1.1	12	1.6	17	1.4	15	3:1
	Intermediate	0.8	9	1.2	13	1.0	11	
	Residential	0.6	6	0.8	9	0.7	8	
Collector[d]	Commercial	0.7	8	1.1	12	0.9	10	4:1
	Intermediate	0.6	6	0.8	9	0.7	8	
	Residential	0.4	4	0.6	6	0.5	5	
Local[d]	Commercial	0.6	6	0.8	9	0.7	8	6:1
	Intermediate	0.5	5	0.7	7	0.6	6	
	Residential	0.3	3	0.4	4	0.4	4	
Alleys	Commercial	0.4	4	0.6	6	0.5	5	6:1
	Intermediate	0.3	3	0.4	4	0.4	4	
	Residential	0.2	2	0.3	3	0.3	3	
Sidewalks	Commercial	0.9	10	1.3	14	1.2	13	3:1
	Intermediate	0.6	6	0.8	9	0.7	8	4:1
	Residential	0.3	3	0.4	4	0.4	4	6:1
Pedestrian Ways and Bicycle Lanes[e]		1.4	15	2.0	22	1.8	19	3:1

SOURCE: From *An Informational Guide for Roadway Lighting*. Copyright © 1984 by the American Association of State Highway and Transportation Officials, Washington, D.C. Used by permission.

a. Average illuminance on the traveled way or on the pavement area between curb lines of curbed roadways.
b. See Table 24.10 for descriptions of various road surface classifications.
c. Both mainline and ramps. Expressways with full control of access are covered separately as freeways.
d. Adapted from *American National Standard Practice for Roadway Lighting*, ANSI/IESNA RP-8-83 (1983). Published by the Illuminating Engineering Society of North America, 120 Wall Street, 17th Floor, New York, NY 10005. Used here by permission.
e. This assumes a separate facility. Facilities adjacent to a vehicular roadway should use the illuminance or luminance levels for that roadway.

Table 24.12   Average Life and Lumens per Watt for Typical Light Sources

Source	Average Life (hours)	Initial Lumens/Watt Including Ballast
Incandescent (100–1,500 W)	1,000–3,000	17–23 (No Ballast)
Fluorescent (30–215 W)	20,000	62–63
Mercury (100–1,000 W)	24,000	35–59
Metal Halide (175–1,000 W)	15,000	67–116
High Pressure Sodium (70–1,000 W)	20,000	64–130
Low Pressure Sodium (35–180 W)	18,000	80–150

SOURCE: Center for Design Planning, *Streetscape Equipment Sourcebook 2* (Washington, DC: Urban Land Institute, 1979), p. 22. Reprinted by permission.

Table 24.13   Streetlight Ratios: Selected Cities

Streetlights per 1,000 Population		Streetlights per Mile	
Oak Ridge, TN 167 (1993)	Norfolk, NE 85 (1996)	White Plains, NY 36.5 (1995)	Denton, TX 17 (1998)
Shreveport, LA 143 (1996)	Brentwood, TN 76 (1996)	Newburgh, NY 34.5 (1991)	West Valley City, UT 16.0 (1996)
Danville, KY 130 (1997)	Glendale, AZ 71 (1998)	Rockville, MD 34.4 (1997)	Norfolk, NE 15.3 (1996)
Norfolk, VA 125 (1998)	Lubbock, TX 71 (1998)	Chandler, AZ 31.2 (1999)	Lubbock, TX 14.8 (1998)
Little Rock, AR 111 (1997)	Ann Arbor, MI 60 (1999)	Charlottesville, VA 27.3 (1997)	Gainesville, GA 14.1 (1999)
Gainesville, GA 105 (1999)	Durham, NC 53 (1998)	Greenville, SC 27.2 (1991)	Flagstaff, AZ 13.4 (1998)
Greenville, SC 104 (1991)	Flagstaff, AZ 47 (1998)	Ann Arbor, MI 24.0 (1999)	Greeley, CO 12.0 (1999)
Kansas City, MO 104 (1997)	Greeley, CO 47 (1999)	Germantown, TN 23.9 (1995)	Brentwood, TN 10.0 (1996)
White Plains, NY 102 (1995)	Norman, OK 46 (1998)	Glendale, AZ 22.4 (1998)	Little Rock, AR 9.5 (1997)
Chandler, AZ 98 (1999)	Avon, CT 45 (1997)	Smyrna, GA 21.6 (1996)	Avon, CT 7.0 (1997)
Charlottesville, VA 92 (1997)	West Valley City, UT 42 (1996)	Kansas City, MO 21.1 (1997)	Blacksburg, VA 6.5 (1997)
Prairie Village, KS 87 (1996)	Blacksburg, VA 39 (1997)	Shreveport, LA 18.3 (1996)	Sterling Heights, MI 6.4 (1996)
Smyrna, GA 86 (1996)	Sterling Heights, MI 17 (1996)	Prairie Village, KS 17.8 (1996)	

Table 24.14   Streetlight Service Responsiveness: Selected Cities

PROMPT RESPONSE TO STREETLIGHT REQUESTS

Orlando, FL
Percentage of streetlight requests investigated within 1 week: 100% (1991)

Winston-Salem, NC
Percentage of streetlight requests processed within 10 days: 95% (1992)

Charlotte, NC
Targets: Investigate and respond to (a) 95% of individual requests within 30 days; (b) 95% of neigh-
borhood requests within 60 days
Actual: (a) 95%; (b) 80% (1987)

Milwaukee, WI
Percentage responded to within 30 days: 83% (1991), 90% (1992), 87% (1993)

PROMPT RESPONSE TO STREETLIGHT MALFUNCTIONS

Lubbock, TX
Percentage of damaged streetlight orders cleared within 1 hour of receipt: 85% (1997)
Percentage of streetlight maintenance orders cleared within 24 hours of receipt: 88% (1997)

Philadelphia, PA
Response time for minor streetlight repair: 1 day
Response time for pole repair: 3 days
Response time for major repair of luminaries: 12 days (1998)

Sunnyvale, CA
Targets: Replace at least 95% of all burned-out streetlights within 24 hours of notification; correct at
least 95% of all minor electrical problems within 24 hours; correct at least 90% of all major electrical problems within 72 hours (1994)

Palo Alto, CA
Percentage of streetlight failures responded to within 2 working days of notification: 100% (1997)

Glendale, AZ
Streetlights repaired within 48 hours: 90% (1998, estimated)

Oak Ridge, TN
Target: Respond to at least 90% of streetlight complaints within 48 hours

Tallahassee, FL
Malfunctioning streetlights repaired within average of 2 days from time reported (1991)

Charlotte, NC
Service restored in average of 2.2 days (1987)

Kansas City, MO
Percentage of isolated streetlight outages responded to within 48 hours: 40% (1995)
Percentage of pole knockdowns replaced within 10 days: 20% (1995)

Cambridge, MA
Percentage repaired within 72 hours: 80% (1996, estimated)

(continued)

Table 24.14   Streetlight Service Responsiveness: Selected Cities (continued)

Overland Park, KS
  Target: Repair routine outages within 3 days of notification; restore knockdowns to service within 7
    workdays (weather permitting)

Chandler, AZ
  Target: Repair or initiate underground repairs for arterial streetlights within 5 working days
  Actual: 96% (1998)

Boston, MA
  Percentage of citizen complaints addressed within 5 days: 84% (1992), 77% (1996)

San Jose, CA
  Percentage of streetlight repairs within 7 days: 58% (1996, estimated)

Reno, NV
  Target: 100% repaired within 10 days (1999)

New York, NY
  Percentage of malfunctioning streetlights repaired within 10 days: 97.5% (1998)

Milwaukee, WI
  Percentage of circuit troubles repaired within 24 hours: 98% (1991), 98.7% (1992), 99% (1993)
  Percentage of unit troubles repaired within 30 days: 85% (1991), 91.5% (1992), 92% (1993)

to 18,000 square feet per custodian would fit comfortably within the range occupied by most of the cities in that table.

In Palo Alto, California, the quality and responsiveness of custodial services are monitored by tracking how quickly reported deficiencies are corrected. During one recent year, 92% of all reported deficiencies involving contractual custodial services were corrected within one scheduled servicing and 100% of those involving in-house custodians were corrected by the next scheduled servicing (City of Palo Alto [CA], *1998-99 Budget Adopted* [Vol. 1], p. 337).

Building Maintenance

Many public works departments are responsible for the maintenance, repair, and occasional remodeling of city facilities. Some have developed systems for evaluating building conditions and the quality of their work. Savannah, Georgia, for example, uses the following 4-point scale to rate the condition of municipal facilities:

Table 24.15 Production Rate Recommendations for Custodial Services

Custodial Service	Time Standard
Basic Service	
Trash removal	2.50 minutes per 1,000 square feet
Clean ashtrays	1.85 minutes per 1,000 square feet
Dust/spot clean horizontally exposed surfaces	2.00 minutes per 1,000 square feet
Spot clean vertical surfaces (e.g., walls, light switches, doors, door frames)	1.10 minutes per 1,000 square feet
Clean glass door (both sides)	2.75 minutes per unit
Clean drinking fountain	1.15 minutes per unit
Clean sink, faucets, and adjacent surface	2.00 minutes per unit
Wash 4′ × 4′ lunchroom table	3.00 minutes per unit
Dust venetian blinds	1.50 minutes per unit
Polish wood desk	5.50 minutes per unit
Restroom Service	
Machine scrub/disinfect restroom floors (150–250 square foot average area)	15 minutes per room
Wash restroom partitions	5.00 minutes per stall
Thoroughly clean and restock	2.50 minutes per fixture (sinks, toilets, urinals)
Floor Service (Carpets)	
Spot vacuum (upright)	3.00 minutes per 1,000 square feet
Traffic lane area vacuum (upright)	5.00 minutes per 1,000 square feet
Detail vacuum (corners, edges, knee wells, etc.)	8.00 minutes per 1,000 square feet
Full vacuum—all exposed carpet (upright)	10.5 minutes per 1,000 square feet
Clean carpet (rotary)	60 minutes per 1,000 square feet
Floor Service (Hard and Resilient)	
Dust mop	
Unobstructed	5.00 minutes per 1,000 square feet
Obstructed	10 minutes per 1,000 square feet
Broom sweep (24″ push broom)	7.00 minutes per 1,000 square feet
Damp mop	
Unobstructed	12.65 minutes per 1,000 square feet
Obstructed	14.50 minutes per 1,000 square feet
Stair Service	
Sweep stairs and clean handrails	6.00 minutes per floor
Vacuum carpeted stairs and clean handrails	11.5 minutes per floor

SOURCE: Excerpted from *BSCAI Production Rate Recommendations*, pp. 1–3. Copyright © 1992 Building Service Contractors Association International. All rights reserved. Reprinted with permission of the Building Service Contractors Association International, 10201 Lee Highway, Suite 225, Fairfax, Virginia 22030.

Table 24.16 Production Ratios for Custodial Services: Selected Cities

City	Square Footage Maintained per Custodian
Reno, NV	21,000 (1999 target)
Santa Ana, CA	20,000 (1995)
Fayetteville, AR	17,913 (1997)
Tempe, AZ	17,637 (1997)
New York, NY	14,400 (1997), 14,200 (1998)
Nashville–Davidson County, TN	13,652 (1997)
Chesapeake, VA	13,646 (1991), 12,938 (1997)
Peoria, AZ	13,500 (1991)
San Antonio, TX	11,000 (1991)

I. All structural components of the building are in excellent condition—only preventive maintenance work is required.

II. All components are structurally sound; however, minor maintenance work is required.

III. Structure shows indications of slight deterioration in several component parts of the building. Maintenance work is required within 6 months.

IV. Structure shows severe deterioration. Immediate major maintenance is required. (*1992 Program of Work*, p. 129)

The building maintenance operation in Savannah has as its objective ensuring that at least 95% of all city facilities meet condition II or better.

Several cities have established performance targets for prompt response to maintenance requests (Table 24.17). The trick is to achieve a reasonably high level of responsiveness without doing so through overstaffing.

A concern of local governments is that their buildings and facilities are accessible and hospitable to persons with disabilities. Many building maintenance operations have become involved in retrofitting existing properties for compliance with provisions of the federal ADA. A sample of building accessibility standards is provided in Table 24.18. Sample requirements for parking spaces are listed in Table 24.19.

Another concern of most building maintenance operations is the unfortunate need to repair the work of vandals. Some cities have even established performance standards for the prompt removal of graffiti from city facilities (Table 24.20). Scottsdale, Arizona; Philadelphia, Pennsylvania; and Ann Arbor, Michigan, for example, strive to remove graffiti within 24 hours.

Table 24.17   Prompt Response to and Completion of Building Maintenance
Requests: Selected Cities

City	Performance Target	Actual
QUICK RESPONSE		
Fayetteville, AR		Average response time of 2 hours to general service requests (1991)
Santa Ana, CA		100% of emergencies responded to within 2 hours (1995)
Orlando, FL	Respond to service calls within 8 hours	90% within 8 hours (1991)
Savannah, GA	Respond to emergencies in average of 6.5 hours; address all maintenance requests within 3 working days	Average emergency response time: 6.5 hours (1998, estimated)
Winston-Salem, NC	Respond to emergency requests within 24 hours	Responded to 100% of emergency requests within 24 hours (1998, midyear)
Oakland, CA	Respond to service requests within 2 days of receipt	100% within 2 days
Reno, NV	Respond to calls for emergency building repairs within 1 hour; to calls for routine repairs within 3 working days	
PROMPT COMPLETION		
Boston, MA		99% of HVAC breakdowns corrected within 4 hours of being reported
Nashville–Davidson County, TN	Complete at least 86% of corrective maintenance calls same day	74% of corrective maintenance calls completed same day (1997)
Alexandria, VA		Average number of days to complete a work order: 2.0 (1990), 2.0 (1991)
Santa Ana, CA		100% of work orders (minor jobs) completed within 3 days (1995)
Boca Raton, FL	Complete air conditioning work orders within 5 days; carpentry within 7 days; electrical within 5 days; plumbing within 3 days; other trades within 5 days (1999)	Air conditioning work completed within 14 days following work order; carpentry, 21 days; electrical, 13 days; plumbing, 8 days; other trades, 10 days (1998, estimated)
Sunnyvale, CA	Complete 90% of emergency repairs within 1 day of notification; complete 80% of nonemergency repairs within 1 week	
Ann Arbor, MI	Complete at least 90% of routine work orders within 7 days; at least 90% of emergency repair orders within 24 hours	95% of routine work orders completed within 7 workdays; 95% of emergency work orders within 24 hours (1998)
Greensboro, NC	Complete at least 95% of all priority/emergency work orders within 5 days of receipt	

Table 24.17   Prompt Response to and Completion of Building Maintenance
Requests: Selected Cities (continued)

Hurst, TX	Complete 70% of work orders within 7 days of receipt	Completed within 7 days: 50% (1991, estimated), 70% (1997)
Fort Collins, CO	Complete more than 90% of service calls within 10 days	88% completed within 10 days (1991)
Cambridge, MA		Average number of days to close request for service: 14.96 (1995), 12.72 (1996, estimated)
Palo Alto, CA	Complete all maintenance service requests within 3 weeks	97% (1997)

Table 24.18   Selected Accessibility Requirements Under the Americans With
Disabilities Act (ADA) of 1990

Automobile Parking
Accessible parking spaces shall be at least 96 inches wide. Parking spaces and access aisles shall be level, with surface slopes not exceeding 2% in all directions.

Drinking Fountains
Where only one drinking fountain is located on a given floor, there shall be a drinking fountain which is accessible to individuals who use wheelchairs and one accessible to those who have difficulty bending or stooping. Where more than one fountain or water cooler is provided on a floor, at least 50% of these shall be accessible to such persons.

Emergency Warning Systems
If emergency warning systems are provided, they shall include both audible alarms and visible alarms.

Wheelchair Accommodations
The minimum clear width for single wheelchair passage shall be 32 inches at a point and 36 inches continuously.

The space required for a wheelchair to make a 180-degree turn is a clear space of 60 inches diameter or a T-shaped space.

The minimum clear floor or ground space required to accommodate a single, stationary wheelchair and occupant is 30 inches by 48 inches.

Ground and Floor Surfaces
Ground and floor surfaces along accessible routes and in accessible rooms shall be stable, firm, and slip-resistant.

Changes in level between one-fourth and one-half inch shall be beveled with a slope no greater than 1:2. Changes in level greater than one-half inch shall be accomplished by means of a ramp.

The maximum pile thickness for carpeting shall be one-half inch.

Water Closet Height
The height of water closets shall be 17 to 19 inches, measured to the top of the toilet seat.

Height of Tables or Counters
The tops of accessible tables and counters shall be from 28 to 34 inches.

SOURCE: U.S. Architectural and Transportation Barriers Compliance Board (Access Board), *Accessibility Guidelines for Buildings and Facilities*, as amended through January 1998 (Washington, DC: Author, 1998).

Table 24.19   Required Number of Handicapped Parking Spaces Under the Americans With Disabilities Act (ADA) of 1990

Total Parking in Lot	Required Accessible Spaces
1 to 25	1
26 to 50	2
51 to 75	3
76 to 100	4
101 to 150	5
151 to 200	6
201 to 300	7
301 to 400	8
401 to 500	9
501 to 1,000	2% of total
Greater than 1,000	20 plus 1 for each 100 over 1,000

SOURCE: U.S. Architectural and Transportation Barriers Compliance Board (Access Board), *Accessibility Guidelines for Buildings and Facilities,* as amended through January 1998 (Washington, DC: Author, 1998), Section 4.1.2.

Table 24.20   Prompt Removal of Graffiti: Selected Cities

Scottsdale, AZ
Percentage of identified graffiti locations abated within 24 hours: 100% (1999)

Philadelphia, PA
Percentage of graffiti removed from city-owned buildings within 1 day: 100% (1998)

Ann Arbor, MI (Parks)
Target: Remove graffiti same day as discovered
Actual: 90% within 2 days (1998)

Oakland, CA
Target: Respond to graffiti complaints by the next day (1997)

Denver, CO
Response time for graffiti removal: 2 working days (1998)

Reno, NV
Target: Remove at least 90% of gang or tagger graffiti within 48 hours; 100% within 1 week (1999)

San Jose, CA
Target: Removal within an average of 3 days

San Diego, CA
Percentage of graffiti service requests completed within 4 days: 92% (1998)

Palo Alto, CA
Percentage of graffiti removed within 5 days of request: 83% (1998, estimated)

Radio Maintenance _____

Many municipal departments—especially those in the public safety field—depend heavily on two-way radio communications. Radio malfunctions can severely damage service effectiveness. Extended downtime is intolerable.

Radio maintenance, whether performed by contract or in-house, must be prompt and reliable. Several cities have established ambitious performance targets and have achieved admirable performance records in the prompt repair of radio equipment (Table 24.21).

Note _____

1. These nine factors were identified by Marc A. Rubin (1991, p. 54) for their potential to influence service efforts and accomplishments indicators. Output indicators recommended by Rubin are number of street miles cleaned, percentage of street miles receiving regular street sweeping, and tons of refuse collected. Recommended outcome indicators are percentage of street sweepings not completed on schedule, average customer satisfaction rating, and percentage of streets rated acceptably clean. Recommended efficiency indicators are cost per mile of street cleaned and cost per ton of refuse collected.

Table 24.21 Prompt Repair of Radios: Selected Cities

GENERAL RADIO REPAIR

Rock Island, IL
Average downtime: 2.16 hours (1990), 1.23 hours (1991)

Raleigh, NC
95% of repairs completed in 4 hours or less (1997)

Portland, OR
Percentage of time portable radios are returned to service within 8 hours: 98% (1991)

Greensboro, NC
Percentage of repairs to communication and security equipment completed within 8 hours of request: 80% (1997)
Percentage of vehicular installations completed within 72 hours of request: 85% (1997)
Percentage of after-hours emergencies responded to within 1 hour: 90% (1997)

Milwaukee, WI
Percentage restored within 24 hours: 98% (1991), 98.2% (1992), 98% (1993)

Savannah, GA
Percentage of mobile radios repaired within 24 hours of request: 94.2% (1997)
Percentage of portable radios repaired within 72 hours of request: 95.1% (1997)
Percentage of mobile installations completed within 96 hours of request: 89.1% (1997)

Palo Alto, CA
Percentage of system support equipment returned to service within 24 hours: 99% (1997)
Percentage of mobile radio equipment returned to service within 3 days: 97% (1997)
Percentage of portable radio equipment returned to service within 5 days: 96% (1997)

Corpus Christi, TX
Targets: Respond to problem within 2 hours; resolve problem within 48 hours; complete at least 90% of service requests on time

San Jose, CA
Percentage of mobile radios repaired within 3 days of request: 91% (1995)
Percentage of other communication equipment repaired/aligned within 1 day of request: 100% (1995)

REPAIR OF PUBLIC SAFETY RADIOS

Portland, OR
100% of police, water, and maintenance mobile radios returned to service within 1 hour; 95% of fire mobile radios returned to service within 2 hours (1991)

Milwaukee, WI
Percentage restored within 24 hours: 100% (1991), 100% (1992), 99% (1993)

Cincinnati, OH
73.3% of police/fire portables repaired within 5 workdays (1993); 77.1% of other public safety equipment repaired within 20 workdays (1993)

25

Purchasing and Warehousing

S trategies that centralize most purchasing and warehousing functions are adopted by many municipalities for three primary reasons:

- To avoid duplication of purchasing and storage efforts across the various departments that acquire common supplies
- To secure economies of scale through bulk purchases
- To derive the benefits of purchasing expertise developed through specialization

In these cities, a department needing a particular item might find it already on hand in the central warehouse or may request its acquisition through the purchasing department, which will then contact vendors to secure the most favorable price.

Typically, a requisition prepared by the requesting department initiates the purchasing process. The purchasing department may seek telephone quotes from suppliers, written quotations, or formal bids, depending in most cases on the anticipated price of the item. Telephone quotations may be obtained quickly and often are deemed sufficient for low-cost items. Securing formal bids takes much longer but is the more appropriate option for high-priced items.

In San Diego, California, for example, the purchasing department processes requisitions valued below $5,000 within 5 days, compared to 7 days for those in the $5,000 to $10,000 range, 21 days for those in the $10,000 to $50,000 range, 30 days for those valued between $50,000 and $1 million, and 60 days for those valued at more than $1 million (City of San Diego [CA], *Fiscal Year 2000 Budget* [Vol. 2], p. 78). Emergency requisitions in San Diego are processed within 24 hours.

Rules of Thumb for Purchasing

To derive maximum value from a centralized purchasing operation, the purchase of equipment and supplies needed by different departments should be well coordinated for quantity discounts. Furthermore, multiple vendors should compete on major transactions, and the cost of administering the centralized purchasing program should be relatively small, compared either with the dollar value of all transactions or with documented savings. An authority in the field has offered these rules of thumb:

> Dividing the total dollar value of all transactions by the total number of all transactions [yields the] average value of each purchase. If the average value is less than $1,000, [the] overall process of buying [should be reviewed].
>
> [The] total of purchase orders issued should not exceed 25% of the total transactions. Purchase orders are expensive, and should be used only when nothing else will do. Change orders and cancellations are indications of not having things together at the outset.
>
> A large number of bids indicates a lack of scheduling, and a large number of re-bids indicates a lack of initial preparation.
>
> Sole source buying can reach as high as 5 percent of total purchases, but should seldom exceed 1 percent. A high percentage of this type buying, even when buying for libraries, points to insufficient bid list development, and will lock [the municipality] into the initial purchases of major systems.
>
> Local purchasing or bid awards should always be done without local preference consideration. If local merchants are partners in pre-bid conferences, they will win 75 to 80% of all bids anyway simply because they will know more about what [is desired]. If [the city] is small [it] may not have local access to all [of its] needs, but in medium to large cities 75 to 80% should be average. Local preference should never be allowed; competition is still the best way for everyone to get what they need.
>
> There is no rule, or average, on how many employees make up a good purchasing department, but if the total budget exceeds 1% of the total dollar value of all transactions, [the department] may not be working at [its] most efficient. (Butler, 1985, p. 43; reprinted with permission of *Texas Town & City,* the official publication of the Texas Municipal League)

Corroborating the reasonableness of the last of these rules of thumb, Calgary, Alberta, reported that operating expenditures by its purchasing department were the equivalent of only 0.7% of the value of all purchasing transactions during a recent year—well below the 1% rule of thumb (Table 25.1). Fort Collins, Colorado, reported an even lower rate of 0.5%. By comparison, a recent study by the ICMA (1999b) reported 1998 median rates of 0.86% ($116.25 in purchases per $1 of central purchasing operating expenditures)

Table 25.1   Cost-of-Purchasing Ratios: Selected Cities

PURCHASING OPERATION EXPENDITURES AS A PERCENTAGE OF THE TOTAL DOLLAR VALUE OF ALL TRANSACTIONS

Fort Collins, CO
  0.5% (1996)

Calgary, Alberta
  0.880% (1994)
  0.749% (1995)
  0.719% (1996)

PURCHASING OPERATION EXPENDITURES AS A PERCENTAGE OF THE ENTIRE MUNICIPAL BUDGET

Charlottesville, VA
  Target: Purchasing budget of less than 1% of total city budget
  Actual: 0.07% (1998)

Santa Ana, CA
  Cost of purchasing operation as a percentage of total city dollars expended: 0.13% (1995)

among participating local governments serving populations of 100,000 or greater and 1.14% among smaller units (pp. 265, 301). Of course, cities comparing the cost of their purchasing operation to the total municipal budget, including personnel and other expenses, can report even lower percentages.

Normally, an aggressive centralized purchasing operation can outperform decentralized arrangements in securing competitive prices for goods and services. Even when consolidation of orders is impractical and the transaction value does not justify formal bids, telephone quotes may be secured from several vendors. Armed with a comprehensive list of potential suppliers, an aggressive purchasing office normally can claim a high percentage of competitive purchases—regardless of the mix of formal bids and informal quotes.

How well do various cities do in securing competitive purchases? Boston reported an average of 6.37 bidders per contract in 1996, up from 3 bidders per contract just 4 years earlier. Rockville, Maryland, reported an average of 6 bidders for solicitations exceeding $7,500. Fort Collins, Colorado, reported 4.7 qualified bidders per bid.

Benchmarks

How long does it take a proficient purchasing department to move from receipt of a requisition to the issuance of a purchase order? Generally, the quickest departments can advance to that stage in a few days for items not requiring formal bids. Although many cities take less than a week to process routine purchase orders, the requirement of formal bids for some items slows the process substantially, often requiring 60 days or more (Table 25.2). Such delays can be frustrating to operating departments anxious to receive the needed items. Steps to streamline the process should be taken where practical, but formal bidding for major purchases safeguards the municipality's integrity and promotes competition. A 1993 survey of 138 municipalities conducted by the National Institute of Governmental Purchasing (1994) indicated that almost half the responding cities were able to move from requisition to award of contract in less than 20 days and that relatively few required more than 30 days (Table 25.3). As a point of comparison, during a recent year Boston, Massachusetts, reported that an average of 24.7 days elapsed between receipt of a requisition by the purchasing department and delivery of the goods to the client department.

Many cities with centralized purchasing operations also maintain a warehouse where items can be secured and held in advance of need. A carefully managed warehouse can be a money saver for a municipality, but one that is poorly managed can be costly.

A good warehouse operation will stock supplies needed on a recurring basis but will not laden its shelves with seldom used items. The decision to place a particular item in the warehouse inventory should be based on the anticipated need for that item in the future. Once such a decision is made, proper inventory management calls for periodic restocking to ensure availability. Out-of-stock ratios of less than 10% are common among leading municipalities (Table 25.4).

A city wishing to avoid high out-of-stock ratios could simply overstock all items to minimize the chance of running out, but adopting such a strategy would be poor inventory management because it would unnecessarily tie up municipal resources in idle inventory and would commit warehouse space unnecessarily. Much like well-managed libraries, well-managed warehouses experience high inventory turnover. They have carefully chosen stock that is "popular," and they manage their supply for maximum advantage. Among cities examined for this volume and reporting such statistics, inventory turnover rates ranged from once per year to six times annually—that is, the value of items issued from the warehouse during a single year was at least as great as the average value of the inventory and in one case was six times its average value (Table 25.5).

Other aspects of warehouse management include prompt issuance of requested items (Table 25.6), prompt processing of newly received warehouse items, and proper inventory control both for accuracy and for minimum loss or theft (Table 25.7). Getting newly received items on the shelves is important. Sunnyvale, California, expects at least 95% of such items to be available within 2 working days of receipt. San Jose, California, processes most items

(text continues on page 374)

Table 25.2   Prompt Purchase: Performance Targets and Actual Experience of
               Selected Cities

City	Issuance of Purchase Order
Palo Alto, CA	Target: Process purchase requests of less than $500 within 1 workday of receipt Actual: 85% (1997)
Fayetteville, AR	Average processing time when sealed bids are required: 45 days (1998) Average processing time when sealed bids are not required: 2 days (1998)
Oak Ridge, TN	Average time required to process purchase order: 2 days (1991)
Ann Arbor, MI	Percentage of incoming purchase orders reviewed within 2 workdays: 95% (1998) Percentage of bid documents prepared and mailed within 30 days of receipt of purchase request: 95% (1998)
Lubbock, TX	Processing time for requisitions: 3 days (1991), 1 day (1992, projected)
Sterling Heights, MI	Average time from receipt of requisition to issuance of purchase order: 3 days (1995)
Ames, IA	Percentage of requisitions processed within 3 workdays: 98% (1997)
Bellevue, WA	Percentage of requisitions processed within 3 days: 98% (1998)
Shreveport, LA	Percentage of requisitions processed within 3 days: 95% (1994)
Boca Raton, FL	Target 1: Prepare specifications and process requisitions greater than $10,000 for competitive sealed bid within 45 days Target 2: Process requisitions of $2,500–$6,000 within 14 days Actual: 90% (1997) Target 3: Process requisitions of less than $2,500 within 3 days Actual: 90% (1997)
College Station, TX	Percentage of purchase orders processed within 3 days of requisition receipt: 90% (1996, estimated)
Chandler, AZ	Target 1: Routine requests of less than $2,000 completed within 3 workdays Actual: 79.6% (1995), 87% (1998) Target 2: Priority requests for purchases of less than $2,000 completed within 8 work hours Actual: 99% (1995)
Denton, TX	Targets: 3 days for standard requests, 30 days for requests requiring formal bids Actual: 4 days, 48 days (1997)
Hurst, TX	Percentage of purchase orders processed within 4 days: 96% (1997)
Orlando, FL	Percentage of sealed bids awarded within 60 days: 92% (1991) Percentage of nonbid purchase requests awarded within 5 workdays: 100% (1991)
Tempe, AZ	Target: Turnaround time of 90 days or less for purchase orders requiring formal bids Actual: 80.4 days (1997) Target: Turnaround time of 5 days or less for purchase orders not requiring formal bids

Reno, NV	Percentage of formal bids opened within 6 weeks: 96% (1990) Percentage of informal quotes secured within 3 weeks: 97% (1990) Percentage of requisitions processed to purchase orders within 5 days: 93% (1990)
Raleigh, NC	Percentage of purchase orders processed within 5 days of receipt of requisition: 95% (1997), 92% (1998)
Athens–Clarke County, GA	Percentage of purchase orders issued within 5 days: 91.6% (1998)
Long Beach, CA	Target 1: Award at least 92% of formal bids within 90 days Actual: 99% (1991), 90% (1992) Target 2: Award at least 92% of informal bids within 30 days Actual: 92% (1991), 90% (1992) Target 3: Process at least 90% of requisitions within 5 days Actual: 85% (1991), 87% (1992)
San Diego, CA	Percentage of requisitions processed within 5 working days: 87% (1998)
Tallahassee, FL	Percentage of requisitions greater than $500 processed within 5 days: 79.9% (1991)
High Point, NC	Percentage of purchase requests processed within 5 workdays: 50% (1997), 65% (1998) Average turnaround time for processing informal quotes: 3 weeks (1998) Average turnaround time for processing formal quotes: 6 weeks (1998)
St. Petersburg, FL	Average days from requisition to purchase order: 10.5 (1995) Average days from requisition to finalization: 56.2 (1995)
Fort Collins, CO	Elapsed time from requisition to purchase order: 64.6 days when sealed bids (1991); 11 days without sealed bids (1992, estimated)
Irving, TX	Target: Issue at least 96% of purchase orders within 15 days when sealed bid is not required
Portland, OR	Percentage of requisitions to purchase orders within 15 days: 90% (1998)
San Antonio, TX	Average time to process formal bid: 62 days (1997) Average time to process informal bid: 17 days (1997)
Cambridge, MA	Target 1: Except during the final 2 months of fiscal year, confirm and issue purchase orders (a) within 24 hours for purchases made independently by departments; (b) within 48 hours for purchases made against existing city/state contracts Estimate: (a) 90%; (b) 30% (1996) Target 2: Receive informal bid/quotes for evaluation within 3 weeks of receipt of requisition Estimate: 90% (1996)
Lubbock, TX	Days from requisition to contract award (avg.): 29.5 (1994), 26.3 (1995)
Tucson, AZ	Percentage of contracts awarded within 120 days: 73% (1997) Percentage of purchase orders awarded within 28 days: 88% (1997)
Chesapeake, VA	Average time to issue purchase orders: 30 days (1997)

Table 25.3   Prompt Award of Contracts: National Survey of Cities

	Average Time From Receipt of Requisition to Award of Contract				
	51 Calendar Days or More	31–50 Calendar Days	26–30 Calendar Days	20–25 Calendar Days	Less than 20 Calendar Days
Percentage of responding municipalities (N = 138)	5%	16%	15%	18%	46%

SOURCE: *1993 Procurement Practices Survey,* National Institute of Government Purchasing, Inc., Reston, VA. Used by permission. (See National Institute of Government Purchasing, 1994.)

Table 25.4   Adequacy of Warehouse Inventory: Selected Cities

City	Adequacy of Inventory
Charlottesville, VA	Target: Fill at least 95% of orders on demand with available inventory Actual: 98% (1997)
San Diego, CA	Percentage of requested items in stock: 97% (1998)
Tallahassee, FL	Percentage of items in stock when requested: 92.5%, vehicle repair parts; 97%, other (1991)
New York, NY	Stock outs: 4.7% (1991), 3.8% (1992)
Lubbock, TX	Orders filled on demand: 96.1% (1991) Office supplies requested that are in stock: 95% (1991) Required vehicle repair parts not in stock: 5% (1991)
St. Petersburg, FL	Target: Out-of-stock ratio of less than 4%
Sunnyvale, CA	Target: Outage of 4% or less in inventoried items
Tucson, AZ	Target: 96% of requisitioned items filled from stock
Hurst, TX	Target: Inventory stock sufficient to meet 95% of needs Actual: 95% (1990)
Peoria, AZ	Stock outs: 5% (1991)
Oakland, CA	Target: Outage of less than 6% in inventoried items
Orlando, FL	Percentage of requested fleet parts filled from stock: 92% (1991)
Winston-Salem, NC	Percentage of orders filled on demand for items advertised as available: 90% (1992)

within 3 days of receipt and delivers these items to requesting departments within 1 day of processing.

If warehouse bins are empty because newly received items remain unpacked or because inventory records incorrectly report that supplies are still available, the projects of operating departments can be delayed. Some cities report high rates of inventory accuracy. Rockville, Maryland, for example, re-

Table 25.5   Warehouse Inventory Turnover Rate: Selected Cities

City	Inventory Turnover Rate
Plano, TX	Actual: 6 (1997)
Savannah, GA	Target: 3.00 Actual: 3.60 (1991), 3.93 (1992), 3.6 (1997)
Largo, FL	Actual: 3.1, central stores; 3.8, central automotive parts (1991)
Amarillo, TX	Actual: 3.1 (1991)
San Diego, CA	Target: 3.5 Actual: 3.1 (1998)
San Jose, CA	Actual: 3 (1995)
Charlottesville, VA	Target: 2.00Actual: 2.99 (1998)
St. Petersburg, FL	Actual: 2.7 (1995)
Winston-Salem, NC	Actual: 2.01 (1997), 1.95 (1998)
Lubbock, TX	Actual: 1.85 (1994), 2 (1995)
College Station, TX	Actual: 1.8 (1995), 1.13 (1998)
Kansas City, MO	Actual: 1.7 (1997)
Chandler, AZ	Actual: 1.6 (1995), 1.72 (1998)
Palo Alto, CA	Target: 3 Actual: 1 (1997)

ported an error rate of only 0.06% between its system count and the physical inventory. Savannah, Georgia, reported a stock item deviation of only 0.09%.

Warehouse losses can be a serious matter to municipalities and also a concern to auditors. In some cases, such losses are accounting problems—that is, items may have been removed from the warehouse and used in legitimate city projects, but not properly recorded. Such a problem is serious and should be corrected, but it is less serious than the problem of warehouse theft. Inadequate security measures can result in significant losses.

No amount of losses should be considered acceptable, but what level of shrinkage is reasonable? Even with minimal loss due to theft, some bookkeeping errors and failures to properly check out emergency items are virtually inevitable. Oakland, California, has an inventory shrinkage target of less than 2%. Savannah, Georgia's target for warehouse shrinkage is 1.5% of inventory or less.

Once an item has been ordered and received, the purchasing department typically is involved in authorizing payment of the vendor. Prompt authorization by the purchasing department following notification of receipt from the operating department facilitates effective cash management and increases the likelihood of securing prompt-payment discounts offered by some vendors (Table 25.8).

Table 25.6   Prompt Issuance of Requested Items: Selected Cities

In-Stock Items (office supplies, etc.)	*Special Order Items*
Palo Alto, CA Target: Fill 100% of stores requisitions within one workday (1999)	**Orlando, FL** Percentage of special order parts for city fleet that are filled within 24 hours: 97% (1991), 100% (1996)
Winston-Salem, NC Percentage of warehouse requisitions filled within 24 hours: 99% (1998)	**Tallahassee, FL** Percentage of vehicle repair parts available within 48 hours (including warehouse and special order): 95.1% (1991)
Reno, NV Percentage of items requested from central stores that are delivered within 36 hours: 98% (1990)	**Cincinnati, OH** Target 1: Special order items within 5 days Actual: 96% (1993)
Peoria, AZ Target: At least 95% within 2 days	Target 2: Emergency items within 24 hours Actual: 100% (1991)
Chandler, AZ Target: Within 2 workdays Actual: 94.4% (1998)	**Ann Arbor, MI** Percentage of special order office supplies received within 5 days of request: 90% (1998)
Portland, OR Target: At least 67% within 48 hours Actual: 69% (1991)	**Chandler, AZ** Target: Fill requests for noninventory materials within 7 workdays Actual: 84% (1991), 56% (1995)
Sunnyvale, CA Target: Within 2 workdays	
Cincinnati, OH Target: Within 72 hours Actual: 100% (1993)	

Table 25.7 Odds and Ends in Purchasing and Warehousing

PROMPT PREPARATION OF BID SPECIFICATIONS

Kalamazoo, MI
Preparation time for bid documents: 17 days (1997)

Orlando, FL
Percentage prepared and processed within 30 days: 98% (1991)

Ann Arbor, MI
Target: Prepare and mail bid documents within 30 days for 90% of all purchasing requests

WELL-DEVELOPED SPECIFICATIONS

Savannah, GA
Percentage of formal bids repeated due to inadequate specifications: 0.95% (1997)

VENDOR LIST

Corpus Christi, TX
Target: Enter new vendors and changes into the system within 2 working days of notification (1999)

Oklahoma City, OK
Target: Enter vendor registrations into the system within 2 days of receipt
Actual: 95% (1999)

Boca Raton, FL
Target: Add vendor applications into the system within 5 days of receipt

MULTIPLE BIDDERS

Boston, MA
Average number of vendors bidding per contract: 6.37 (1996)

Rockville, MD
Average number of bids received per solicitation exceeding $7,500: 6.0 (1997)

Fort Collins, CO
Qualified bidders per bid: 4.7 (1991)

Albuquerque, NM
Average number of offers per solicitation: 4 (1998)

COMPETITIVE BIDDING RATIO

Decatur, IL
Target: Seek competitive bids for 70% of all purchases (dollar volume)
Actual: 50% (1991)

BID VARIANCE

Rockville, MD
Average difference between low and high bid: 77.3% (1997)

SAVINGS

Charlottesville, VA
Target: Save 10% of costs of items processed through the purchasing department
Actual: 16.9% (1998)

INTEGRITY OF PROCUREMENT PRACTICES

College Station, TX
Percentage of bids that are protested: < 2.5% (1996, estimated)

(continued)

Table 25.7   Odds and Ends in Purchasing and Warehousing (continued)

Tucson, AZ
Vendor protests sustained by hearing officer, as a percentage of all contracts: 0% (1997)

PROMPT PROCESSING OF MATERIALS
RECEIVED INTO WAREHOUSE INVENTORY

Sunnyvale, CA
Target: At least 95% available within 2 working days of receipt (1994)

San Jose, CA
98% processed within 3 days of receipt; 99% delivered to departments within 1 day of processing (1995)

MINIMIZE ERRORS IN ISSUING INVENTORY ITEMS

Savannah, GA
Target: Limit recording errors to 0.25% of lines issued or less
Actual: 0.18% (1990), 0.20% (1992), 0.21% (1997)

ACCURATE INVENTORY

Rockville, MD
Error between inventory physical and system counts: 0.06% (1997)

Savannah, GA
Deviation: 0.09% (1997)

Palo Alto, CA
Target: Annual inventory within 3% of accounting record
Actual: 0.16% (1997)

INVENTORY SHRINKAGE

Savannah, GA
Target: Limit warehouse losses to 1.5% of inventory value
Actual: 0.14% (1992), 0.10% (1997)

Oakland, CA
Target: Less than 2% inventory shrinkage

Table 25.8   Prompt Authorization for Payment: Selected Cities

City	Prompt Authorization for Payment of Invoices
Lubbock, TX	Percentage of payment authorizations processed in one day: 98% (1991)
Oak Ridge, TN	Target: Process at least 95% of payment authorizations within one day after receipt of invoice Actual: 75%
Milwaukee, WI	Average number of days until payment authorization: 7.8 (1991)
Shreveport, LA	Percentage of invoices processed within 8 days: 90% (1991)
Savannah, GA	Target: Take advantage of all prompt-payment discounts (1999)
Boston, MA	Percentage of invoices processed within 20 days: 57% (1992)

26

Risk Management[1]

Even a cautious local government stands vulnerable to losses arising from accidents, illnesses, and acts of nature. Add to these the possibility of negligence or misconduct by government officials in today's litigious environment and the magnitude of this vulnerability can soar. Escaping all vulnerability is impractical, if not impossible, but prudent steps can be taken to minimize a local government's risk. These steps include efforts to spot any practices or conditions that increase the likelihood of loss, to influence improvements wherever practical, and to minimize the financial ramifications when losses do occur.

Aggressive risk management programs examine local government workplaces to be sure they are safe, promote safe practices in the operation of government vehicles and equipment, encourage employee participation in practices that enhance wellness, and endorse operating procedures that reduce vulnerability to liability claims. They make prudent insurance decisions, keeping loss exposure as well as premiums in mind as they seek to minimize the "cost of risk."[2]

Outstanding risk managers see that their own operations function efficiently, investigating accidents promptly and thoroughly, handling claims quickly, and keeping their coverages up to date. Furthermore, they aggressively protect the local government's resources, defending against inappropriate claims and collecting for damage to local government property from responsible parties or their insurers whenever possible.

Seeking Benchmarks

Energetic risk managers typically can recite the array of activities that consume their time and that of their coworkers. What they often *do not know* is whether the service expectations they hold and the results they achieve compare favorably with those of their counterparts in other local government risk management programs. In most cases, they have few benchmarks against which to assess their own performance expectations and results.

When it comes to reporting performance, most risk management programs in local government rely on workload measures, also known as output measures. They tabulate the numbers of incidents reported, claims processed, meetings conducted, buildings insured, inspections performed, and reports prepared. Raw numbers like these are of little value to another city or county seeking an external peg for judging its own risk management operation. Of greater value as benchmarks are measures of efficiency, service quality, and effectiveness—performance characteristics more relevant for comparison purposes. Although in shorter supply, measures of efficiency, quality, and effectiveness can be found. Many such measures are reported in this chapter.

When more local governments begin to report their performance using higher-order measures rather than raw workload counts, a clearer picture will emerge regarding what constitutes truly outstanding performance in risk management. Until that time, the cities and counties now reporting the efficiency, quality, and effectiveness of their programs at least provide an outline of the state of the art, as well as a target for others to achieve and exceed. Consider this merely the first round of risk management benchmarks.

Performance Benchmarks

Although some cities report unit costs for various risk management activities, using such figures as benchmarks is often problematic. Elements included as a component of unit costs in one local government (e.g., overhead and various indirect costs) are sometimes excluded from unit cost calculations in others. Furthermore, costs per claim might be defined as "processing costs only" in one jurisdiction, but include settlement costs in another. Such variations in cost-accounting rules and performance measurement definitions help explain wide swings ranging, for instance, from $313 per general liability claim in one city in this review of performance documents to $1,500 in another. Some of the interjurisdictional variation may be reduced by taking broader cuts at standardized costs—focusing not on the cost of an individual unit of service but instead on an entire category or on risk management as a whole. Norman, Oklahoma, for example, attempts to keep workers' compensation costs below 5% of payroll costs and reported achieving a 2.5% mark in 1998. Although not entirely unencumbered by cost-accounting considerations, these figures provide more useful targets for others.

The safest guideline regarding the use of unit costs as benchmarks is simple: *unit costs should be used as benchmarks only after ensuring consistency of cost-accounting rules and performance measurement definitions.* Because the task of ensuring consistency requires the matching of rules and definitions across jurisdictions, it is sometimes deemed impractical.

Table 26.1   Asset Inventory and Coverage

Accuracy of Inventory	*Timely Renewal*
Fairfax County, VA Target: Variance of 10% or less between insurable assets reported to insurer and insurable assets reported by agencies during inventory Actual: 10% (1998)	**Decatur, IL** Average number of days insurance renewals and enrollments are completed before deadlines: 8 days (1991) **Fairfax County, VA** Target: At least 90% of assets inventoried and insured within 10 days of policy renewal (2000) Actual: 88% (1998)

Many higher-order measures are less vulnerable than unit costs to different cost-accounting rules and, therefore, are more easily used as benchmarks. These include measures that reflect prompt service, quick turnaround times, accuracy, service quality, and effectiveness. Examples featured in the various tables of this chapter provide benchmarks for other risk management programs to strive to achieve or surpass.

Asset Inventory and Insurance Coverage

Good risk management programs have a firm grasp of their local government's coverage needs. Asset inventories in these communities are thorough and values are accurate. Although some variance is likely to exist between agency inventories and insurance listings, the disparity should be small. Fairfax County, for example, attempts to keep its variance below 10% (Table 26.1). Good programs also are diligent in meeting insurance renewal deadlines and keeping coverage listings up to date.

Responsiveness

Being responsive has many connotations in risk management. In Ann Arbor, Michigan, and Lubbock, Texas, local performance expectations include responsiveness to prospective claimants (Table 26.2). The standard in Lubbock is to answer at least 85% of all claim questions within 2 days. Ann Arbor attempts to respond to prospective claimants by providing filing instructions within 24 hours of an inquiry.

Fairfax County, Virginia, expects to respond promptly to risk assessment requests and to complete its reports within 10 days of an inspection. Corpus

Table 26.2   Responsiveness in Risk Management

Responsiveness to Inquiries	Risk Assessment Requests	Prompt Resolution of Risk Management Issues
Ann Arbor, MI Target: Distribute instructions for filing a claim: 95% of prospective claimants within 24 hours of inquiry	**Fairfax County, VA** Percentage of risk assessments conducted within 3 days of request: 65% (1998) Percentage of reports disseminated within 10 days of inspection: 86% (1998)	**Palo Alto, CA** Percentage of loss control and insurance issues resolved within 24 hours: 85% (1997)
Lubbock, TX Target: 85% of claim questions answered within 2 days Actual: 85% (1998)		**Corpus Christi, TX** Target: Respond to risk management issues within 72 hours

Christi, Texas, and Palo Alto, California, emphasize prompt response and resolution of risk management issues.

Claims Processing

Risk management programs report claims-processing proficiency in a variety of ways. Although some isolate particular types of claims (e.g., health
claims or workers' compensation claims) and report the percentage processed
within a targeted time frame, others report targets and performance statistics
for claims in general (Table 26.3). Ann Arbor, Michigan, attempts to process
at least 95% of all citizen claims within 48 hours of receipt. In Asheville,
North Carolina, and Danville, Kentucky, the targets are 72 and 96 hours, respectively. Charlottesville, Virginia, reports that 30% of its claims were settled
within 1 week, 95% within 1 month, and 100% within 6 months in 1998.

Winston-Salem, North Carolina, attempts to provide health claim benefits
within 10 days of notification. The average turnaround in San Antonio, Texas,
is 5 days. Danville, Kentucky, processed 100% of its workers' compensation
claims within 24 hours in 1997.

Some programs report claim closure ratios, emphasizing the importance of
closing enough cases to offset the number of new cases entering the system. In
Santa Ana, California, closed claims exceeded new claims by 33% in 1995.

Vehicular Accidents

Many risk management programs focus part of their energy on the reduction of accidents involving local government drivers. Scottsdale, Arizona, tries
to limit vehicle liability accidents to no more than 10 per million miles driven

Table 26.3 Claims Processing

GENERAL

Ann Arbor, MI
Target: Process 95% of all citizen claims within 48 hours of receipt

Asheville, NC
Target: 75% of liability claims closed within 3 days (2000)

Danville, KY
Percentage of liability claims processed within 96 hours: 100% (1997)

Charlottesville, VA
Percentage of claims settled within 1 week: 30%; within 1 month: 95%; within 6 months: 100% (1998)

Fairfax County, VA
Percentage of claims processed within 30 days: 85% (1998)
Average claims processing time: 22 days (1998)

Washoe County, NV
Average processing time for general and auto liability claims: 45 days (1998)

San Diego, CA
Percentage of claims investigated within 60 days: 80% (1998)

Shreveport, LA
Target: Close at least 65% of all claims within 90 days

San Antonio, TX
Percentage of auto liability claims closed within 12 months: 55% (1995)
Percentage of general liability claims closed within 12 months: 72% (1995)
Percentage of tort claims closed within 12 months: 88% (1997)

HEALTH

San Antonio, TX
Average turnaround time for claims paid, pended, or denied: 5 days (1997)

Winston-Salem, NC
Percentage of claims for which benefits were provided within 10 days of notification: 97% (1998)

Lubbock, TX
Percentage of health claims processed within 14 days: 75% (1998)

Jacksonville, FL
Percentage of medical payments processed within 45 days: 99.9% (1994)

WORKERS' COMPENSATION

Danville, KY
Percentage of workers' compensation claims processed within 24 hours: 100% (1997)

Ann Arbor, MI
Target: Distribute 95% of workers' compensation accident reports to claims service within 24 hours

Hurst, TX
Percentage of workers' compensation claims filed within 3 days of receiving injury report: 98% (1997)

(continued)

Table 26.3   Claims Processing (continued)

Corpus Christi, TX
Target: Process workers' compensation claims within 3 days (1999)

Cincinnati, OH
Percentage of workers' compensation and injury-with-pay claims reviewed within 5 days of receipt:
    100%

San Clemente, CA
Target: Review and process at least 95% of all workers' compensation claims within 5 days
Actual: 100% (1998)

Knoxville, TN
Average turnaround for processing workers' compensation claims: 7 days (1991)

Jacksonville, FL
Percentage of indemnity payments processed within 14 days: 100% (1994)

CLOSURE RATIO

Santa Ana, CA
Closed claim–to–new claim ratio: 133% (1995)

San Diego, CA
Target: Maintain a closed claim–to–new claim ratio of 1.00 or greater
Actual: 0.99 (general); 0.84 (workers' compensation) (1998)

Table 26.4   Vehicle Accident Rate per Million Miles

Jurisdiction	Vehicle Accidents per Million Miles
San Antonio, TX	16.5 (1997)
Scottsdale, AZ	Target: Limit vehicle liability accidents to no more than 10 per million miles and vehicle physical damage accidents to no more than 27 per million miles Actual: 6.43 vehicle liability accidents and 24.98 vehicle physical damage accidents per million miles (1999)
Montgomery County, MD	29.5 (1998)
Flagstaff, AZ	Public works accident rate per million miles: 45.5 (1 per 22,000 miles) (1997)

and vehicle physical-damage accidents to no more than 27 per million miles (Table 26.4). Scottsdale was successful on both counts in 1999. San Antonio, Texas, reported a vehicle accident rate in 1997 of only 16.5 accidents per million miles.

Some local governments choose to isolate the accident rates of different categories of vehicles, recognizing that the accident experience of some categories will be higher than others because of the nature of the work being performed. Public works vehicles, for example, are likely to be involved in more accidents per million miles, as confirmed by the statistics of Flagstaff, Arizona.

Occupational Injuries and Illnesses

Occasionally, broad-based statistics relevant to a particular aspect of local government operations are available for use as benchmarks. Such is the case for programs attempting to curb the incidence of occupational injuries and illnesses. The U.S. Bureau of Labor Statistics regularly compiles data on rates of occupational injuries and illness among local government workers in the states that report these statistics (Table 26.5). Median rates in 1997 were 7.8 incidents and 2.5 lost-time incidents per 100 local government workers. Local governments with lower incidence rates deserve acclaim for their accomplishment.

Program Effectiveness

Benchmarks of effectiveness focus squarely on program results. Although often related to program effectiveness, the volume of activity and the pace of the risk management staff are not the issues here. The question instead is whether risk management efforts—from inspections to instruction to claims defense—are achieving favorable results.

The objective of a safety inspection program is not merely to conduct numerous inspections. The objective is to identify hazards and get them corrected. In 1997, Albuquerque, New Mexico, recorded an impressive 98% success rate in this regard by correcting 483 of the 493 hazards identified through its inspections (Table 26.6).

Among performance reports examined for this volume, many of the local governments that reported accident and injury statistics incurred higher rates than the median experience reported by the Bureau of Labor Statistics in Table 26.5. However, two cities that were competitive with leaders in the field were Fort Worth, Texas, and Durham, North Carolina.

The cities of Scottsdale, Arizona, and Savannah, Georgia, report claims and losses relative to total operating expenditures and property value. Scottsdale, for instance, strives to incur no more than one general liability claim for every $1 million of city operating expenditures.

San Diego, California, focuses on thorough review of workers' compensation medical expenses and judges program effectiveness in part by the savings achieved. In 1998, such reviews accounted for savings of 25%.

Table 26.5   Occupational Injury and Illness Rates for Local
Government, 1997

State	Incidence Rate per 100 Workers	
	All Incidents	Incidents With Days Away From Work
North Carolina	5.8	1.6
Iowa	6.1	1.8
Alaska	6.7	2.9
Maine	6.8	2.1
South Carolina	6.8	2.0
Vermont	6.8	2.5
Tennessee	6.9	2.6
Oregon	7.0	2.1
Utah	7.1	2.2
Indiana	7.2	1.8
Kentucky	7.2	2.7
Minnesota	7.5	2.1
Virginia	7.8	2.5
New Mexico	8.4	2.2
Arizona	8.5	1.8
Maryland	9.0	3.4
Michigan	9.7	3.0
Washington	9.7	3.3
Nevada	9.8	2.3
Wisconsin	9.8	3.1
California	10.0	3.4
New York	11.1	6.1
New Jersey	12.4	4.9
Connecticut	13.6	4.6
Hawaii	15.7	7.7
Median	7.8	2.5

SOURCE: U.S. Department of Labor, Bureau of Labor Statistics, "Incidence Rates of Nonfatal Occupational Injuries and Illnesses by Industry and Selected Case Types," Survey of Occupational Injuries and Illnesses in cooperation with participating State agencies. Posted at www.bls.gov/oshstate.htm.
Note: Local government rates were reported only by the 25 states listed here. Incidence rate is calculated as incidents per 200,000 work hours.

Successful defense against claims is another sign of program effectiveness, one that can take several forms. Some jurisdictions report favorable outcomes (92% for Eugene, Oregon) or favorable settlements (90% for San Diego), whereas others report claims closed without payment (70% for Fort Collins,

---

Table 26.6   Effectiveness of Risk Management Programs

HAZARDS CORRECTED

Albuquerque, NM
Percentage of identified hazards and deficiencies corrected within 60 days: 98% (1997)

Fairfax County, VA
Percentage of identified deficiencies corrected as a result of safety inspections: 76% (1998)

ACCIDENT/INJURY RATE

Fort Worth, TX
Injury Incident Rate per 100 employees: 5.06 (1997)

Durham, NC
Rate of disabling injuries per 200,000 hours:[a] 6.3 (1998)

Scottsdale, AZ
Target: Fewer than 50 workers' compensation accidents per million hours worked
Actual: 45.69 (1999)

San Antonio, TX
Personal injuries per 100 employees: 23.9 (1997)
Lost-time injuries per 100 employees: 10.0 (1997)

Montgomery County, MD
Injury incidents per 100 employees: 11.1 (1998)

Cary, NC
Workers' compensation claims per 100 employees: 13.4 (1999)

Flagstaff, AZ
Public works employee time lost due to accidents: 0.16% (1997)

Phoenix, AZ
Job accidents per 100 employees: 19 (1993)

San Jose, CA
Target: Not more than 27 claims per 100 employees (1997)

CLAIMS/LOSSES

Scottsdale, AZ
Target: 1.0 or fewer general liability claims per $1 million of city operating expenditures
Actual: 0.49 (1999)

Target: Less than $6,000 general liability losses per $1 million of city operating expenditures
Actual: $4,251 (1999)

Savannah, GA
Target: Property loss of 0.0020% or less of property value
Actual: 0.0007% (1997)

BILL REVIEW

San Diego, CA
Percentage of workers' compensation medical expenses saved due to bill review: 25% (1998)

CLAIM DEFENSE

Eugene, OR
Percentage of legal/formal resolutions producing favorable outcome for local government: 92%
(1997)

(continued)

Table 26.6   Effectiveness of Risk Management Programs (continued)

San Diego, CA
 Percentage of cases seeking monetary damages settled favorably to the city: 90% (1998)

Denver, CO
 Percentage of claims cases in which the city prevailed: 85% (1998)

Bellevue, WA
 Target: At least 75% of contested claims successfully defended
 Actual: 70% (1998)

Fort Collins, CO
 Percentage of liability claims closed without payment: 70% (1998)

Reno, NV
 Target: At least 70% of general liability claims closed with no payment (1999)
 Target: Not more than 10% of general liability claims resulting in lawsuit (1999)

Albuquerque, NM
 Target: 96% of claims closed without litigation (1999)
 Actual: 97.9%, workers' compensation; 96%, others (1998)

College Station, TX
 Percentage of claims filed against the city resolved without litigation: 99% (1998)

RETURN ON INVESTMENT

Decatur, IL
 Target: Maintain annualized self-insurance program savings of at least 175% of Risk Management
  Division costs
 Actual: 251% (1995)

a. 200,000 hours approximates the work years of 100 employees.

Colorado) or claims closed without litigation (97.9% for workers' compensation and 96% for all others in Albuquerque, New Mexico).

At least one local government attempts to summarize its program's effectiveness by documenting "return on investment." Decatur, Illinois, compared the costs of the Risk Management Division in 1995 to the annualized savings of that city's self-insurance program and declared a 151% return—in other words, savings equal to 251% of program expenditures.

Identifying and Collecting From
Responsible Parties or Their Insurers

The city loses when it is at fault in an accident. It also loses, even when it is not at fault, if it fails to collect from the responsible party or that party's insurer. Reno, Nevada, strives to identify the responsible party and initiate collection proceedings in at least 85% of all incidents of city property damage (Table 26.7). College Station, Texas, subrogated 100% of all third-party losses

Table 26.7   Identifying and Collecting From Responsible Parties or Their Insurers

IDENTIFYING RESPONSIBLE PARTIES

Reno, NV

Target: Identity responsible party (for damage to city property) and initiate collection proceedings in at least 85% of all incidents (1999)

RECOVERIES

Alexandria, VA

Incidents in which city losses were recovered from third parties, as a percentage of all incidents of city property damage caused by third parties: 81.8% (1994), 100.0% (1995), 100.0% (1996), 70.6% (1997), 68.3% (1998)

College Station, TX

Percentage of third party losses subrogated:[a] 100% (1995), 100% (1998)
Percentage of funds recovered: 81.15% (1995), 57% (1998)

Bellevue, WA

Percentage of self-insurance losses recovered: 79% (1997), 69% (1998)

Chandler, AZ

Target: At least 80% collection on recoverable claims
Actual: 77% (1991)

San Antonio, TX

Target: Subrogation[a] amount or value recovered of at least 67% of total value (1999)
Actual: 73% (1995), 52% (1997)

Washoe County, NV

Success rate on subrogation claims:[a] 60% (1998)

---

a. *Subrogation* is the substitution of one creditor for another, as when collection is made from the responsible party's insurer.

in 1995 and recovered 81% of those losses. Alexandria, Virginia, recovered more than two thirds of all third-party losses during each year of a 5-year span from 1994 through 1998, capturing 100% during two of those years.

Cost of Risk

Among risk managers, the phrase "cost of risk" is taken to mean the combined costs of the risk management program, other expenditures in support of risk management, and insurance premiums, plus all losses incurred by the local government. Some risk management programs set targets that attempt to minimize the cost of risk relative to the overall local government budget (Table 26.8). The cities of Scottsdale and Chandler, Arizona, for instance, have been successful in their attempts to hold the cost of risk below 2% of the local government budget. San Antonio, Texas, has an even more ambitious

Table 26.8   Cost of Risk[a]

COST OF RISK AS A PERCENTAGE
OF THE LOCAL GOVERNMENT BUDGET

San Antonio, TX
  Target: 1.00% or less (1999)
  Actual: 0.92% (1997)

Scottsdale, AZ
  Target: Risk management budget that equals 2% or less of the local government's
    operating budget
  Actual: 1.04% (1999)

Chandler, AZ
  Target: 2% or less
  Actual: 1.5% (1998)

a. A community's *total* cost of risk includes insurance premiums, risk management program costs, and other expenditures in support of risk management, as well as losses incurred in accidents, acts of nature, lawsuits, and other incidents.

target of 1% or less. Although vulnerable to some of the same issues that plague unit cost comparisons, these problems are reduced for this measure by the more broadly accepted dictum that "cost of risk" includes program costs, premiums, and losses, and by the fact that the denominator is the local government's entire budget.

Initial Volley

How good are the best local government risk management programs? Unfortunately, we cannot answer with absolute certainty. The best programs may very well include the ones featured in this chapter, but some outstanding programs undoubtedly were missed. This search for benchmarks did not include all local governments, but focused instead on a carefully selected portion of them. The performance reports examined were those provided by a select group of local government risk managers regarded by peers and experts as leaders in the field,[3] supplemented by performance reports from other jurisdictions noted for their use of performance measures. Even among this select group, the search for benchmarks was hampered by the persistent tendency of most risk management programs—including many that perhaps are excellent—to report only raw workload or output numbers rather than measures of efficiency, service quality, and effectiveness. In the benchmarking arena, these jurisdictions remain on the sidelines.

So, how good are the best local government risk management programs? To be the best, a program has to be at least as good as the ones featured in this chapter. Because these programs were willing to provide performance data that revealed their performance targets and accomplishments, they have fired the first volley in the campaign to discover how good *the best of the best* really are. Now we await the second volley from local governments that are prepared to raise the performance bar.

Notes

1. This chapter was adapted from David N. Ammons, *Performance Benchmarks in Local Government Risk Management* (Fairfax, VA: Public Entity Risk Institute, 2000). Adapted by permission.

2. A local government's cost of risk includes insurance premiums, risk management program costs, and other expenditures in support of risk management, as well as losses incurred in accidents, acts of nature, lawsuits, and other incidents.

3. David N. Ammons, "Taking the Lead in Risk Management," *American City & County, 115* (July 2000), pp. 46–50.

27

Social Services

For many years, some cities have been direct producers of social services that many of their counterpart municipalities left to state and county governments, nonprofit organizations, and others. Times have changed and so has the municipal role in the provision of social services in many of these communities. For some, the old role has expanded; for others, a new role has emerged.

Benchmarks

The range of social services offered by municipalities varies from one community to another. A few categories of service, however, are fairly common. The performance targets and results reported for these services by different cities offer potential benchmarks for others.

Employment Assistance

Many cities have established programs designed to provide job training and employment assistance to their citizens (Table 27.1). Some are directed toward adults and others toward youth. Several cities examined for this volume report impressive results from their programs. Alexandria, Virginia, for example, reported that 97% of its adult job-training clients one recent year were placed in jobs and were still employed 52 weeks later.

To assess local performance in employment assistance, Cincinnati, Ohio, compares its statistics on key indicators with statewide averages (Table 27.2). The city program beat the statewide average in every reported category.

Table 27.1   Employment Assistance: Selected Cities

Alexandria, VA
  Percentage of adult job training clients who entered employment: 77% (1995), 96% (1998)
  Percentage of adult clients who were employed 13 weeks after terminating job training: 67.9% (1995)
  Percentage of adult job training clients placed in jobs and still employed 13 weeks later: 90% (1995),
    95% (1998); 26 weeks later: 90% (1995), 96% (1998); 52 weeks later: 89.5% (1995), 97% (1998)

San Antonio, TX
  Percentage of clients who increased skills or income: 85% (1997)

Winston-Salem, NC
  Percentage of welfare recipients completing JTPA[a] program, placed and still working after 13 weeks
    of employment: 83% (1998)
  Percentage of Youth Service Corps earning general equivalency diploma (GED): 23% (1998)
  Percentage of Youth Service Corps entering unsubsidized employment: 43% (1998)

Greensboro, NC
  Percentage of adult clients finding and retaining unsubsidized employment for 90 days or more:
    72% (1997)
  Percentage of dislocated workers entering unsubsidized employment: 80% (1997)

Portland, OR
  Percentage of youth participants placed in jobs: 78% (1998)

Oklahoma City, OK
  Percentage of youth participants (16–21) employed after training: 75%
  Percentage of adult participants over 21 employed after 13 weeks: 77%
  Percentage of welfare participants over 21 employed after 13 weeks: 82%
  Percentage of dislocated workers over 21 employed at program completion: 78%
  Percentage of older workers employed: 63% (1999)

Durham, NC
  Percentage of adult enrollees placed and retained in jobs after 13 weeks: 70% (1998)

Santa Ana, CA
  Percentage of youth participants entering employment: 50% (1995)
  Percentage of adult participants entering employment: 70% (1995)

New York, NY
  Adult participants working at 90 days: 47.0% (1996), 46.3% (1998)
  Youth participants placed remaining on the job at 30 days: 60% (1996), 48.3% (1998)

Raleigh, NC
  Percentage of youths applying for job programs who are placed in jobs: 43% (1998)

a. JTPA stands for the Job Training Partnership Act.

Child and Youth Services

Child and youth services range from protective services for victims of child
abuse to foster care, adoption services, after-school programs, and programs
designed to divert young people from the criminal justice system. Relatively

Table 27.2   Comparing Local and Statewide Employment Assistance Statistics:
Cincinnati

	City of Cincinnati	State of Ohio
Adult Standard		
Percentage employed after 90 days	60.1%	47.38%
Weekly earnings after 90 days	$280.00	$214.98
Welfare recipients employed after 90 days	56.1%	44.3%
Welfare recipients weekly earnings after 90 days	$265.00	$205.51
Youth Standard		
Employed at termination	54.46%	42.9%
Employability enhancements (e.g., high school diploma, GED)	56.12%	30.37%

SOURCE: City of Cincinnati (OH), *Measures of Success: 1995* (July 1996), p. 17.

few of the cities examined for this volume report effectiveness statistics for
their child and youth service programs, but those statistics that are reported
offer potential benchmarks for others to meet or exceed (Table 27.3)

Housing Assistance

Housing assistance services offered by municipalities come in different
forms. Many cities operate, or support financially, shelters for homeless individuals. Some cities provide *family* housing assistance with the objectives of
meeting short-term needs and helping families move to permanent housing
(Table 27.4). Some also have programs that investigate fair housing complaints or that attempt to resolve tenant-landlord disputes (Table 27.5).

Miscellaneous Social Services

A variety of other programs offer social services of one kind or another.
Table 27.6 lists performance statistics for several of these programs. Social service caseloads reported by selected cities are listed in Table 27.7.

Table 27.3   Social Services for Children and Youth: Selected Cities

PROMPT RESPONSE: CHILD PROTECTIVE SERVICES (CPS)

New York, NY
 Percentage of responses within 1 day following report to the state central register: 99.1% (1996), 98.6% (1997), 97.0% (1998)

Alexandria, VA
 Percentage of child abuse or neglect cases in which investigation was initiated within 24 hours of receipt of report: 72% (1995), 79% (1998)
 Percentage of cases without subsequent CPS complaints during the month: 95% (1995), 96% (1998)

FOSTER CARE

Alexandria, VA
 Percentage of foster care children who were stabilized after 18 months in foster care: 75% (1994), 41% (1995), 61% (1998)

New York, NY
 Average length of foster care: 4.42 years (1996), 4.28 years (1997)

ADOPTION

Alexandria, VA
 Percentage of children who were adopted or who achieved an alternate goal: 75% (1995), 66% (1998)
 Average number of months that children placed were waiting for adoption: 10.5 months (1995), 12.2 months (1998)

New York, NY
 Average length of time to complete adoptions: 3.3 years (1996), 3.4 years (1997)

JUVENILE AFTERCARE

Bellevue, WA
 Percentage of youth clients showing improvement in academic achievement and school behavior: 77% (1997)

New York, NY
 Percentage who improved in school attendance by program completion: 44% (1997)

Raleigh, NC
 Percentage of youth in tutorial program with improved grades: 43% (1997), 36% (1998)

ACADEMIC BOOST

White Plains, NY
 Percentage of day campers with pre- to posttest improvement in reading and math: 100% (87 of 87) (1995)

Alexandria, VA
 Percentage of at-risk high school program participants enrolling in college upon graduation: 85% (1995), 88% (1998)

YOUTH DIVERSION FROM JUSTICE SYSTEM

San Antonio, TX
 Percentage of youth diverted from juvenile justice system: 95% (1995)
 Recidivism rate for youth served: 3% (1997)

Table 27.4 Family Housing Assistance: Selected Cities

Raleigh, NC
Percentage of persons transitioned out of incentive shelter: 86% (1997), 8% (1998)

Boston, MA
Percentage of clients placed in housing or on waiting lists: 75% (1996)

San Antonio, TX
Percentage of Dwyer Center families that move to permanent housing: 20% (1995), 46% (1997)
Average number of days to locate permanent housing: 100 days (1997)
Percentage of clients losing housing and returning to emergency shelters: 10% or less (target)

Alexandria, VA
Target: 30 days or less average length of stay for families
Actual: 36.1 days (1995), 43 days (1998)
Percentage of families discharged from the shelter who obtained stable housing: 35.1% (1995), 35% (1998)

New York, NY
Average days in temporary housing (all families): 223 (1996), 265 (1997), 306 (1998)

Table 27.5 Prompt Response to Fair Housing and Tenant-Landlord Complaints: Selected Cities

San Antonio, TX
Initial response time: 3 days (1995)

Greensboro, NC
Percentage of fair housing complaints investigated within 100 days: 63% (1997)

Kansas City, MO
Average days to complete fair housing investigation: 123 days (1995)

Table 27.6 Social Services Odds and Ends: Selected Cities

CASE PROCESSING TIME

Orlando, FL
Average processing time for EEO cases: 188 days (1994)
Average processing time for fair housing cases: 158 days (1994)

Alexandria, VA
Percentage of food stamp applications for severe hardship cases processed within 5 days: 99.5% (1995), 98% (1998)
Percentage of routine food stamp applications processed within 30 days: 98.4% (1995), 96% (1998)
Percentage of family assistance applications processed within 45 days: 72.9% (1995), 66% (1998)

Boston, MA
Percentage of disability discrimination cases resolved within 30 days: 36% (1996)

MEDICAID

Chesapeake, VA
Percentage of Medicaid applications processed within 45 days: 97.3% (1997)

INCOME SUPPORT PROGRAMS

New York, NY
Payment error rate (ADC only): 5.89% (1996), 7.25% (1997)

ATTENDANT CARE EFFICIENCY

San Antonio, TX
Hourly cost as a percentage of average private agency charge: 46% (1995), 48% (1997)

SENIOR CITIZENS

Boston, MA
Percentage of senior citizen requests for medical trips fulfilled: 98% (1996)

PROMPT RESPONSE TO ABUSE OF THE ELDERLY CASES

Alexandria, VA
Percentage of new cases investigated within 5 days: 99% (1995), 99% (1998)

HOMELESS SHELTER

Raleigh, NC
Percentage of persons seeking shelter who were housed (winter operation): 100% (1997), 100% (1998)

Boston, MA
Percentage of homeless population accessing shelters: 55.5% (1996)

LITERACY PROGRAMS

San Antonio, TX
Percentage of students with increased level of literacy: 95% (1995)

LEVERAGING GRANTS AND PRIVATE FUNDS

San Antonio, TX
Grant funds received per $1 city match: $19.03 (1995), $11.87 (1997)

Table 27.7   Social Service Caseloads: Selected Cities

San Antonio, TX
Average monthly cases per youth worker: 135 (1995), 42 (1997)

New York, NY
Average child welfare protective worker caseload: 24.1 (1996), 16.8 (1997), 13.7 (1998)
Average field office family service worker caseload: 13.5 (1996), 12.2 (1997), 12.7 (1998)

Orlando, FL
Annual caseload per EEO investigator: 36 (1994)
Annual caseload per community/housing assistance investigator: 15 (1994)

28

Solid Waste Collection

In most American communities, the municipal role in the removal of household refuse has grown considerably since the 18th and early 19th centuries, when that task, according to one authority, was left to "unreliable contractors, scavengers, and the pigs" (Yates, 1977, p. 70). Although refuse collection is a popular candidate for intergovernmental and private contracting, even cities choosing one of those options remain heavily involved. They still specify service scope and quality, monitor contract compliance, and retain ultimate responsibility for services.[1] In either of the two predominant modes of municipal involvement—direct service delivery or contracting—the municipality has need for a system of performance monitoring.

Performance Measures

Recommended performance measures for solid waste collection emphasize the amount of refuse collected, the efficiency with which it is collected (e.g., unit costs), collection reliability, community cleanliness, and citizen satisfaction (Rubin, 1991). Measures recommended for the related functions of solid waste recycling and disposal typically address the percentage of households participating in a recycling program, the amount of waste recycled (either as a percentage of all solid waste or as an average quantity per household), unit costs, net recycling income, tons of waste disposed in landfills or by other means, cubic yards of landfill used, unit costs for disposal, and the degree of local compliance with state and federal environmental standards governing solid waste disposal.

Refuse Collection Benchmarks

Understandably, officials in communities where municipal crews collect the garbage often are interested in production ratios for sanitation employees.

398

Table 28.1 Production Ratios for Sanitation Employees: Selected Cities

STOPS PER COLLECTION EMPLOYEE

Scottsdale, AZ
Target: 2,600 residences serviced per employee per week
Actual: 2,450 (1999)

Shreveport, LA
1,145 stops per week per employee (1994)

Long Beach, CA
Stops per worker hour: 28.2 (1994), 29.3 (1995), 36 (1996)

Corpus Christi, TX
58,777 stops per year per employee (1992)

LABOR RATIOS

College Station, TX
Labor hours per ton of household garbage: 0.64 (1995), 0.60 (1998)

Victoria, TX
Tons of solid waste per worker hour: 1.51 (1996)
Tons of recyclables per worker hour: 0.12 (1996)

Peoria, AZ
230.45 tons per month per operator (1991)

Long Beach, CA
Tons per worker hour: 0.7 (1994), 0.8 (1995), 0.8 (1996)

College Park, MD
Target: 1,000 pounds per labor hour

Dunedin, FL
845 pounds per labor hour (1992)

Charlotte, NC
Tonnage per sanitation staff member: 625 (1993)

Chapel Hill, NC
Pounds of residential refuse per collector per week: 18,503 (1995), 18,417 (1996), 19,334 (1999)

They want assurances that their own crews are performing a reasonable amount of work at an appropriate pace. Although some guidance in that regard may be provided by production ratios reported by other municipalities (Tables 28.1 and 28.2), such ratios should not be applied indiscriminately. Different types of refuse collection equipment, modes of operation, work rules, collection frequency (e.g., once a week or twice a week), collection location (e.g., curbside, alley, or backdoor), terrain, distance between stops, and even the prevalence of on-street parking can affect production ratios. A municipal official whose local production ratio is deemed unacceptable relative to others should consider those factors and possible changes in mode of operation, as well as modifications in equipment size and type, before leaping to conclusions regarding the diligence of refuse collection employees.

Table 28.2   Tons Collected per FTE Employee: Statistics From North and South
Carolina Comparative Performance Projects, 1998

	Average Tons of Residential Solid Waste Collected per Year per FTE	Pacesetter
7 North Carolina cities of 63,000–260,000 population	886.8 tons	Greensboro (2,385.2 tons)
12 North Carolina cities of 2,000–62,000 population	847.9 tons	Gastonia (1,539.0 tons)
11 South Carolina cities of 10,000–49,000 population	1,123.2 tons	North Augusta (4,737.0 tons)

SOURCES: *Interim Report on Services for Seven Cities* (Chapel Hill, NC: Institute of Government/University of North Carolina, 1998), pp. 6–21; *Performance and Cost Data: Phase III City Services* (Chapel Hill, NC: Institute of Government/University of North Carolina, 1999), pp. 10–37; *South Carolina Municipal Benchmarking Project: 1998* (Columbia, SC: Institute of Public Affairs/University of South Carolina, 1999), pp. 247–270. Reprinted with permission of the Institute of Governent, The University of North Carolina at Chapel Hill.

A common concern among city officials who are disturbed by unfavorable rates of program efficiency is the problem of employee absenteeism among sanitation personnel. A benchmark for assessing that concern was provided by a federally sponsored study of service delivery among Los Angeles–area cities in the mid-1980s that revealed an average absenteeism rate of 10.6% for sanitation employees (Stevens, 1984, p. 173). The rate was 13.4% in communities using city crews for refuse collection and 7.9% for cities using contractors. Among cities examined for this chapter, New York City reported average annual sick leave rate of 11.0 days per sanitation employee in 1998. Savannah, Georgia, reported average annual sick leave rate per sanitation employee ranging from a high of 6.9 days to a low of 5.4 days from 1988 through 1991.

Municipal officials attempting to sort out the relevance of equipment size for refuse collection productivity will be interested in the production ratios specified in a City of Philadelphia study prepared in 1991. That report called for the collection of an average of 12.66 tons per 8-hour shift when using a 32-cubic-yard compactor, 11.84 tons per 8-hour shift when using a 20-cubic-yard compactor, and an average of 9.21 tons per 8-hour shift when using a 9-cubic-yard compactor (p. 4).

The quality of refuse collection services often is measured by collection reliability relative to announced schedules, the extent to which collection stops are occasionally missed, and the satisfaction of residents with collection services, often gauged by the number of complaints received. Citizens expect their garbage to be picked up according to the announced collection schedule, and some cities have an extremely low tolerance for deviation (Table 28.3). Savannah, Georgia, for example, reported that 99.9% of its residential solid waste collections in 1997 were completed on schedule. Charlottesville, Vir-

Table 28.3   Collections Completed on Schedule: Selected Cities

GENERAL

Savannah, GA
Percentage of residential units' refuse collected on schedule: 99.9% (1997)

Charlottesville, VA
99% of curbside collections made on schedule (1997)

College Park, MD
Target: 98% (2000)

Durham, NC
On-schedule collection of roll-out carts: 96% (1998)

Philadelphia, PA
95.5% (1998)

Calgary, Alberta
95% (1992–1996)

LEAF ROUTES

Winston-Salem, NC
Percentage completed on schedule (once every 2 weeks): 63% (1997), 51% (1998)

BRUSH ROUTES

Winston-Salem, NC
Percentage completed on schedule (once per week): 92% (1997), 88% (1998)

BULKY ITEMS

Winston-Salem, NC
Percentage completed on schedule: 100% (1997), 100% (1998)

Savannah, GA
99.9% (1997)

Charlottesville, VA
95% of large-item-collection appointments worked within 4 weeks of original call (1997)

RECYCLING

Philadelphia, PA
98.0% on schedule (1998)

ginia, completed 99% of its curbside collections on schedule. College Park, Maryland, targets 98% schedule compliance or better. Philadelphia, Pennsylvania, reported a decline from 94% on-time collection in 1995 to 81.7% in 1996, attributable to record snowfall (Figure 28.1). Except for those snow days, Philadelphia's on-time rate was 95% in 1996 and climbed to 95.5% in 1998.

Similarly, sanitation officials attempt to keep missed collections to a minimum because, in addition to taking their toll in public relations, each such instance imposes costs in equipment and labor when it becomes necessary to return to the site to correct the oversight. Some "missed collections" are bum raps, the result of a resident depositing garbage at the collection location after

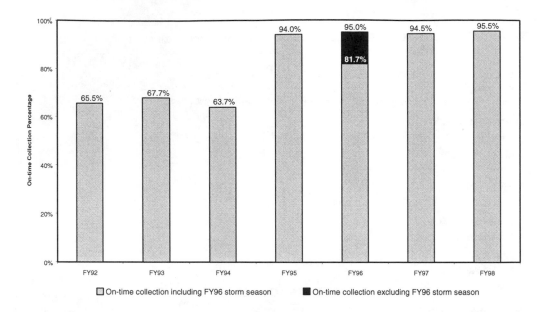

Figure 28.1. On-Time Refuse Collection Service in Philadelphia
SOURCE: City of Philadelphia (PA), *Mayor's Report on City Services* (1998), p. 36.

the crew has already passed. Even in such instances, the sanitation department has little to gain by arguing the point. Most cities reporting statistics on missed collections experienced 5 to 23 misses per 10,000 scheduled stops, with a low of 1.6 per 10,000 stops reported by Rockville, Maryland (Table 28.4).

Statistics on refuse collection complaints are reported in a variety of ways. The Los Angeles–area study noted earlier (Stevens, 1984), for example, reported approximately 21 complaints per year for each 1,000 households (Table 28.5). Alternatively, some municipalities report complaints per 10,000 persons or per 10,000 collections. Charlotte, North Carolina, sets its performance target at minimizing complaints per refuse collection crew, with each crew striving to receive no more than 5 valid complaints per month.

Responses to service requests and complaints are swift in many cities. White Plains, New York, responded to resident requests in a recent year in an average of 30 minutes. Boca Raton, Florida, resolved complaints in an average of 2 hours. Several cities report high percentages of complaints responded to or resolved the day of the complaint, within 24 hours, or in some cases within 48 hours.

Benchmarks for Recycling Programs _____

Many cities have developed programs to collect recyclable materials. By diverting these materials from the waste stream, the life of a landfill is extended and the environment benefits.

Table 28.4 Missed Collections: Selected Cities

City	Missed Collections per 10,000 Scheduled Stops
Rockville, MD	Missed homes per 10,000 due to collector error: 1.6 (rate of 0.016%) (1997)
Ocala, FL	2.52 (1996)
Greensboro, NC	5 (1997)
Oak Ridge, TN	10 (1992)
Shreveport, LA	10 (1994)
Winston-Salem, NC	11.8 among household collections; 20 among businesses (1998)
College Station, TX	23 (rate of 0.23%) (1998)

Table 28.5 Refuse Collection Complaints: Selected Cities

COMPLAINTS PER YEAR PER 1,000 HOUSEHOLDS

Study of Los Angeles–area cities, 1984[a]
20.7 among cities with municipal crews
21.2 among cities with contractors

Raleigh, NC
21 (1997), 30 (1998)

Corpus Christi, TX
27.9 refuse collection complaints per 1,000 households (1997)

Boston, MA
31.1 (1996)

COMPLAINTS PER 10,000 POPULATION

Plano, TX
5.1 legitimate complaints per year (1997)

Greenville, SC
6.3 per year (1990)

Denton, TX
31.2 per year (1991)

Chandler, AZ
61.4 complaints about missed pickups per year (1998)

COMPLAINTS PER 10,000 COLLECTIONS

Alexandria, VA
0.21 (1995), 0.12 (1998)

San Diego, CA
Actual: Less than 1 per 10,000 service stops (1998)

OTHER

Charlotte, NC
Target: No more than 25% of crews receiving more than 5 valid complaints per month, and no more than 25% of supervisors having more than one crew exceeding the 5-complaint limit

a. Statistics from Los Angeles–area study reported in Barbara J. Stevens (Ed.), *Delivering Municipal Services Efficiently: A Comparison of Municipal and Private Service Delivery* (prepared by Ecodata, Inc., for the U.S. Department of Housing and Urban Development, June 1984), p. 161.

Table 28.6   Average Participation and Diversion Rates Among Recycling Programs in 158 Surveyed Cities in 1996

	Mean Rate of Participation	Mean Rate of Waste Diversion
All programs	72.8%	33.1%
Mandatory programs	80.0%	36.4%
Voluntary programs	65.3%	29.9%
Curbside pick-up	68.8%	30.4%
Drop-off only	50.4%	28.2%

SOURCE: Adapted from David H. Folz, "Recycling Policy and Performance: Trends in Participation, Diversion, and Costs," *Public Works Management & Policy, 4* (October 1999), p. 136. Reprinted by permission of Sage Publications.

A 1996 survey of 158 cities with recycling programs revealed average participation rates of 50% to 80%, depending on program type (Table 28.6). On average, these programs diverted 28% to 36% of their community's solid waste from the waste stream (Folz, 1999, p. 136). Participation rates as high as 90% and waste diversion rates of as much as 60% for specific customer categories were reported in the performance reporting documents of individual cities examined for this volume (Table 28.7).

Note

1. Although most municipalities either provide refuse collection services directly through city crews or contract with another government or a private firm for that service, some cities assume a lesser role—either regulating the private market through franchise arrangements or refraining from any involvement whatsoever.

Table 28.7 Recycling Participation and Waste Diversion Rates: Selected Cities

Participation Rates for Recycling Programs	Waste Diversion Rates
Bellevue, WA Percentage of households participating: 90% (1997) **Chandler, AZ** Curbside recycling participation rate: 85% (1995) **Portland, OR** Percentage of households participating: 83% (1998) **Overland Park, KS** Residents participating in curbside program: 82% (1997) **San Jose, CA** Curbside participation rate: 80% of households (1995) Yard waste participation rate: 81% of households (1995) **College Station, TX** 73% (1998) **Winston-Salem, NC** 68.22% (1998) **Raleigh, NC** 64.2% (1997), 62.9% (1998) **Albuquerque, NM** 52% (1998) **Plano, TX** Average weekly set-out rate: 50% (1997) **Denver, CO** 45% (1998) **Corpus Christi, TX** Set-out rate: 34.1% (1997) **San Antonio, TX** Percentage in-house participation weekly: 32.5% (1995), 29.0% (1997) **Macon, GA** Percentage of households participating: 31% (1995) **Oklahoma City, OK** Set-out rate for curbside recycling: 23.7% (1999)	**Bellevue, WA** Percentage of waste recycled: Multifamily and business, 23%; single-family residential, 60% (1998) **Cary, NC** 46% (1999) **San Diego, CA** 44.8% (1998) **Portland, OR** 40% (1998) **Rockville, MD** 35% (1997) **Plano, TX** 32% (1997) **Aiken, SC** 30.3% (1998) **Orlando, FL** Percentage of total waste stream recycled: 30% (1996) **Cambridge, MA** 28.5% (1997) **San Clemente, CA** 27.4% (1998) **Oakland, CA** 27% (1996) **Scottsdale, AZ** Percentage of residential waste stream diverted from the landfill through curbside recycling: 27% (1999) **Durham, NC** 24.6% (1997) **College Station, TX** 24.5% (1998) **Blacksburg, VA** Percentage of waste stream recycled (curbside program): 22% (1997) **Largo, FL** Percentage recycled: 21.4% (1994) **Tempe, AZ** Residential diversion rate: 20.1% (1997) **New York, NY** Percentage of residential waste stream recycled: 16.3% (1998) **Gainesville, GA** Percentage of waste stream recycled: 14.0% (1998) **San Antonio, TX** Percentage of waste diverted from landfills: 13.11% (1995), 12.19% (1997)

29

Streets, Sidewalks, and Storm Drainage

Street maintenance is a major element of municipal public works—so major, in fact, that it is allocated a separate chapter in this volume. Typically, street maintenance operations are also responsible for sidewalk maintenance and storm drainage in their community.

Streets

The bottom line for judging a municipal streets program is the condition of community roadways and the resulting quality of ride for motorists. Some cities focus primarily on pavement condition, visually rating streets for potholes, surface cracking, and other signs of deterioration. Other cities attempt to gauge the more difficult dimension of street quality, smoothness and ride quality, through the use of mechanical devices that record the degree of roughness of city streets. Among the most sophisticated of this genre is the laser scanning profilometry technology used in New York City (Center on Municipal Government Performance, 1998). Derived from the aircraft industry's work with runways, this approach records the incidence of bumps or jolts as a vehicle traverses a roadway. The score in New York City? The study recorded an average of 9.49 jolts per mile in 1998 (p. 10).

More commonly, cities that seek independent ratings or that rate their own streets focus mostly on pavement characteristics rather than primarily on the riding comfort of motorists. A popular method of assessing pavement condition is through the use of a simple scale called the *present serviceability rating* (PSR), with scores ranging from 0 for *very poor* to 5 for *very good* (Table 29.1). A community may rate itself and track its progress in upgrading the condition of local roadways as measured by the change in average PSR or by change in the percentage of street miles with favorable ratings.

Table 29.1   Pavement Condition Rating: Federal Highway Administration

PSR^a	Verbal Rating	Description
4.01–5	Very good	Only new (or nearly new) superior pavements are likely to be smooth enough and distress free (sufficiently free of cracks and patches) to qualify for this category. Most pavements constructed or resurfaced during the data year would normally be rated in this category.
3.01–4	Good	Pavements in this category, although not quite as smooth as those described above, give a first class ride and exhibit few, if any, visible signs of surface deterioration. Flexible pavements may be beginning to show evidence of rutting and fine random cracks. Rigid pavements may be beginning to show evidence of slight surface deterioration, such as minor cracks and spalling.
2.01–3	Fair	The riding qualities of pavements in this category are noticeably inferior to those of new pavements, and may be barely tolerable for high-speed traffic. Surface defects of flexible pavements may include rutting, map cracking, and extensive patching. Rigid pavements in this group may have a few joint failures, faulting and/or cracking, and some pumping.
1.01–2	Poor	Pavements in this category have deteriorated to such an extent that they affect the speed of free-flow traffic. Flexible pavement may have large potholes and deep cracks. Distress includes raveling, cracking, and rutting, and occurs over 50% of the surface. Rigid pavement distress includes joint spalling, patching, cracking, and scaling, and may include pumping and faulting.
0.0–1.0	Very poor	Pavements in this category are in an extremely deteriorated condition. The facility is passable only at reduced speeds, and with considerable ride discomfort. Large potholes and deep cracks exist. Distress occurs over 75% or more of the surface.

SOURCE: U.S. Department of Transportation, Federal Highway Administration, *Highway Performance Monitoring System Field Manual* (Washington, DC: Government Printing Office, December 1987), Table IV-2. Available online at www.fhwa.dot.gov/ohim/hpmsmanl/hpms.htm (25 January 1999).
a. Present Serviceability Rating (PSR) established by the American Association of State Highway Transportation Officials (AASHTO).

Street ratings by consultants or other outside entities may use similar systems, though often with different scales. Charlotte, North Carolina, for instance, reported an overall street condition rating of 90 ("good") in 1998, as assessed by North Carolina State University's Institute for Transportation Research and Education. Winston-Salem, North Carolina, reported that 79% of its streets had a pavement condition rating of 85 or better.

Some cities have devised their own street condition standards that differ somewhat from the PSR. Cincinnati, Ohio, for example, has used an "Infrastructure Management" rating system since 1992. In 1996, 60% of the city's lane miles were rated as being in good or excellent condition. Savannah, Georgia, considers a street to be in "standard condition" despite relatively minor flaws. A street meets the Savannah standard even with the following conditions:

- Minor defects less than 12 inches across may exist
- Minor bumps less than 3.5 inches high may exist
- Rideability is slightly less than perfect (City of Savannah (GA), *Neighborhood Quality Benchmarks,* p. 166).

Service areas in Savannah are prioritized according to the number of defects per mile, and maintenance resources are directed accordingly. In 1996, Savannah's streets averaged 1.36 defects per mile.

The condition of city streets is partly a result of construction standards, soil conditions, and climate, and partly a product of maintenance practices. Many street construction factors, too numerous to be covered here, can influence the quality and durability of a roadway, but once constructed, the roadway's condition becomes the responsibility of the street maintenance program.

Local officials and citizens who are concerned with the condition of city streets often focus on the number and depth of potholes, the promptness of pothole repair, plans for resurfacing of city streets (especially the choice of particular streets to be included or excluded from those plans), and the diligence of street crews in pursuit of the community's objectives. When citizens express dismay over the productivity level of local street crews, they almost invariably mention what they perceive to be examples of deteriorated street conditions and instances of municipal laborers just "leaning on their shovels."

Everyone who has ever wielded a shovel—even as a weekend amateur—knows that some shovel leaning provides a necessary breather in hard labor. But how much is too much? When does the frequency or duration of "breathers" begin to interfere with, rather than sustain, good performance? Knowing the answer requires a determination of what constitutes reasonable production in various tasks.

Among the performance documents examined, several cities reported production ratios and work crew configurations for various tasks (Table 29.2). Charlotte, North Carolina, for example, reported about 0.4 staff hours per square yard of skin patching and 0.2 staff hours per square yard for base failure repair. Dunedin, Florida, reported 0.62 feet of curb repair per labor hour and 0.73 feet of sidewalk repair per labor hour. Additional sets of crew size and production standards are presented in Tables 29.3 and 29.4. Still further guidance on standard labor output may be found in a manual (updated annually) produced by R. S. Means Company for construction contractors.

Responsiveness is also an important aspect of street maintenance. High-quality street maintenance programs will not allow potholes and other road hazards to remain unrepaired for extended periods, for example. Among the

Table 29.2   Production Ratios for Street Maintenance Operations in Selected Cities

ASPHALT PAVEMENT REPAIR

Charlotte, NC
  Targets: 0.43 staff hours per sq. yd. of skin patching; 0.47 staff hours per sq. yd. of base failure repair;
    2.5 staff hours per sq. yd. of pothole repair
  Actual: 0.44 staff hours per sq. yd. of skin patching; 0.20 staff hours per sq. yd. of base failure repair;
    1.86 staff hours per sq. yd. of pothole repair (1987, 5-month record)

Dunedin, FL
  Surface repair: 15.2 sq. ft. per labor hour (1.7 sq. yd. per labor hour) (1992)
  Surface holes: 1.2 sq. ft. per labor hour (1992)

Wichita, KS
  Sq. yds. of asphalt pavement repairs (permanent) per labor hour: 1.1 (1990), 1.6 (1991)
  Blocks of temporary pavement repairs per labor hour: 2.9 (1990), 3.0 (1991)

ASPHALT REPAVING

Pasadena, CA
  Sq. ft. of asphalt repaving per person hour: 30.8 (1986)
  Tons per person hour: 0.176 (1986)

Victoria, TX
  Worker hours per square yard of streets rebuilt: 0.51 (1996)
  Worker hours per ton of cold mix: 4.62 (1996)

CONCRETE PAVEMENT REPAIR

Wichita, KS
  Sq. yds. of concrete pavement repair (permanent) per labor hour: 1.0 (1990), 1.3 (1991)

Little Rock, AR
  Square yards of concrete replaced per day in concrete pavement repairs: 30 to 70 sq. yds.
    (7-person crew)

Lubbock, TX
  Average worker hours per paving cut and base failure repaired: 7.71 (1991)

CURB REPAIR

Dunedin, FL
  0.62 ft. per labor hour (1992)

Charlotte, NC
  Target: 1.2 staff hours or less per linear foot of curb and gutter replaced
  Actual: 1.58 staff hours per linear foot (1987, 5-month record)

SIDEWALK REPAIR

Dunedin, FL
  0.73 ft. per labor hour (1992)

Santa Ana, CA
  Sidewalk/curb forms built and cu. yds. of concrete poured and finished per direct worker hour: 0.122
    (1987)

(continued)

Table 29.2 Production Ratios for Street Maintenance Operations in Selected Cities (continued)

MISCELLANEOUS TASKS

Little Rock, AR
Blocks of unimproved road prepared for chip sealing per day (i.e., ditches pulled, debris removed, roadway graded): 2 to 4 blocks (6-person crew)

Wichita, KS
Crack sealing of asphalt/concrete streets—lineal feet per labor hour: 93 (1990), 65 (1991)

Little Rock, AR
Sq. yds. chip sealed: 10,000 to 16,000 sq. yds. per 10-hour day for 19-person crew

Little Rock, AR
Alley blocks graded and surface material placed per day: 3 to 5 blocks of alley (4-person crew)

Little Rock, AR
Guardrail installed per 8-hour day: 25 to 75 linear feet (3-person crew)
Guardrail repaired per 8-hour day: 50 to 100 linear feet (3-person crew)
Guardrail painted per 8-hour day: 200 to 300 linear feet (3-person crew)

Little Rock, AR
Shoulder blocks of right-of-way moved: 20 to 40 shoulder blocks per day (4-person crew)

Little Rock, AR
Alley blocks cleared of brush and debris: 1 to 3 alley blocks per day (4-person crew)

examined performance reporting documents, several performance targets called for pothole repairs within 1 or 2 days of notification (Table 29.5). These targets are consistent with the standards established by the LCC (1994), calling for repair within 24 hours by high-service departments and within 48 hours by medium-service departments (p. 43). Similarly, top operations establish ambitious performance targets for the repair of utility cuts and other general street and sidewalk maintenance. Average response times for a variety of street maintenance functions in Corpus Christi, Texas, are reported in Table 29.6.

Many street maintenance operations are responsible for restoring city streets to passable conditions during winter onslaughts of ice and snow. Reno, Nevada's performance target is to plow or sand 60% of all priority 1 and 2 routes within 8 hours after the end of a snow or ice event; 75% within 16 hours; and 90% within 24 hours. Cincinnati, Ohio, tries to treat all priority 1 streets within 2 hours following a snowfall of 2 inches or more. Rock Island, Illinois, expects its crews to have snow removed from city streets within 48 hours after the end of a storm. In Sterling Heights, Michigan, the target is snow removal within 30 hours after snowfall ceases.

Fairly elaborate plans for snow removal are not unusual. Overland Park, Kansas, adopted a priority system for restoring passable conditions to its streets:

Table 29.3   Production Standards for Selected Maintenance Operations:
                   Fayetteville, Arkansas

Maintenance Operation	Crew Size	Average Daily Production
Right-of-way mowing	4	20 lane miles (2 passes)
Pesticide spraying	1	75 to 100 gallons applied
Pothole patching (hot mix)	3	3 tons
Asphalt milling	4	1,000 square yards
Spot surface asphalt repair/replacement	8	450 square yards
Asphalt overlay	14	4,000 square yards
Asphalt surface treatment	4	5,000 square yards
Paving fabric application	5	4,000 square yards
Crack sealing	5	7,500 square yards
Street cut repairs	3	3 cuts repaired
Base repair	4	125 square yards
Shoulder repair	5	750 lineal feet
Curb and gutter installation/repair	3	25 lineal feet
Guardrail installation/repair	4	200 lineal feet
Street sweeping	1	30 curb miles
Clean catch basins and other drainage structures	4	15 structures
Wash drain tiles	4	150 lineal feet
Mechanical ditch cleaning	5	750 lineal feet
Install/replace French drain	4	300 feet
Install drain tile	4	100 feet
Concrete construction/repair	3	1.5 cubic yards

SOURCE: City of Fayetteville (AR), *Public Works Maintenance Management System: System Manual* (January 1989).

Priority I. Bare pavement on all heavily traveled thoroughfares . . . within 2 hours of snow or ice storm ending between hours of 7 a.m. and 9 p.m.

Priority II. Secondary thoroughfares on traffic routes to schools . . . will have been plowed, sanded, or salt and sand mix applied as conditions require within 2.5 hours of storm ending. Sand or salt/sand mix will be applied at a rate of 700 to 800 pounds per two-lane mile.

Priority III. Collector and industrial commercial streets . . . will have been plowed, sanded, or salt and sand mix applied within 3.5 hours of storm ending. Sand or salt/sand mix will be applied at a rate of 700 to 800 pounds per two-lane mile.

Priority IV. All other public streets . . . receiving 4- to 8-inch snowfall will be plowed and sand or salt/sand mix will be applied on steep grades and at intersections within 18 hours of snow ending. Application rate of abrasives, where required, will be 700 to 800 pounds per two-lane mile. (*1996 Budget,* p. 5.165)

Table 29.4   Street Maintenance Standards

Title	Recommended Crew Size	Standard Labor Time	Hourly Crew Output
Joint and crack sealing	7	00.00182 labor hours/ pound	862.0 (pounds)
Slurry seal	4	00.3024 labor hours/ 1000 lane feet	13.2 (1000 lane ft.)
Premix patching	4	03.69 labor hours/ton	1.08 (tons)
Reshape shoulders	4	01.988 labor hours/ 1000 lane feet	2.08 (1000 lane ft.)
Remove/install concrete pavement	7	07.56 labor hours/ cubic yard	0.925 (cubic yards)
Remove/install bituminous pavement	8	00.1544 labor hours/ton	51.8 (tons)
Machine sweeping	1	00.240 labor hours/ lane mile	4.16 (lane miles)
Machine flushing (streets)	1	00.275 labor hours/ lane mile	3.64 (lane miles)
Other patching and paving	As required	—	—
Premix patching (miscellaneous)	2	01.044 labor hours/ton	1.92 (tons)
Regrade, gravel or dirt street	1	00.497 labor hours/ 1000 linear feet	2.01 (1000 linear ft.)
Chip seal	6–8	00.529 labor hours/ 1000 linear feet	13.23 (1000 linear ft.)
Oil surface	1	00.145 labor hours/ 1000 linear feet	6.89 (1000 linear ft.)
Clean and reshape ditches	4	29.16 labor hours/ 1000 feet of ditch	0.137 (1000 ft. of ditch)
Clean drainage structures	2–3	00.0764 labor hours/ structure cleaned	26.18 (structure cleaned)
Manual policing	2	01.18 labor hours/ cubic yard (of debris)	1.69 (cubic yards)
Machine mowing	1	00.046 labor hours/ 1000 square feet	21.74 (1000 square ft.)
Stump removal	2	00.666 labor hours/stump	3.00 (stumps)
Other roadside and drainage	As required	—	—
Remove/install concrete curb	7	02.80 labor hours/ cubic yard	2.5 (cubic yards)
Remove/install concrete sidewalk	5	06.09 labor hours/ cubic yard	0.821 (cubic yards)

Remove/install miscellaneous concrete structures	7	—	—
Signal maintenance	2	00.73 labor hours/signal	2.74 (signals)
Pavement marking	2	02.53 labor hours/1000 linear feet	0.790 (1000 linear ft.)
Sign fabrication	1	00.931 labor hours/sign	10.7 (signs)
Sign replacement	2	00.732 labor hours/sign	2.73 (signs)
Miscellaneous barricade, sign, and signal work	1	—	—
Spread sand or salt	2	00.16 labor hours/lane mile	12.5 (lane miles)
Control slick spots	2	—	—
Plow snow	1	01.61 labor hours/lane mile (normal)	4.97 (lane miles—normal)
		02.36 labor hours/lane mile (heavy)	3.39 (lane miles—heavy)
Setup (instructions, tools)	—	0.220 hours/day/worker	—
Travel and material handling	—	00.054 hours/mile/worker	—
Cleanup	—	00.192 hours/day/worker	—

SOURCE: Excerpted from Earl J. Ferguson, "Street Maintenance," in George J. Washnis (Ed.), *Productivity Improvement Handbook for State and Local Government* (New York: John Wiley, 1980), p. 682. Reprinted by permission.
NOTE: Indiscriminate use of these labor standards may result in significant errors. They are included as a guide only.

Sidewalks, Curbs, and Gutters

Sidewalks, curbs, and gutters are also elements of the street maintenance program. The City of Sunnyvale, California (1993), uses a standard decision rule to determine whether to include particular sidewalks, curbs, and gutters in the upcoming year's repair schedule. Repair within 1 year is justified for a sidewalk when the differential between walking surfaces is greater than three fourths of an inch. For curbs and gutters, repair within 1 year is justified when displacement of the gutter lip is three fourths of an inch or greater and impounds water more than 10 feet up the gutter swale (p. 7).

Storm Drainage

As in the case of street maintenance, the quality of a municipal storm drainage program may be judged ultimately on its ability to achieve its objectives.

(text continues on page 416)

Table 29.5   Prompt Road Repairs: Performance Targets and Experience of Selected
                   Cities

Prompt Repair of Potholes and Road Hazards	*Prompt Patching of Utility Cuts*	*General Maintenance*

Kalamazoo, MI
 Percentage of potholes patched
  within 24 hours of report:
  100% (1997)

Norman, OK
 Targets: (1) Patch potholes
  smaller than 1 cubic foot
  within 24 hours of notification;
  (2) Schedule maintenance on
  larger potholes immediately
 Actual: (1) 100%; (2) 80%
  (1998)

Fort Collins, CO
 Target: Fill potholes within
  1 business day of notification
 Actual: 98% (1991)

Cary, NC
 Percentage of potholes repaired
  within 24 hours of notification:
  95% (1999)

Overland Park, KS
 Targets: Chuckholes greater
  than 3 inches deep or with
  subbase exposed will be
  repaired within 24 hours;
  holes or overlay/surface
  failures will be repaired within
  1 week

Greensboro, NC
 Percentage of potholes repaired
  within 24 hours: 87% (1997)

Boston, MA
 Percentage of potholes filled
  within 1 day: 87% (1996)

Raleigh, NC
 Target: Repair asphalt failures
  (e.g., potholes) within
  24 hours of notification
 Actual: 81% (1997), 66% (1998)

Santa Ana, CA
 Percentage of street damage
  greater than $3/4''$ variance
  repaired within 24 hours of
  notification: 80% (1995)

Tallahassee, FL
 Utility service cut repair time:
  1 day (1991)

Charlotte, NC
 Target: Close 95% of utility cuts
  within 24 hours of completion
  by utility company
 Actual: 96% (1987)

Raleigh, NC
 Percentage repaired within
  24 hours: 34% (1997),
  50% (1998)

Denton, TX
 Percentage patched within
  48 hours: 95% (1991)

Corpus Christi, TX
 Target: Within 3 days (1999)
 Actual: 2-workday average
  (1996)

Cary, NC
 Target: Within 5 days (2000)

Peoria, AZ
 Target: Within 7 days

Savannah, GA
 Target: 7 to 14 days

Rock Island, IL
 Percentage repaired within
  30 days: 89% (1990), 100%
  (1991)

Alexandria, VA
 Percentage of repairs completed
  within 30 days:
  95% (1998)

Corpus Christi, TX
 Target: Repair base failure and
  cracked pavement within
  2 weeks (1999)
 Actual: 6-workday average
  (1996)

San Antonio, TX
 Street repair within 7 days: 35%;
  within 8 to 30 days: 55%
  (1995)

Reno, NV
 Target: Routine pavement
  repairs within 2 weeks

Orlando, FL
 Target 1: Complete asphalt
  street patches within
  15 workdays
 Actual: 100% (1991)
 Target 2: Complete brick
  street repair requests within
  5 workdays
 Actual: 100% (1991)
 Target 3: Complete concrete
  repair requests within
  15 workdays
 Actual: 100% (1991)
 Target 4: Repair broken/
  displaced sidewalks within
  30 workdays of notification

Cambridge, MA
 Average number of days to
  close request for unscheduled
  service (e.g., sidewalk/street
  repair): 19.61 days (1995)

Savannah, GA
 Target: Priority III maintenance
  (street defect, tree root
  damage, curb defect, ROW[a]
  mowing, overgrown lane,
  unpaved street grading, sidewalk repair, alligator cracking,
  shoulder repair) performed
  within 14 to 50 days of
  request
 Actual: 69% (1990), 100% (1991)

San Jose, CA
Target: At least 75% of repairs completed within 24 hours of request

Cincinnati, OH
Target: Repair at least 80% of potholes within 24 hours of request from citizen
Actual: 60% (1995)

Kansas City, MO
Arterial defects repaired within 24 hours: 20% (1995)

Greeley, CO
Percentage of potholes repaired within 48 hours: 94% (1997)

San Antonio, TX
Percentage of potholes repaired within 2 workdays: 85% (1995), 91.7% (1997)

Corpus Christi, TX
Target: Repair potholes within 3 days (1999)
Actual: 2-workday average (1996)

Philadelphia, PA
Number of workdays to repair pothole following notification during peak season: 3 workdays; during non–peak season: 6 workdays (1998)

Milwaukee, WI
Percentage response (patching) within 4 days: 100% (1993)

Oklahoma City, OK
Target: Respond to pothole complaints within 5 working days (2000)

Macon, GA
Percentage of repairs within 7 days: 85% (1996, estimated)

Palo Alto, CA
Percentage of potholes repaired within 10 working days: 80% (1997)

New York, NY
Percentage of potholes repaired within 30 days: 26% (1997), 44% (1998)

Milwaukee, WI
Percentage of sidewalk replacement requests serviced same year: 85% (1992), 54% (1993)

---

a. ROW stands for the publicly defined right-of-way.

Table 29.6   Responsive Street Maintenance: Corpus Christi

Service in Response to Complaints/Notification (first 7 months of 1995)	Average Response Time
Repaired 406 potholes	3 days
Repaired 78 base failures	8 days
Leveled 52 standing water areas	10 days
Leveled 37 low gutters	3 days
Repaired 35 cave-ins	1 day
Repaired 34 utility cuts	1 day
Repaired 33 cracked pavements	5 days

SOURCE: Susan Cable, John Longoria, and Ed Martin (City of Corpus Christi [TX]), "Performance Management: An Essential Foundation for Reengineering Government," p. 7. Presented at SAM International Management Conference on Reengineering: Businesses and Business Schools, March 14–16, 1996. Reprinted by permission.

Minimal flooding and rapid dissipation of standing water are signs of a successful program. In Norman, Oklahoma, for example, the local performance target is for new street surfaces to have no ponded stormwater runoff deeper than one eighth of an inch 2 hours after the end of a storm (*Budget: Fiscal Year Ending June 30, 2000*, p. 186). Norman reported an 80% success rate in 1998. In St. Petersburg, Florida (1992), the local performance standard calls for intersection drainage within 30 minutes following a normal rainfall. Observation of 336 intersections in 1990 revealed 333—or 99%—draining within the prescribed time (p. 243).

Diligent work crews, conscientious maintenance, and responsive services contribute to service proficiency and good public relations. Few cities report production ratios for storm drainage crews, but one that does is Little Rock, Arkansas (Table 29.7). More often reported are maintenance cycles and responsiveness of storm drainage operations to complaints and requests—emergency and nonemergency in nature (Tables 29.8 and 29.9).

Table 29.7 Production Targets for Storm Drain and Ditch Maintenance: Little Rock, Arkansas

Number of catch basins rebuilt per 8-hour day:
  0.6 (3-person crew)

Number of catch basins repaired per 8-hour day (i.e., patching of front facing of catch basin top):
  2 to 5 (3-person crew)

Number of catch basins vacuumed per day:
  15 to 20 (2-person crew)

Linear feet of storm sewer removed and replaced per day:
  30 to 40 linear feet (6-person crew)

Linear feet of storm sewer and culverts cleaned per day, when cleaning by hand:
  100 to 150 linear feet (4-person crew)

Linear feet of street drainage ditch cleared by hand per day:
  600 to 1,000 (4-person crew)

Linear feet of off-street drainage ditch cleared by hand per day:
  50 to 100 (4-person crew)

Linear feet of drainage ditch dug out by backhoe per day:
  300 to 500 (5-person crew)

Linear feet of ditch cleared per day using heavy equipment:
  100 to 250 (6-person crew)

Square yards of riprap placed per day to stabilize ditch banks:
  30 to 50 square yards (6-person crew)

SOURCE: City of Little Rock (AR), *Management Information Systems Manual,* 1990. Updated per electronic message from David Wooley, Little Rock Public Works, February 1999.

Table 29.8   Storm Drainage Maintenance Cycles and Service Responsiveness: Selected Cities

Routine Maintenance Cycle	*Prompt Responses to/Resolution of Drainage Emergency Complaints/Requests*

Lubbock, TX
Targets: Check/clean all storm sewer inlets at least once every 2 months; all "high-risk" inlets, once per week

Reno, NV
Target: Clean catch basins twice per year
Actual: 67% (1989–1990)

Fort Collins, CO
Target: Clean and/or check catch basins twice per year

Orlando, FL
Target: Maintain all drainage wells once each year; all open drainage facilities three times per year

Tallahassee, FL
Targets: Enclosed system maintenance cycle, 4.7 years; detention facilities, 7.3 years; roadside ditches, 11 years; outfall ditch dredging, 9 years

Lee's Summit, MO
Target: 6-year cycle

San Clemente, CA
Percentage of responses to working hour emergencies within 15 minutes: 100% (1998)
Percentage of responses to non–working hour emergencies within 45 minutes: 100% (1998)

Lubbock, TX
Average response time on emergency calls: 15 minutes (1991)

Orlando, FL
Percentage of emergency responses within 2 hours: 100% (1991)
Percentage of citizen inquiries/complaints investigated within 72 hours: 100% (1991)

Palo Alto, CA
Percentage of responses within 4 hours to emergency flooding calls: 95% (1997)
Percentage of service requests investigated within 5 working days: 98% (1997)
Percentage of responses within 1 working day to incidents of illegal disposal to storm drain and sanitary sewer system: 100% (1997)
Percentage of compliance directives issued within 1 working day in confirmed cases of illegal discharge: 100% (1997)
Percentage of responses within 1 working day to reported nonhazardous/nonemergency storm drain contamination: 100% (1997)
Percentage of nonhazardous spills within the public right-of-way contained/cleaned up within 4 hours: 95% (1997)

Overland Park, KS
Target: Curb inlet guards that are a hazard to children's safety will be repaired within 4 hours of notification

Denton, TX
Percentage of responses to drainage complaints/ inquiries within 24 hours: 41% (1991)

Cincinnati, OH
Target: respond to complaints/requests within 48 hours
Actual: 85% (1992), 96% (1993)
Investigations of drainage claims completed within 2 weeks: 90% (1993)

TABLE 29.9 Responsiveness to Routine Storm Drainage/Catch Basin Maintenance Requests

New York, NY
Average time from catch basin complaint to completion of maintenance/repair: 15.9 days (1997)
Percentage of catch basin complaints resolved within 30 days: 85.9% (1997)

Corpus Christi, TX
Target: Repair damaged curb and gutter, inlets, manholes, and culverts within 4 weeks of notification (1999)

30

Traffic Engineering and Control

Pity the poor traffic engineer. Sometimes it seems that everyone who has ever driven a car thinks he or she has the answers to the community's traffic problems. Too bad those answers are rarely the same as the ones offered by the city's official traffic engineer!

A stop sign here, a traffic signal there, and a left-turn arrow with a turning lane over there should do the job nicely. With three simple moves, the problems plaguing one particular motorist could be solved. The speeding that occurs on the neighborhood's streets could be reduced, lengthy delays in making a difficult left-turning maneuver from the neighborhood onto a major thoroughfare could be cut, and an equally difficult left turn across traffic into the office complex could be simplified. Any traffic engineer who cannot grasp these simple solutions must not be trying to understand! Or could it be that the traffic engineer has more than a single motorist in mind?

Like so many other public servants, the traffic engineer is invisible when things are running smoothly. But let a few traffic snarls develop, inconvenience a few drivers for the sake of traffic flow, or have the community experience a rash of accidents—perhaps a tragic fatality—and the traffic engineer suddenly might be thrust into the eye of a storm.

Standards

Traffic engineers strive to design and regulate transportation arteries to allow safe movement with minimum delay. Professional guidance is available in the form of standards on the use of uniform traffic control markings and guidelines identifying the conditions that call for the use of traffic control devices. Often, the "solutions" that seem so obvious to amateur traffic engineers impose a host of problems that are apparent to their professional counterparts—for example, the stop sign or traffic signal that eases conditions for a few motorists may cause travel delays for many others. Add enough stops along a given route and traffic flow is seriously impaired.

The *Manual on Uniform Traffic Control Devices for Streets and Highways* (U.S. Department of Transportation, 1988) provides standards that govern the size of traffic control signs, their shape and color, the size of their letters or figures, their placement relative to roadways and intersections, and their height. The manual also provides guidance in identifying the circumstances that warrant various traffic control devices. The installation of a traffic signal, for instance, may be warranted by a single factor, or by a combination of any of 11 factors:

- *Warrant 1—Minimum vehicular volume.* This warrant is reached when the hourly traffic volume, sustained for 8 hours of an average day, reaches 500 vehicles traveling either direction on a major street and 150 vehicles traveling in one direction on a minor, intersecting street, if each is a two-lane street with one lane of traffic moving in each direction.

- *Warrant 2—Interruption of continuous traffic.* This warrant is met, for example, when 75 vehicles per hour traveling a single direction on a minor street (one lane each direction) attempt to enter or cross a major street (two lanes each direction) carrying 900 vehicles per hour.

- *Warrant 3—Minimum pedestrian volume.* This warrant is met when 190 or more pedestrians attempt to cross a major street during any single hour or when 100 or more cross during each of 4 hours.

- *Warrant 4—School crossing.* This warrant is met when traffic engineering studies indicate inadequate gaps in the traffic stream.

- *Warrant 5—Progressive movement.* This warrant is met, for example, when the nearest signals are too far apart to effect vehicle platooning and speed control.

- *Warrant 6—Accident experience.* This warrant is met when five or more accidents of a type that could be avoided with a traffic signal occur within a 12-month period, the location experiences substantial vehicular and pedestrian volume (Warrants 1, 2, and 3), other remedies have proven ineffective, and signal installation will not seriously disrupt traffic flow.

- *Warrant 7—Systems warrant.* In some cases, traffic signal installation may be warranted to encourage concentration and organization of traffic flow networks.

- *Warrant 8—Combination of warrants.* Even in cases where no single warrant is met, meeting Warrants 1 and 2 at the 80% level may be sufficient to justify signal installation.

- *Warrant 9—Four-hour volumes.* High volumes of vehicles through an intersection, in various major street–minor street combinations during each of 4 hours of an average day, may justify signal installation.

Table 30.1   Minimizing Traffic Congestion: Selected Cities

San Jose, CA
  Target: At least 90% of vehicles at peak traffic times traveling at speeds greater than 50% of
    free-flow speed
  Projected: 88% (1996)

Savannah, GA
  Target: To maintain the average travel speed for all major corridors at 23 miles per hour (mph)
    (1999)

Scottsdale, AZ
  Percentage of intersections with average delay greater than 40 seconds in peak hours: 56% (1997),
    47% (1998)

- *Warrant 10—Peak hour delay.* This warrant is met when extreme delays
  are experienced by drivers on a minor street attempting to enter or cross
  a major street during any single hour of an average weekday.
- *Warrant 11—Peak hour volume.* Like Warrant 10, this warrant addresses
  the difficulties experienced by drivers at a high-volume intersection during rush-hour traffic. (pp. 4C.3–4C.8)

Despite the availability of professional guidelines, the degree to which these
conditions are actually required prior to signal installations differs from community to community. Emotion and politics—as well as professional expertise—often weigh heavily in such decisions.

Other Possible Benchmarks

Responsible traffic engineers strive for traffic flow and safety. In that regard,
some cities declare precise traffic flow targets (Table 30.1). For example, San
Jose, California, tries to maintain at least 50% of free-flow speed even during
peak traffic times. Savannah, Georgia, has established a system for grading
traffic flow (Table 30.2) and set a target of maintaining average travel speed
for all major corridors at 23 miles per hour. Winston-Salem, North Carolina,
reported average travel time within coordinated signal sections of 2.42 minutes per mile during peak times and 2.34 minutes per mile during off-peak
times in a recent year. Scottsdale, Arizona, monitors congestion by tracking
the percentage of intersections with average delays greater than 40 seconds
during peak traffic hours.

Traffic safety can be reported in several ways. Portland, Oregon, defines as
"high-accident intersections" those with 20 or more accidents in a 4-year pe-

Table 30.2   Grading Traffic Flow in Savannah

Level of Service	Description	Average Overall Speed (mph)
A	Free flow (relatively—some stops will occur)	30
B	Stable flow (delays not unreasonable)	20
C	Stable flow (delays significant, but acceptable)	15
D	Approaching unstable flow (delays tolerable)	10
E	Unstable flow (congestion—intolerable delay)	below 10
F	Force flow (jammed)	stop and go

SOURCE: City of Savannah (GA), *Neighborhood Quality Benchmarks: 1996 Report to the Community*, p. 196.

Table 30.3   Traffic Injuries and Fatalities: 1997 National Statistics

	per 100 Million Vehicle Miles Traveled	per 100,000 Population
Injured Persons	133	1,270
Fatalities	1.6	15.68

SOURCE: U.S. Department of Transportation, National Highway Traffic Safety Administration, *Traffic Safety Facts 1997* (November 1998).

riod and periodically reports the number of such intersections. More common methods are to report accidents per 1,000 or 100,000 population or accidents per 1 million vehicle miles (Tables 30.3 and 30.4).

Professional conduct for traffic engineers requires the conscientious application of professional norms to local traffic considerations, but it also implies responsiveness and diligence in the conduct of various traffic engineering duties (Table 30.5). The traffic engineering department in Orlando, Florida, for example, reported providing complete responses to most citizen requests within 1 week and the completion of most major studies within 30 days.

Good traffic flow depends not only on sound engineering but also on the installation and maintenance of appropriate traffic control devices. Traffic control signs, for instance, should conform to uniform standards in appearance and placement. Pavement markings should be properly designed and maintained. A panel assembled by the LCC (1994) to consider local government performance standards regards the repainting of centerlines, curb markings, crosswalks, and legends on a 9- to 18-month cycle to be the mark of high-level service; an 18- to 30-month cycle to reflect medium service; and less frequent repainting to indicate a low service level (p. 43). Many cities examined for this volume attempt to repaint pavement markings at least annually (Table 30.6).

Table 30.4   Accident Rates: Selected Cities

REPORTED ACCIDENTS PER 1,000 POPULATION

Long Beach, CA
13 (1996)
Tucson, AZ
28.6 (1997)
Corpus Christi, TX
34.3 (1997)
San Antonio, TX
44.6 (1995)

ACCIDENTS PER MILLION VEHICLE MILES

Sunnyvale, CA
Targets: 2.97 accidents per 1 million miles traveled; 0.14 bicycle accidents per 1 million miles traveled;
0.06 pedestrian accidents per 1 million miles traveled (1994)
Winston-Salem, NC
3.08 accidents per 1 million vehicle miles; 1.61 accident injuries per 1 million vehicle miles;
0.01 accident deaths per 1 million vehicle miles (1992)
Charlotte, NC
Target: Not to exceed 6.44 traffic accidents per million miles traveled
Actual: 4.75 (1994, 5-month record)

INJURY ACCIDENTS PER 1,000 POPULATION

San Jose, CA
4.9 (1995)
Calgary, Alberta
6.37 (1996)
Long Beach, CA
Traffic injuries per 1,000 population: 6.6 (1996)
Duncanville, TX
Injury accidents per 1,000 population: 7.3 (1998)
Wichita, KS
9.95 (1998)
Tucson, AZ
10.02 (1997)
National Average[a]
12.7 traffic injuries per 1,000 population (1997)

TRAFFIC FATALITIES PER 1,000 POPULATION

Calgary, Alberta
0.049 (1996)
Raleigh, NC
0.08 (1998), 0.09 (1999)
Corpus Christi, TX
0.11 (1997)
National Average[a]
0.157 (1997)

PEDESTRIAN ACCIDENTS PER 1,000 POPULATION

Raleigh, NC
0.72 (1998), 0.74 (1999)

a. National averages for traffic injuries and fatalities reported in U.S. Department of Transportation, National Highway
Traffic Safety Administration, *Traffic Safety Facts 1997* (November 1998).

Table 30.5   Traffic Engineering: Performance Targets and Actual Experience of Selected Cities

Prompt Investigation of Traffic Problems	Prompt Completion of Traffic Studies	Prompt Review of Plans
Orlando, FL Target: Provide complete response to citizen requests for information/assistance within 1 week Actual: 97% (1993), 100% (1994)	**Long Beach, CA** Percentage of traffic engineering studies completed within 3 weeks: 60.0% (1991), 65.6% (1992)	**Lubbock, TX** Percentage of subdivision plats checked within 1 workday: 100% (1991)
Rock Island, IL 11.7 days (1990), 26.2 days (1991)	**Cincinnati, OH** Target: Within 30 days	**Raleigh, NC** Average review time for development plans: 2.0 days (1997)
Reno, NV Target: Within 14 days Actual: 65% (1990)	**Orlando, FL** Percentage of major studies completed within 30 days: 98% (1993), 100% (1994)	**Norman, OK** Target: Review subdivision plats, construction traffic control plans, traffic impact statements, and other transportation improvement plans within 7 days (1999)
Lubbock, TX Percentage of citizen requests resolved within 15 workdays: 66% (1991)	**Norman, OK** Target: Within 45 days (1999)	
Tacoma, WA 29 days (1990), 30 days (1991)	**Chandler, AZ** 7.5-week average (1995)	**Reno, NV** Percentage of development applications reviewed within 10 days: 95% (1990)
Raleigh, NC Traffic signal requests: 35 days; traffic sign requests: 4 days; individual streetlight requests: 0.94 days (1991)		**Long Beach, CA** Percentage reviewed within 3 weeks: 95.0% (1991), 91.0% (1992)
Corvallis, OR Target: Evaluate and respond to neighborhood traffic/parking concerns within 60 days Actual: 90% (1991), 100% (1992)		**Cincinnati, OH** Target: Within 30 days for zoning changes, building permits, PUD[a] projects, streetscape projects

a. PUD = planned unit development.

When stop or yield signs are damaged or when traffic signals malfunction, prompt response is important. Many cities report a response time target of 4 hours or less for the repair or replacement of damaged stop or yield signs (Table 30.7). Because of their customary placement at hazardous, high-volume intersections, targeted response times for malfunctioning traffic signals are even quicker. Several cities reported performance consistent with the findings of a federally sponsored study of Los Angeles–area cities, where city crews had an average response time to malfunctioning traffic signals of 32.5

Table 30.6   Pavement Marking Cycles: Selected Cities

City	Pavement Marking Cycle
Fort Collins, CO	Target: Repaint most crosswalks and traffic lane symbols once per year, more frequently on arterials and collectors; repaint traffic lanes on arterials 4 times per year, on collectors 3 times, and residentials once per year
Lubbock, TX	Target: Repaint thoroughfares twice per year; marked collector streets and crosswalks, annually Actual: Thoroughfares, 68%; collectors and crosswalks, 100% (1995)
Overland Park, KS	Target: Twice per year (unless thermoplastic) (1999)
Phoenix, AZ	Target: Repaint major streets and intersections twice per year
Milwaukee, WI	Average marking frequency per year for lanes and centerlines: 1.3 (1991), 1.5 (1992), 1.2 (1993) Average marking frequency per year for crosswalks and other markings: 0.6 (1991), 0.9 (1992)
Cincinnati, OH	Target: Repaint centerlines, lane lines, and edge lines throughout the city at least once a year
Grand Prairie, TX	Percentage of centerline miles maintained/restriped annually: 90% (1996)
Shreveport, LA	Percentage of streets striped annually: 70% (1994)
Sunnyvale, CA	Target: 12-month maintenance cycle for markings on arterials; 24-month cycle off arterials
Plano, TX	School crosswalks repainted annually: 100% (1997) Intersection markings maintained annually: 85% (1997) Thoroughfare markings maintained annually: 35% (1997)
Rock Island, IL	Target: Repaint center and lane lines annually; curbs on a 2-year cycle; stop bars, crosswalks, and lane use markings at intersections on a 2-year cycle
Tallahassee, FL	Target: 2-year cycle

minutes and on average completed the repairs 77 minutes after arrival (Stevens, 1984, p. 279). When contractor crews were involved, the average response time was 43 minutes and the average repair time was 37 minutes.

Proper staffing for the maintenance of traffic control devices balances the need for quick response with the desire to avoid overstaffing. The prudent administrator, as well as the wily budget officer, is likely to want statistics on the incidence of traffic signal problems, the normal rate of traffic control sign repair and replacement, and relevant production ratios for maintenance personnel. Such statistics are collected by many municipalities. Tallahassee, Florida, for example, had 9.34 trouble reports per signal during one recent year, compared with 8.8 and 15 trouble calls per signalized intersection in Nashville–Davidson County, Tennessee, and Raleigh, North Carolina, respectively. Ames, Iowa, reported an average of 17 service calls per signalized intersection during a recent year. Kansas City, Missouri, and Scottsdale, Arizona, reported 42 and 39 traffic signals per maintenance employee, respectively.

Several cities report production ratios for a variety of traffic control activities (Table 30.8). Although a given city's own production ratios might not match those reported here precisely, substantial variation may suggest the need for review.

Table 30.7   Prompt Response to Damaged Traffic Control Signs and Malfunctioning
Traffic Signals: Selected Cities

Response to Damaged Traffic Control Signs

High-Priority Signs	General	Response to Malfunctioning Traffic Signals
Bellevue, WA Percentage of traffic sign emergency calls responded to within 45 minutes: 95% (1997) **Palo Alto, CA** Percentage of responses to missing stop sign reports within 1 hour: 100% (1997) **Plano, TX** Percentage of stop or yield signs repaired within 1 hour: 100% (1997) **Lubbock, TX** Service calls on stop/yield signs responded to within 1 hour: 99% (1997) **Denton, TX** Responded within 1 hour to 80% of service calls (1991) **Norman, OK** Target: Respond to high-priority sign damage (stop or yield signs) within 1 hour (1999) **Reno, NV** Target: Respond to calls regarding damaged stop or yield signs within 1 hour **Orlando, FL** Target: Damaged regulatory signs replaced within 2 hours at hazardous intersections **Overland Park, KS** Target: Respond within 2 hours (1999) **Tucson, AZ** Target: Within 2 hours **Charlotte, NC** Percentage of damaged stop or yield signs repaired/replaced within 3 hours of notification: 100% (1994, 5-month record)	**Lubbock, TX** Service calls on traffic control signs (other than stop or yield) responded to within 1 workday: 98% (1997) **College Station, TX** Average replacement time for lower priority regulatory signs (i.e., other than stop signs): within 24 hours (1998) **Norman, OK** Target: Respond to high-priority sign damage (stop or yield signs) within 1 hour; lower priority signs within 1 day; street name signs within 2 weeks (1999) **Orlando, FL** Targets: Damaged regulatory signs replaced within 2 hours at hazardous intersections; damaged informational signs replaced within 24 hours **Reno, NV** Targets: Respond to calls regarding damaged stop or yield signs within 1 hour; other traffic control signs within 1 workday; street name signs within 2 weeks **Fort Collins, CO** Average time for repair/replacement of signs: 2 workdays (1991) **Tempe, AZ** Percentage of responses to nonemergency sign service calls within 3 days: 95% (1997) **Overland Park, KS** Target: Within 3 workdays for "routine" problems and within 5 workdays for "minor" problems (1999)	**Winston-Salem, NC** 18-minute average (1998) **Shreveport, LA** Target: Respond to traffic signal trouble calls within 30 minutes Actual: 97% (1994) **Tempe, AZ** Percentage of emergency signal service calls responded to within 30 minutes: 95% (1997) Percentage of nonemergency responses within 2 days: 100% (1997) **St. Petersburg, FL** Target: Respond to traffic signal trouble calls within 30 minutes Actual: 93.5% (1995) **Lubbock, TX** Percentage of emergency service calls responded to within 30 minutes: 92% (1997) **San Jose, CA** Target: At least 90% within 30 minutes of notification **Palo Alto, CA** Percentage of responses within 4 hours of notification: 100% (1997) Average response time: 0.6 hour (1997) Percentage restored within 8 hours: 100% (1997) **Oklahoma City, OK** Percentage of signal trouble calls completed within 1 hour: 76% (1999) **Plano, TX** Emergency service responses within 1 hour: 100% (1997)

College Station, TX
Missing stop sign replacement time: within 4 hours (1998)

Chandler, AZ
Percentage repaired/replaced within 1 workday: 100% (1998)

Tempe, AZ
Percentage of responses to emergency sign service calls within 24 hours: 100% (1997)

Corpus Christi, TX
Target: Same day for stop signs (1999)
Actual: 5-day average (1997)

New York, NY
Percentage repaired or replaced within 9 days: 100% (1998)

Palo Alto, CA
Percentage of missing street name signs replaced within 5 working days of notification: 85% (1997)

Corpus Christi, TX
Target: Within 3 days for speed limit and defaced signs; within 14 days for street name, other signs (1999)
Actual: 6-day average for "other" signs (1997)

Tucson, AZ
Target: High-priority within 2 hours of notification; lower priority, within 1 week

Boston, MA
Target: Repair or replace non–public safety signs (i.e., lower priority) within 15 days
Actual: 50% (1992)

San Diego, CA
Percentage of requests responded to within 1 hour: 77% (1998); within 1 working day: 97% (1998)

Raleigh, NC
Average response time for traffic signal complaints: 61 minutes (1997), 60 minutes (1998)

Corpus Christi, TX
Target: Within 1 hour for signal out and for physical damage; within 3 hours for flash (1999)

Norman, OK
Target: Within 1 hour (1999)

Reno, NV
Target: Respond to at least 90% of emergency traffic signal calls within 1 hour during working hours; within 2 hours during non–working hours (1999)

Macon, GA
Percentage of emergency repairs completed within 2 hours: 98% (1995)

College Station, TX
Response time to after-hours calls: < 2 hours (1998)

Overland Park, KS
Target: Correct or alleviate every priority malfunction within 2 hours working time

Kansas City, MO
Target: Within 60 minutes on priority calls
Actual: 132-minute average (1995)

Philadelphia, PA
Percentage of responses within 24 hours of notification: 100% (1996)

New York, NY
Target: Repair 99% of malfunctioning traffic signals within 48 hours of notification
Actual: 98.7% (1998)

Table 30.8　　Production Ratios: Traffic Control Operations in Selected Cities

SIGN PRODUCTION

San Antonio, TX
Traffic and other signs produced per labor hour: 2.04 (1995)
Number produced per FTE: 2,842 (1997)

Duncanville, TX
Average labor hours to make up and install new sign: 1.71 (1997), 1.69 (1998)

STREET NAME SIGN INSTALLATION

Little Rock, AR
Street name markers installed per day: 5 to 8 if new posts required (2-person crew); 15 to 20
   if mounted on existing posts (2-person crew)

STREET NAME SIGN REPAIR

Little Rock, AR
Street name markers repaired/replaced per day: 30 to 50 (2-person crew)

LANE MARKINGS

Raleigh, NC
Staff hours per mile of lane markings: 4.2

Wichita, KS
Lineal feet of pavement marking per worker hour: 832 feet (1990), 608 feet (1991)

Santa Ana, CA
Lineal feet of pavement marking per direct worker hour: 648 feet (1987)

LEGEND/ARROW INSTALLATION (THERMOPLASTIC)

Little Rock, AR
Legends/arrows installed in traffic lanes per day: 4 to 6 locations (2-person crew)

LEGEND/ARROW PAINTING

Little Rock, AR
Legends/arrows painted in traffic lanes per day: 10 to 12 locations (2-person crew)

RAISED PAVEMENT MARKING

Little Rock, AR
Number of raised pavement markings per day: 30 to 50 (2-person crew)

LANE LINE INSTALLATION (THERMOPLASTIC)

Little Rock, AR
Linear feet installed per day: 3,600 feet (4-person crew)

LANE LINE PAINTING

Cincinnati, OH
Linear feet of lane lines painted per day: 100,000 (2-person crew)

Little Rock, AR
Linear feet of lane lines painted per day: 20,000 to 60,000 (4.2-person crew)

Norman, OK
Worker hours per gallon of traffic paint used: 0.8 (1999, projected)

Winston-Salem, NC
Worker hours per gallon of traffic paint used: 0.84 (1992)

MAINTENANCE OF TRAFFIC CONTROL SIGNS

Cincinnati, OH
Number of traffic control signs repaired (straightened, replaced, etc.) per day: 40 (2-person crew)

Little Rock, AR
Number of traffic control signs and posts installed per day: 14 to 18 (2-person crew)
Number of traffic control signs and posts removed per day: 30 to 50 (2-person crew)
Number of traffic control signs repaired (straightened, replaced, etc.) per day: 30 to 50 (2-person crew)

Dunedin, FL
Signs repaired per worker hour: 1.3 (1992)
Signs replaced per worker hour: 1.3 (1992)
Posts replaced per worker hour: 1.1 (1992)
Posts maintained per worker hour: 1.1 (1992)
Posts removed per worker hour: 1.8 (1992)

San Clemente, CA
Target: Install, remove, repair 1.2 signs per worker hour (1999)

Santa Ana, CA
Signs fabricated, installed, replaced, or repaired per direct worker hour: 0.965 (1987)

Winston-Salem, NC
Worker hours per sign maintenance activity: 1.03 (1992)

MOUNTING OVERHEAD SIGNS

Little Rock, AR
Number of overhead signs mounted on traffic signal equipment per day: 4 locations (2-person crew)

TRAFFIC SIGNAL CONSTRUCTION

Fort Collins, CO
Total work hours needed to construct traffic signal: 312 (1991)

TRAFFIC SIGNAL REPAIR/MAINTENANCE

Winston-Salem, NC
Worker hours per signal failure correction: 0.45 (1992)

Wichita, KS
Average worker hours per traffic signal repair: 1.29 (1990), 1.08 (1991)

Duncanville, TX
Average worker hours to maintain traffic signal (field): 1.47 (1998)

Little Rock, AR
Traffic signals repaired per day: 2 locations (2-person crew)

SIGNS/SIGNALS PER EMPLOYEE

Kansas City, MO
Number of traffic signals maintained per FTE employee: 42 (1997)

Scottsdale, AZ
Number of signs maintained per employee: 10,190 (1999)
Number of traffic signals per maintenance employee
  Target: 34 to 38
  Actual: 39 (1999)

31

Utilities Business Office

Field maintenance crews and plant operators are important to a public utility's success, but so are the meter readers and the people in the utilities business office who receive the telephone calls, serve walk-in utility customers, prepare utility bills, process payments, and maintain accounting records. Unless meter reading and office personnel perform their jobs well, customer relations—not to mention the utility's balance sheet—will suffer.

Benchmarks

By comparing its own performance with relevant operating statistics of counterparts, the utilities business office and meter reading operation for a natural gas, electric, water, or sewer utility may gauge the adequacy of local performance in the context of what others are able to achieve.

Office Operations

A 1991 survey of gas and electric utilities revealed a high level of administrative and clerical attention given both to customer relations and to the diligent collection of utility revenues (Table 31.1). The average utility customer had to wait only 37.5 seconds to speak to a company representative by telephone and 3 minutes in person. Among utility leaders, uncollectible expenses represented less than one half of 1% of total revenues. Among the 78 utilities providing information on the problem of delinquent accounts, one out of every five reported at least 5% of its accounts to be more than 90 days delinquent. At the other end of the continuum, a similar portion reported less than 1% of the accounts to be delinquent for more than 90 days (Mercer Management Consulting, 1992, p. 7; used by permission).

Employees in a utilities business office typically deal with many customers throughout the day. Wichita, Kansas, for example, reported 16.28 inquiries

Table 31.1   Selected Performance Marks of Gas and Electric Utility Leaders and
Average Performers

Measurement	Leaders	Average Performers
Waiting time for customer to speak to company representative on telephone	< 20 seconds	37.5 seconds
Abandoned telephone calls	< 3%	7%
Waiting time for customer to speak to company representative in person	< 1 minute	3 minutes
Uncollectible expenses as a percentage of total revenue		
Electric	< 0.20%	0.42%
Gas	< 0.45%	1.24%
Combination	< 0.15%	0.44%

SOURCE: Excerpted from Mercer Management Consulting, Inc., *Benchmarking the Quality of Utility Customer Services: 1991 Survey Results* (Lexington, MA: Author, 1992), p. iii. Reprinted by permission.
NOTE: *N* = 109 utilities.

Table 31.2   Utilities Business Office Production Ratios: Selected Cities

Tempe, AZ
  Utility service accounts per customer services employee: 1,997 (1997)
  Utility service accounts per customer services representative: 6,657 (1997)
Gainesville, GA
  Ratio of customers to office staff (Customer Service & Billing): 3,766:1 (1998)
Plano, TX
  Accounts per clerk:[a] 3,634 (1997)
  Customer transactions per clerk:[a] 3,551 (1997)
  Phone calls answered per clerk per month: 497 (1997)

a. Ratio includes clerks in addition to those at front counter.

handled per telephone representative per hour in the early 1990s, compared with an average of approximately 12 per hour reported in a survey of 191 gas and electric utilities (Mercer Management Consulting, 1992, pp. 3–4; used by permission). Selected production ratios in three cities, indicating the number of utility accounts per employee and similar statistics, are shown in Table 31.2.

A variety of indicators of business office effectiveness, ranging from low error rates and high collection rates to prompt customer service, have been reported by municipal utility offices and can serve as useful points of comparison for counterparts seeking benchmarks (Table 31.3).

(text continues on page 437)

Table 31.3   Indicators of Utilities Business Office Effectiveness: Selected Cities

MINIMAL CASHIER ERRORS

College Station, TX
  Target: Error-free daily balance at least 95% of time

Sunnyvale, CA
  Target: Daily cash no more than $1 out-of-balance at least 95% of time

MINIMAL TRANSACTION ERRORS

Savannah, GA
  Target: Complete at least 99.8% of all customer transactions correctly, as requested
  Actual: 99% (1997)

ON-TIME MAILING OF BILLS

Duncanville, TX
  Percentage of bills mailed within 1 day of target date: 95.0% (1996), 100% (1997)

Cary, NC
  Percentage of bills mailed within 1 day of target date: 95% (1999)

Lubbock, TX
  Percentage of electric accounts billed within 3 days of meter reading: 99.6% (1997)

Charlottesville, VA
  100% of bills mailed within 5 working days of meter reading (1998)

San Jose, CA
  Percentage of bills mailed within 5 workdays of meter reading: 98% (1995)

BILLING ACCURACY

Rockville, MD
  Rebillings due to error: 0.052% (1997)

Grand Prairie, TX
  Percentage of accounts billed correctly: 99.96% (1996)

Plano, TX
  Adjustments as a percentage of all billings: 0.29% (1997)

San Clemente, CA
  Target: Achieve at least 95% accuracy on monthly utility bills
  Actual: 95% (1998)

PROMPT PROCESSING OF RECEIPTS

Duncanville, TX
  Percentage of same-day deposits: 100% (1998)

Lubbock, TX
  Percentage of electric payments posted on day of receipt: 98.4% (1997)

Raleigh, NC
  Percentage of same-day deposits of utility bill payments: 96% (1998)

Hurst, TX
  Percentage of payments processed and deposited within 1 day of receipt: 100% (1997)

Plano, TX
  Percentage of funds deposited within 24 hours: 100% (1997)

San Diego, CA
Percentage of payments accurately processed within 1 working day: 92% (1998)

COLLECTION RATE

Raleigh, NC
99.5% (1997)

Danville, KY
99% (1997)

Shreveport, LA
99% (1994)

Winston-Salem, NC
Percentage of billed water/sewer use collected: 99% (1991), 98% (1997), 96% (1998)

Savannah, GA
Target: Collect at least 99.5% of water, sewer, and refuse billings
Actual: 98.2% (1997)

DELINQUENT ACCOUNTS

Albuquerque, NM
Accounts receivable delinquency rate: 0.50% (1997)

Rockville, MD
Percentage of accounts delinquent: 0.6% (1997)

Boca Raton, FL
Target: Maintain a delinquency rate of less than 1% on utility billings more than 60 days past due

San Clemente, CA
Target: Limit utility receivables over 30 days old to 20% or less; limit write-offs to 1% of total billings

BAD DEBTS

Ft. Collins, CO
Bad debts written off as a percentage of the total billed: 0.26% (1991)

College Station, TX
0.39% (1998)

Fayetteville, AR
Percentage of utility revenue dropped due to bad debt: 0.50% (1997)
Percentage of final utility bills dropped due to bad debt: 13.89% (1997)

Fort Smith, AR
0.5% (1997)

LOBBY WAITING TIME

Denton, TX
5-minute average (1991)

Houston, TX
10-minute average (1992)

Long Beach, CA
Target: 10 minutes or less
Average wait during peak hours: 10 minutes (1991), 10 minutes (1992)

(continued)

Table 31.3   Indicators of Utilities Business Office Effectiveness: Selected Cities
(continued)

TELEPHONE SERVICE

Oklahoma City, OK
Percentage of calls answered within 15 seconds: 78% (1999)

High Point, NC
Call abandon rate (target < 5%): 3.4% (1998)
Percentage of calls answered by fourth ring: 87.2% (1998)

Palo Alto, CA
Target: Maintain an average time under 20 seconds for customer calls placed on hold
Actual: 15-second average (1997)

Tucson, AZ
Average wait time a customer spends holding for a customer service representative: 41 seconds
   (1997)

Tempe, AZ
Percentage of calls answered in less than 2.5 minutes: 91% (1997)

PROMPT UTILITY SERVICE

Lubbock, TX
Percentage of electric utility service orders (e.g., turn-ons, turn-offs, transfers) worked day of
   receipt: 100% (1997)

College Station, TX
Actual: Provided on date requested 99.9% of time (1998)

Tempe, AZ
Percentage of service initiations or terminations completed by promised date and time: 99% (1997)

Duncanville, TX
Percentage of service orders worked within 24 hours: 98% (1996), 99% (1997), 99% (1998)

San Jose, CA
Percentage of customer service requests responded to within 48 hours: 99% (1995)

Palo Alto, CA
Percentage of water service orders (turn-ons, turn-offs, reads) processed within 2 working days
   of scheduled date: 81.9% (1997)

TROUBLESHOOTING

Lubbock, TX
Percentage of electric service investigation orders worked the day of receipt: 100% (1997)
Percentage of solid waste, sewer service, and high bill investigations completed within 24 hours
   of work order: 98% (1997)

Plano, TX
Percentage of errors corrected/inquiries answered within 3 workdays: 100% (1997)

Table 31.4   Production Ratios for Meter Readers: Selected Cities

Meters Read per Meter Reader		
per Month	*per Day*	*per Hour*
Plano, TX 9,854 (1997)	**Ames, IA** 684 (water and electric) (1997)	**Fayetteville, AR** 49.8 (1991)
Chandler, AZ 8,246 (1995)	**Santa Ana, CA** 500 (1995)	**Winston-Salem, NC** 33.3 per worker hour (0.030 worker hours per meter) (1998)
Tallahassee, FL 7,608 (1991)	**College Station, TX** 387 (1998)	**Dunedin, FL** 17.7 (1992)
Shreveport, LA 6,000	**Kansas City, MO** 348 (1995)	
	Raleigh, NC 341 (1998)	
	Denton, TX 306	
	Gainesville, GA 290 (1999)	
	Rochester, NY 145 attempted reads; 87 reads completed (1991)[a]	

a. Where meters are located inside buildings, the problem of "lock-outs" may restrict production rates.

Meter Reading

Efficient meter reading helps keep operating costs down. Accurate meter reading enhances customer relations, promotes proper revenue flow, and alerts the utility to system or service abnormalities detectable through individual account records.

Among cities examined for this volume, considerable variation emerged in the number of reporting production ratios for meter readers—that is, some ratios were reported as meters read per month, some as meters per day, and others as meters per hour (Table 31.4). A rough average would be approximately 350 meters per day, with several cities reporting higher rates and several reporting lower rates.

Most of the examined cities report high rates of meter reading accuracy —or low error rates (Table 31.5). Error rates of less than one half of 1% were common among these cities.

Table 31.5   Meter Reading Accuracy: Selected Cities

METER READING ERRORS/ACCURACY

Chandler, AZ
Meter reading error rate: 0.01% (1995)

High Point, NC
Meter reading accuracy: 99.99% (1998)

Ocala, FL
Error rate: 0.01% (1996)

Tucson, AZ
Errors per 10,000 reads: 2.3 (1997)

College Station, TX
Accuracy rate: 99.96% (1998)

Grand Prairie, TX
Percentage of meters read correctly: 99.90% (1996)

Lubbock, TX
Meter reading error rate (electric): 0.17% (1995), 0.17% (1996), 0.13% (1997)

Raleigh, NC
Percentage of meter reading errors and adjustments: 0.15% (1997), 0.14% (1998)

Albuquerque, NM
Meter reading accuracy: 99.84% (1997)

Palo Alto, CA
Meter reading accuracy: 99.7% for electric; 99.2% for water (1997)

San Diego, CA
Accuracy rate: 99.7% (1998)

Gainesville, GA
Accuracy rate: 99.58% (1999)

Hurst, TX
Accuracy rate: 99.5% (1997)

Denton, TX
Error rate: 0.60% (1997)

Winston-Salem, NC
Water meter reading error rate: 0.98% (1997), 0.99% (1998)

Boca Raton, FL
Target: Accuracy rate of at least 95%

32

Water and Sewer Services

Unlike the case of electricity and natural gas, where many of the cities that provide these utilities operate simply as distributors, cities often take on much greater responsibilities in water and sewer services. Many municipalities assume a comprehensive role in the treatment and distribution of potable water and the collection and treatment of wastewater.

Federal and state water and sewer regulations form a set of basic standards governing operations and product quality. Some of these regulations will be noted in this chapter.[1] In addition, professional standards of the water and sewer "industry," as well as the performance targets and actual experience of water and sewer counterparts, can serve as the basis for benchmarks in the various service specialties.

Water Treatment and Distribution

The residents of a community typically know little about the process of purifying water or maintaining distribution lines and may willingly leave the details of formulating standards to local government officials, but that does not mean their expectations are ambiguous. As one authority notes,

> For the public, the goals of water supply and distribution are simple. The supply should be ample and dependable. The water should be tasteless and odorless, without ingredients that are harmful to health. It should be reasonably soft and cool. Its cost should not be excessive.
>
> For the local government, the goals are more detailed. The volume of water must be sufficient not only to satisfy domestic, commercial, and industrial requirements but also for fire protection. The pressure should be maintained at not less than 30 to 40 pounds per square inch (107 to 276 kilopascals) throughout the system. Fire hydrants must be reliable, as must

Table 32.1   Summary Statement on Water Quality

AWWA (American Water Works Association) advocates the principle that water of the highest quality should be delivered to all consumers and encourages each public water supplier to provide water of a quality that is as close to the AWWA goals as is feasible. In addition,

- Water delivered to the consumer for domestic purposes should contain no pathogenic organisms and be free from biological forms that may be aesthetically objectionable. It should be clear and colorless and should have no objectionable taste or odor. It should not contain concentrations of chemicals that may be physiologically harmful, aesthetically objectionable, or economically damaging.
- The water should neither be corrosive nor leave excessive or undesirable deposits on water-conveying structures, including pipes, tanks, and plumbing fixtures.
- The water should be obtained from the best sources available and should be treated as necessary to ensure safety and consistent quality.
- Existing and potential sources of community water supply should be protected from contamination to the highest degree feasible.
- A water utility has the responsibility to be aware of the nature of its source water. Therefore, water utilities should adequately monitor their source waters to determine their quality.
- Care should be taken to avoid and minimize the introduction of unnecessary constituents during the treatment and delivery process.
- In the distribution process, every reasonable effort, including backflow prevention, should be made to protect water from degradation.
- Water utilities should undertake adequate monitoring and periodic sanitary surveys to ensure that water quality objectives are met on a continuous basis.

SOURCE: Reprinted from *American Water Works Association 1992–1993 Officers and Committee Directory*, by permission. Copyright © 1992, American Water Works Association. Also posted at www.awwa.org/govtaff/driwapol.htm (January 2000).

be the valves in the system. Water mains must be free from leaks so that as much as is practicable of the water that has been purified and pumped at rather substantial cost actually will be used by the customers. Water rates must be adequate to make the system economically viable. (William S. Foster, *Handbook of Municipal Administration and Engineering,* Copyright © 1978 by McGraw-Hill, p. 16.20. Reprinted with permission of McGraw- Hill, Inc.)

In contrast to the typical citizen's imprecisely expressed desire for an ample and dependable supply of tasteless and odorless water, the prescriptions of the American Water Works Association (AWWA) embodied in its statement of objectives are more precise and more narrowly focused on water quality (Table 32.1). The AWWA summary statement, addressing matters ranging from water supply contamination to the corrosion of plumbing fixtures, is supplemented by more precise standards of water quality (Table 32.2).

The U.S. Environmental Protection Agency (EPA) weighs in significantly on the subject of water quality. Among the water contaminants regulated by

Table 32.2   Recommended Standards of Water Quality

Element/Characteristic	Recommended Standard
Chloride	Operating level of less than 250 mg/L in finished water; goal of less than 100 mg/L
Chlorine residual	Maintain a detectable free chlorine residual or 1.0 mg/L of combined chorine throughout the distribution system; goal of free chlorine residual of at least 0.5 mg/L or combined chlorine of at least 2.0 mg/L
Color	Less than 5 cobalt platinum units (cpu) true color in finished water; goal of less than 3 cpu
Foaming agents	Sufficiently low to prevent foaming and other aesthetic problems (Environmental Protection Agency prescribes less than 0.5 mg/L)
Manganese	Less than 0.05 mg/L in finished water
Nitrate	Less than 10 mg/L of nitrate as nitrogen in finished water
Phenols	Less than 0.002 mg/L of phenols at the point of chlorination
Taste and odor	Less than 3 threshold odor number (TON) in finished water; taste should be nonoffensive
Turbidity	No more than 0.5 NTU of turbidity in filter plant effluent; goal of no more than 0.2 NTU
Zinc	Less than 5 mg/L in finished water; goal of less than 2 mg/L

SOURCE: Reprinted from *American Water Works Association 1992–1993 Officers and Committee Directory*, by permission. Copyright © 1992, American Water Works Association.

the EPA are those shown in Table 32.3, with maximum allowable contaminant levels noted. Also shown are performance statistics regarding these contaminants compiled from consumer confidence reports issued by 17 water systems across the nation. Median statistics for the group fall comfortably within the EPA standards. Pacesetters for water quality in each category reported remarkably favorable statistics.

Additional water quality standards sometimes are set by individual cities or by other authorities. Portland, Oregon, for example, repeats two of the factors noted previously in the EPA list and adds two others—pH and chlorine residual—in the set of principal indicators it routinely reports (Table 32.4). The Massachusetts Water Resources Authority has established performance targets addressing not only water content, but also line loss and customer complaints (Table 32.5). Statistics compiled on various aspects of water quality from the reports of individual cities round out the pool of potential benchmarks even further (Table 32.6).

Although the significance of many of the AWWA and EPA water quality standards may be more fully understood by chemists and lab technicians than by city managers, elected officials, or general citizens, awareness of their importance even among nonspecialists may be increasing. A few key characteristics

Table 32.3 Water Quality Performance Statistics of 17 Selected Cities Compared to EPA Drinking Water Standards

Contaminant	Maximum Allowable Levels[a]	Range	Mean	Median	Pacesetters[b]
Arsenic	50 ppb	0.3 to 2.1 ppb	1.05 ppb	0.9 ppb	Virginia Beach, VA (0.3 ppb)
Barium	2 ppm	0.01 to 0.20 ppm	0.046 ppm	0.025 ppm	Austin, TX (0.01 ppm) Chicago, IL (0.014 ppm) Dallas, TX (0.015 ppm)
Copper	1.3 ppm (Action Level)	0 to 0.90 ppm	0.202 ppm	0.096 ppm	Austin, TX (none detected) Providence, RI (none detected) Cincinnati, OH (0.002 ppm)
Fluoride	4 ppm	0.39 to 3.1 ppm	1.03 ppm	1.01 ppm	Santa Barbara, CA (0.39 ppm) Dallas, TX (0.4 ppm) Savannah, GA (0.4 ppm, Dutch Island System)
Lead	15 ppb (Action Level)	0 to 29 ppb	7.06 ppb	6.0 ppb	Austin, TX (none detected) Dallas, TX (none detected) Santa Barbara, CA (none)
Nitrate	10 ppm	0.01 to 2.20 ppm	0.556 ppm	0.50 ppm	Denver, CO (0.01 ppm) St. Petersburg, FL (0.04 ppm) High Point, NC (0.127 ppm)
Turbidity	At least 95% of samples at less than 0.5 NTU[c]	0.03 to 2.34 NTU	0.195 NTU	0.125 NTU	Virginia Beach, VA (0.03 NTU) Austin, TX (0.06 NTU) Cincinnati, OH (0.06 NTU)
Total coliform bacteria	Coliform bacteria present in < 5% of monthly samples	0% to 2.8%	0.74%	0.77%	Dallas, TX (none detected) St. Petersburg, FL (none detected) Santa Barbara, CA (0.09%)

SOURCE: Compiled from 1998 and 1999 Consumer Confidence Reports from 17 U.S. cities.

a. These are the maximum levels of contamination that the EPA allows in drinking water. Furthermore, the EPA encourages water utilities to strive for even lower levels of contamination, noting that some water customers may be more vulnerable to contaminants in drinking water than the general population—such as those with suppressed immune systems, some elderly, and infants.

b. In instances when a water system reports results from different treatment plants, the best performing facility was considered in naming pacesetters.

c. All samples must have turbidity of less than 5 NTUs.

Table 32.4   Principal Indicators of Water Quality in Portland, Oregon

	1997–1998	*Standard*
Turbidity maximum (NTUs)	2.44	< 5.00
pH (standard units):		
Minimum	7.3	6.0
Maximum	7.6	8.5
Coliform bacteria (% positive samples)	0.06%	< 5.00%
Chlorine residual (mg/L):		
Minimum	0.10	0.02
Maximum	2.20	4.00

SOURCE: City of Portland (OR) Office of the City Auditor, *City of Portland Service Efforts and Accomplishments: 1997–98* (December 1998), p. x.

Table 32.5   Water System Targets: Massachusetts Water Resources Authority

Targets
- Not more than 5% of all water samples may be total coliform positive (or no more than 1 positive when 40 samples or less are collected in a given month).
  - Residual disinfectant, measured 2 hours downstream of disinfection, shall be at a level of at least 1.4 milligrams per liter (mg/L).
  - Chlorine residuals shall be greater than or equal to 0.2 mg/L.
- Fecal coliform counts should not exceed 20/100 milliliter (a violation occurs when the count exceeds this level 13 or more times in a 6-month period).
- Turbidity of unfiltered water should not exceed 1.0 NTU.
- Unaccounted-for water should not exceed 6%.
- Complaints should be kept at a low level (low: less than 50 complaints per month per 100,000 population; moderate: between 50 and 100 complaints per 100,000 population; high: more than 100 complaints per 100,000 population).

SOURCE: Excerpted from Massachusetts Water Resources Authority, *Management Indicators for June 1997* (Boston: Author, July 23, 1997).

are being reported routinely by a growing number of cities. Turbidity, for example, is a measure of particulate matter in a sample of water and, in effect, refers to the cloudiness of water. Too much turbidity can protect bacteria from the disinfectant effects of chlorine. Increasingly, municipal water systems are reporting turbidity (Figure 32.1)—and occasionally they are noting in performance reports other water qualities such as chlorine residual and the absence of coliform bacteria.

Although a degree of uniformity is provided to water quality standards by the involvement of EPA and AWWA, remarkable variation exists in performance

Table 32.6   Meeting the Test for Drinking Water Quality: Selected
Cities

GENERAL COMPLIANCE

Philadelphia, PA
Percentage of time drinking water met or surpassed state/federal standards:
100% (1998)

San Clemente, CA
Percentage of monitoring samples in full compliance with state and federal
requirements: 100% (1998)

High Point, NC
Percentage of bacteriological tests passing state and federal requirements:
100% (960 tests) (1997), 99.2% (972 tests) (1998)

Grand Prairie, TX
Percentage of total number of water system samples collected that met or
surpassed state standards: 98% (1996)

TURBIDITY

Ocala, FL
0.05 NTU (1996)

Norman, OK
0.06 NTU average; 0.50 NTU maximum (1998)

Philadelphia, PA
0.147 NTU (1995), 0.063 NTU (1998)

Oklahoma City, OK
0.09 NTU (1999)

High Point, NC
Target: Average finished water turbidity of less than 0.5 NTU
Actual: 0.12 NTU (1997), 0.122 NTU (1998)

Fort Worth, TX
Percentage of samples with turbidity of 0.20 NTUs or less: 85% (1997)

Gainesville, GA
Percentage of samples with turbidity of 0.50 NTU or less: 99.67% (1998)

Cary, NC
Percentage turbidity removed: 99.99% (1999)

Raleigh, NC
Percentage turbidity removed: 98.94% (1997), 98.65% (1998)

CHLORINE

High Point, NC
Average chlorine concentration in finished water: 3.9 ppm (1997), 3.2 ppm (1998)

Chandler, AZ
Average free chlorine throughout distribution system: 1.3 ppm (1995)

Ocala, FL
0.60 mg/L (1996)

FLUORIDE

High Point, NC
Average fluoride concentration in distribution: 1.01 ppm (1997), 0.98 ppm (1998)

Ocala, FL
0.79 mg/L (1996)

IRON

Ames, IA
Average iron concentration in system: 0.09 mg/L (1997)

TRIHALOMETHANES (THM)

Oklahoma City, OK
THM average compared to maximum allowable levels: 47/80 (1999)

COLIFORM

Ocala, FL
0/100 ml (1996)

pH

High Point, NC
Average finished water pH (standard: 7.6): 7.60 (1997), 7.52 (1998)

HARDNESS/SOFTNESS

Ames, IA
Average hardness: 9.7 grains/gallon (1997)

Ocala, FL
Hardness: 115 mg/L (1996)

mance targets for other aspects of water operations. Consider, for example, performance targets regarding water storage and system capacity. Charlotte, North Carolina, attempts to maintain a daily treated water capacity exceeding average daily demand by a factor of at least 1.5. Portland, Oregon, strives to maintain in storage the equivalent of three times the average daily demand. Fort Collins, Colorado, attempts to provide sufficient water supply to meet the demand of a once-in-50-year drought and reported a ratio of a water supply to demand of 2.15 in the early 1990s. Interpretations of "adequate" water supply differ from place to place.

Similarly, standards for "good" water pressure vary widely. Not every system subscribes to the water pressure prescriptions of 30 to 40 pounds per square inch (psi) offered above by Foster (1978). Tallahassee, Florida, for example, attempts to maintain system pressure of at least 20 psi. Portland, Oregon, strives for pressure between 20 and 110 psi. Sunnyvale, California, has a system target of 40 to 105 psi, and Houston, Texas, strives for 50 psi.

"Line loss" is a phrase sometimes used interchangeably with "unaccounted water." Actually, the two are not synonymous. Unaccounted water typically

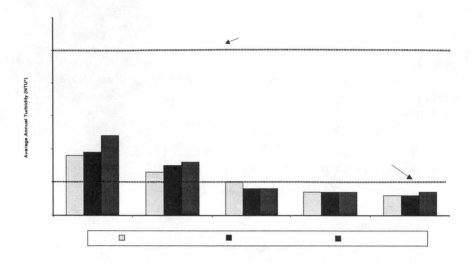

Figure 32.1. Keeping Turbidity Low in Philadelphia's Drinking Water
SOURCE: City of Philadelphia (PA), *Mayor's Report on City Services (Fiscal Year 1998)*, p. 38.

includes more than line loss alone. Unaccounted water is the difference between the amount of treated water entering the distribution system and the
amount metered (and usually billed) for use by individual customers or other
authorized users. Steady leakage and intermittent main breaks (i.e., line loss)
are major culprits, but unaccounted water may also result from inaccurate meters, theft, and unmetered water used for fire fighting, hydrant flushing, street
cleaning, and other legitimate municipal purposes. Urban Institute researchers found the average of unaccounted-for water among 49 major cities to be
15% in 1980 (Peterson, Miller, Godwin, & Shapiro, 1984, p. 36). Highest
rates of line loss were reported in northeastern cities, cities experiencing high
levels of distress, and cities with older housing stock. Urban Institute researchers contended that unaccounted-for water exceeding 15% to 20% is generally
considered to be a problem. AWWA rules of thumb place the acceptable rate
of unaccounted-for water between 5% and 25% (U.S. General Accounting
Office, 1980, p. 28).

Among municipal documents examined for this volume, several reported
performance targets and experience with main breaks and unaccounted water
losses. Winston-Salem, North Carolina, and Ames, Iowa, for example, each
reported the number of water main breaks experienced per 1,000 miles of water main. Winston-Salem had 111.7 main breaks per 1,000 miles of water
main in 1998. Ames had 100.8 breaks per 1,000 miles in 1997. These figures
compare favorably with national data from studies in the 1980s conducted by
the Urban Institute and the U.S. General Accounting Office. In the Urban

Institute study, the average number of water main failures among 34 major U.S. cities was 229 per 1,000 miles of pipe (Peterson et al., 1984). The U.S. General Accounting Office study (1980) reported a nearly identical average of 231 breaks per 1,000 miles of pipe.

Several cities emphasize quick response and prompt repair of main breaks—some striving to maintain an average response time of less than 30 minutes and others targeting completion times within 4 or 5 hours (Table 32.7). Unaccounted water—water sent into the distribution system but not subsequently metered or billed—is reported by many cities, with most experiencing losses of 4% to 15% (Table 32.8).

Water maintenance personnel typically are responsible for installing, repairing, and replacing fire hydrants. Perhaps not surprisingly, cities differ in their expectations regarding hydrant reliability and the speed at which deficiencies should be corrected. New York City reported finding approximately 1% of its hydrants broken or otherwise inoperative in the early 1990s. Savannah, Georgia, discovered not a single unreported dry hydrant at a fire scene between 1988 and 1991. When fire hydrants needed to be repaired, the work was done within 14 days in Savannah. Portland, Oregon, repaired fire hydrants and returned them to service within 5 days. Cary, North Carolina, attempts to repair or replace defective hydrants within 5 days of notification. Lubbock, Texas, reported that in a recent year, it completed such work within 24 hours in every case.

Promptness of water connections is another aspect of service tracked by several municipalities (Table 32.9). Still other maintenance services are reported by individual cities in Table 32.10.

In addition to the work associated with water treatment and maintenance of the water distribution system, another key aspect of water services—especially from the standpoint of system revenues—is the maintenance and reading of water meters. To ensure the system's financial integrity, meters must operate at a high degree of accuracy. To achieve and sustain such accuracy, many municipalities establish regular programs of testing and replacement. Fayetteville, Arkansas, for example, tests at least 6% of its water meters annually. Sunnyvale, California, attempts to test its large meters—2-inch meters or larger—on a 2-year cycle. Decatur, Illinois, replaces any meter exceeding 10 years of age. In Sunnyvale, $5/8$-inch to 1-inch disc meters are on a 15-year replacement schedule and $1\frac{1}{2}$- to 2-inch disc meters are on a 10-year replacement schedule. The meter replacement policy in Hurst, Texas, calls for large commercial meters to be tested annually and replaced when inaccuracy exceeds 5%. Residential meters are replaced every 1 million gallons or every 10 years, whichever occurs first.

Even if meters are registering water flow accurately, meter-reading errors can negate that precision. Therefore, cities often take steps to ensure low error rates by meter readers. Many cities report error rates of less than one half of 1% (see Chapter 31, Table 31.5).

Table 32.7 Water Main Failures: Performance Targets and Experience of Selected Cities

Contending With Main Breaks

Quick Response	*Prompt Restorations*
San Clemente, CA Percentage of responses to working hour emergencies within 15 minutes: 100% (1998) Percentage of responses to non–working hour emergencies within 30 minutes: 100% (1998)	**Norman, OK** Average repair time (water-off to water-on): 2 hours (1997), 1.29 hours (1998)
Denton, TX Average response time: 15 minutes (1991)	**Dayton, OH** Target: Hold 80% of emergency mainline shut-downs to a 2-hour maximum
Flagstaff, AZ Average response time to emergencies: 20 minutes (1997)	**Decatur, IL** 4-hour average repair time (1995)
Portland, OR Target: Response to system emergencies within 30 minutes Actual: 100% within 30 minutes (1991)	**Oklahoma City, OK** Target: Repair main breaks in an average of 4.5 hours or less Actual: 4.0 hours (1999)
Sunnyvale, CA Target: Respond to at least 95% of emergency calls within 30 minutes (1994)	**Palo Alto, CA** Percentage of service disruptions due to main line leak repairs limited to 4 hours or less: 86.9% (1997)
Oklahoma City, OK Target: Average response time of 30 minutes to emergency main breaks Actual: 93% (1999)	**Raleigh, NC** Percentage of water main breaks repaired within 5 hours: 75% (1998)
Charlottesville, VA Percentage of water service emergency calls responded to within 30 minutes: 74% (1998)	**Tucson, AZ** Target: Complete 90% of emergency repairs within 5 hours; no repair exceeding 8 hours if water service interrupted
Danville, KY Target: Respond to after-hours water breaks and service emergencies within 2 hours of notification Actual: 100% (1997)	**Hurst, TX** Target: Average repair time of 5 hours or less for main breaks Actual: 6-hour average (1990); less than 5-hour average (1997)
St. Petersburg, FL Target: Respond to at least 75% of reported damage to the system within 3 hours Actual: 96.6% (1995)	**Rock Island, IL** Target: Restore water service within an average of 12 hours after a main break Actual: 7.2-hour average (1991)
	Philadelphia, PA Average time to complete water main repair: 8.3 hours (1998)

When water systems fail to attain local objectives for water quality, officials may hear the news as quickly from angry citizens as from lab technicians. Foul-tasting water, brown water, or no water at all is sure to prompt several calls. An effective water treatment and distribution system, however, should

Table 32.8 Unaccounted Water in Selected Cities

UNACCOUNTED WATER LOSSES

Wichita, KS
3.1% (1990), 1.7% (1991)

Irving, TX
3.24% (1998)

Addison, TX
Target: 5% or less
Actual: 3.8% to 6% (1988–1991)

Lubbock, TX
4% (1996), 5% (1997)

Rockville, MD
4.4% (1997)

Chandler, AZ
7% (1995), 4.76% (1998)

Oak Park, MI
8.4% (October 1997), 6.4% (November 1997)

Fort Collins, CO
Target: 7% or less
Actual: 7% (1991)

College Station, TX
8.0% (1998)

Ames, IA
8.3% (1997)

Winston-Salem, NC
8.56% (1998)

Corvallis, OR
Target: 5% or less (water treated but not metered)
Actual: 8.88% (1992)

Duncanville, TX
9.0% (1998)

Greensboro, NC
9% (1997)

Boca Raton, FL
9.2% (1997)

San Clemente, CA
Target: 5%
Estimated: 9.5% (1999)

Plano, TX
10% (1997)

Oak Ridge, TN
10.76% (1997)

Albuquerque, NM
Target: 7% or less
Actual: 11.5% (1998)

Danville, KY
12% (1997)

St. Petersburg, FL
12.0% (1995)

Kalamazoo, MI
15.6% (1995), 13.9% (1996)

Victoria, TX
15% (1996)

High Point, NC
20% (1998)

Durham, NC
25% (1998)

Brentwood, TN
27.2% (1996)

Philadelphia, PA
Percentage of water pumped that is not billed to customers: 30.97% (1998)

Kingsport, TN
Target: 20% or less
Actual: 31% (1997), 25% (1998)

hear few such complaints—the rate in Norman, Oklahoma, was 0.144 water quality complaints per 100 million gallons purified and pumped in 1998 (Table 32.11). Despite best efforts, a few complaints are virtually inevitable. When they do occur, top water departments respond quickly (Table 32.12).

Proper staffing for water system efficiency and effectiveness requires balancing the desire to have enough well-qualified employees to ensure sound

Table 32.9   Prompt Water Connections: Selected Cities

Chandler, AZ
Average turnaround time for meter and service installation: 4 days (1998)

College Station, TX
Percentage of new services installed within 5 working days of connection request date: 88% (1995)

Winston-Salem, NC
Percentage of ¾" water connections installed within 10 days of permit issuance: 100% (1997)

Palo Alto, CA
Target 1: Review plans for water service extensions within 10 working days
Actual: 88.5% (1997)
Target 2: Install approved water services within 38 days of receipt of payment
Actual: 100% (1997)

Table 32.10   Water System Odds and Ends: Selected Cities

ADEQUATE WATER PRESSURE

Bellevue, WA
Percentage of customers with at least 1,000 gpm fire flow protection: 92.0% (1997)

HYDRANT MAINTENANCE

Scottsdale, AZ
Percentage of hydrants receiving preventive maintenance annually: 22.0% (1999)

PROMPT REPAIR/REPLACEMENT
OF NONFUNCTIONING HYDRANTS

Savannah, GA
Percentage of nonfunctioning hydrants repaired within 14 days: 100% (1997)

VALVE MAINTENANCE

Scottsdale, AZ
Percentage of values receiving preventive maintenance annually: 7.7% (1999)

WATER METER REPLACEMENT/REPAIR

Lubbock, TX
Target: Repair or replace all malfunctioning meters within 30 days of notification

---

operations and quick response to emergencies with the desire to avoid over-staffing. A good sense of workload, as well as of the human resources needed to match that workload, is necessary to strike the appropriate balance between those desires.

Some cities report overall ratios of employees to an approximation of workload—perhaps water and sewer employees per 1,000 population or wa-

Table 32.11   Water Quality Complaints: Selected Cities

COMPLAINTS PER 100 MILLION GALLONS

Norman, OK
Complaints per 100 million gallons: 0.144 (1998)

Flagstaff, AZ
Water quality complaints per 100 million gallons: 0.27 (1997)

COMPLAINTS PER 1,000 CUSTOMERS

Raleigh, NC
Water quality complaints per 1,000 customers: 1.60 (1998)

Ocala, FL
Water quality/pressure complaints per 1,000 customers: 2.23 (1996)

COMPLAINTS PER 10,000 RESIDENTS

Chandler, AZ
Water quality complaints per 10,000 residents: 19.9 (1995), 16 (1998)

ter distribution employees per 100 pipeline miles. Overall rates vary from one community to another and offer only a rough gauge for judging the adequacy of a system's employment level. More precise calculations of human resource requirements can be made using labor ratios for various water system tasks. The labor ratios reported by several municipalities are shown in Table 32.13.

Sewage Collection and Treatment

The quality of a community's wastewater collection and treatment system is of concern not only to local residents but also to state and federal environmental protection officials. Sludge from the treatment process is considered clean if various pollutants do not exceed allowable concentrations (Table 32.14).

Many cities report compliance with state and federal standards or regulations as evidence of sewage collection and treatment effectiveness. Some report the percentage of permit standards met during a given year. Some calculate the percentage of days in compliance with state and federal regulations. Others report the number of combined sewer overflows, the percentage of sewage bypassing treatment, the percentage of industrial samples in compliance with discharge limits, or the percentage of stream samples in compliance with ambient standards.

The two most common gauges of wastewater plant performance focus on the quality of treated water as it returns to the receiving stream. One measure

Table 32.12   Responsiveness to Water Treatment and Distribution Complaints and
Requests: Selected Cities

City	Response to Complaint or Miscellaneous Service Request
Tempe, AZ	Percentage of service calls responded to within 30 minutes and repaired within 24 hours: 100% (1997)
Tucson, AZ	Percentage of site investigations regarding water cleanliness, clarity, flow, or pressure performed within 3 hours of customer's request: 94% (1998, estimated)
Chandler, AZ	Percentage of responses to service requests and water quality complaints within 24 hours: 100% (1995) Percentage of problems/disputes resolved within 2 days: 99% (1995)
Savannah, GA	Target: Process at least 99% of all water tap and water connection requests within 2 workdays; respond to water quality complaints within 1 workday Actual: 100% of taps processed within 2 workdays; 100% of water quality complaints responded to within 1 workday (1997)
Scottsdale, AZ	Percentage of water quality and pressure complaints responded to within 24 hours: 100% (1999)
Tallahassee, FL	Target 1: Respond to water quality complaints within 24 hours Actual: 100% (1991) Target 2: Respond within 48 hours to requests to identify water line locations Actual: 100% (1991)
Danville, KY	Average response time to customer requests: 24 hours (1997)
Duncanville, TX	Percentage of service orders worked within 24 hours: 96% (1990)
Lubbock, TX	Percentage of work orders completed within 24 hours of receipt: 59% (1997) Percentage of work orders completed within 3 days of receipt: 79% (1997)
Phoenix, AZ	Target: Respond to at least 98% of contractor-requested utility locations, shutdowns, and taps within 2 days of notification
Winston-Salem, NC	Percentage of ¾" water connections and 4" sewer connections installed within 10 days of permit issuance: 100% (1997)
New York, NY	Percentage of leak complaints requiring excavation that are resolved within 30 days: 84.1% (1997)

pertains to biological oxygen demand (BOD) and the other to total suspended solids (TSS) in the water. Neither BOD nor TSS is desired; wastewater treatment is designed to remove as much of each as possible.

High levels of BOD and TSS can be harmful to lakes and streams. BOD is a measure of the oxygen required to decompose organic material. A high level of BOD in a pond or stream threatens the oxygen supply, and therefore the survival, of fish in that body of water. Suspended solids include organic and inorganic particles present in water. High levels of TSS restrict light, coat lake and stream beds, and thereby threaten aquatic life.

Measures of BOD and TSS commonly are reported in one of two ways: (a) the percentage of BOD or TSS removed from the water or (b) the concentra-

Table 32.13 Production Ratios for Water Services: Selected Cities

---

WATER TESTING

Kansas City, MO
General lab tests per associate: 30,000 (1997, estimated)
Samples collected per sample collector: 6,400 (1997, estimated)

Duncanville, TX
Average worker hours per sample collected: 0.2

Wichita, KS
Number of lab tests per worker hour: 5.64 (1991)

WATER MAIN REPAIR

Santa Ana, CA
Total hours per main break repair: 98.6 (1987)

Duncanville, TX
Average worker hours per major leak repair: 20.39

Dunedin, FL
Feet of water main replacement per worker hour: 1.7 (1992)

MINOR LEAK REPAIR

Duncanville, TX
Average worker hours per minor leak repair: 4.65

VALVE MAINTENANCE

Duncanville, TX
Average worker hours per valve stack cleaned and raised: 1.33

FIRE HYDRANT INSTALLATION/MAINTENANCE

Duncanville, TX
Average worker hours per fire hydrant installation: 12.0

Dunedin, FL
Worker hours per hydrant repair/replacement: 10.9 (1992)

Santa Ana, CA
Worker hours per hydrant testing and repair: 2.25 (1987)

SERVICE INSTALLATION

Wichita, KS
Average worker hours per service installation: 15.57 (1991)

Winston-Salem, NC
Worker hours per ¾-inch water connection installed: 15.29 (1991), 16.85 (1992), 15.42 (1997)

CUSTOMER SERVICE TURN ON/OFF

Dunedin, FL
Turn on/offs per worker hour: 3.5 (1992)

METER CHANGE OUTS

Winston-Salem, NC
Work hours per meter changed out: 0.33 (1997)

Table 32.14 Allowable Pollution Concentrations: Sewage Sludge

Pollutant	Monthly Average Concentration (milligrams/kilogram)[a]
Arsenic	41
Cadmium	39
Copper	1,500
Lead	300
Mercury	17
Nickel	420
Selenium	100
Zinc	2,800

SOURCE: *Code of Federal Regulations,* Protection of Environment, 40 § 503.13 1999, Table 3, p. 756.
NOTE: If bulk sewage sludge is to be applied to agricultural land, a forest, a public contact site, or a reclamation site, the concentration of each pollutant in the sludge must not exceed the allowable limits noted above, unless cumulative loading rates fall within allowable maximums as specified in *Code of Federal Regulations,* 40 § 503.13 1999.
a. Dry weight basis.

tion of BOD or TSS remaining in the water following treatment. A high percentage of BOD and TSS removal is desirable, as is a low concentration of BOD and TSS remaining in the effluent following treatment. Most cities that routinely report BOD and TSS removal indicate high percentages removed —often well above 90% (Table 32.15). Targeted concentrations in the treated effluent typically are 20 milligrams per liter or less.

Wastewater systems that strive to improve effectiveness and efficiency focus not only on effluent quality but also on operating costs. A recent benchmarking project sponsored by the Water Environment Research Foundation included an examination of expenditure allocations among 31 public utilities (Olstein et al., 1996). Average expenditures for chemicals and labor, excluding fringe benefits, constituted two thirds of the costs of wastewater plant "wet" operations, with fringe benefits, the cost of power, and other expenses making up the rest (Figure 32.2). Plant operators with high-cost facilities who are searching for the cause of their expenditure problems might compare their own expense allocations to the benchmark average to pick up a few clues.

The treatment of wastewater is only part of the job of sewer departments. Another major part is the maintenance of the sewage collection system. Considerable variation in operating patterns and maintenance demand is evident among the cities reporting performance targets and operating statistics for wastewater collection systems. Some cities report relatively few stoppages or backups per mile of sewer line, whereas others experience much heavier demand for maintenance services. Boise, Idaho, for example, reported only 5.6 stoppages per 100 miles of sewer line—only a tiny fraction of the rate re-

Table 32.15   Common Indicators of Wastewater Treatment Effectiveness:
BOD & TSS in Selected Cities

Biological Oxygen Demand (BOD)		Total Suspended Solids (TSS)	
Percentage of BOD Removed	*Remaining BOD in Effluent*	*Percentage of TSS Removed*	*TSS Remaining in Effluent*
Raleigh, NC 98.3% (1998)	**Rock Island, IL** Target: 20 mg/L or less Actual: 5.1 (plant 1), 0.6 (plant 2) (1990); 4.2 (plant 1), 0.6 (plant 2) (1991)	**Denton, TX** 99% (1991)	**Rock Island, IL** Target: 25 mg/L maximum Actual: 19.8 (plant 1), 0.6 (plant 2) (1990); 16.8 (plant 1), 1.0 (plant 2) (1991)
Denton, TX 98% (1991)		**Corpus Christi, TX** 98% (1991)	
Oak Ridge, TN 98% (1992), 97% (1997)		**San Diego, CA** 96% (1998)	
Greeley, CO 97% (1997)	**Raleigh, NC** Limits: 5 mg/L (summer), 10 mg/L (winter) Actual: 1.30 mg/L (1997); 2.30 mg/L (1998)	**Boca Raton, FL** 95% (1997)	**Ocala, FL** Target: Less than 5.0 mg/L Actual: 1.63 mg/L (1996)
Danville, KY 96.14% (1997)		**Greeley, CO** 95% (1997)	
Corpus Christi, TX 96% (1991)		**Houston, TX** Target: Remove at least 92% of suspended solids Actual: 95% (1992)	**Raleigh, NC** 2.00 mg/L (1997), 2.40 mg/L (1998)
Fort Collins, CO Target: Maintain organic removal percentages above 90% Actual organic removal percentage: 95.0% (1991)	**Ocala, FL** Target: Less than 20.0 mg/L Actual: 2.28 mg/L (1996)		**Corpus Christi, TX** 6 mg/L (1991) 6.4 mg/L (1995)
	Corpus Christi, TX Effluent BOD: 8 mg/L (1991), 6.4 mg/L (1995)	**Danville, KY** 89.36% (1997)	
Portland, OR Target: 85% or greater Actual: 93.8% (1998)			

ported by several other cities (Table 32.16). Such a result is only mildly surprising, for previous studies have shown remarkable variation among cities in the number of sewer breaks and backups, a variation attributed less to differences in the age of pipeline than to "problems of root infiltration, ground movement and poor bedding soils, faulty pipe installation, and inadequate protection from surface weights" (Peterson et al., 1984, p. 28).

Top performing sewer maintenance operations respond quickly to sewer emergencies, as well as to miscellaneous complaints (Table 32.17). They are also good at the more routine elements of the operation (Tables 32.18, 32.19 and 32.20).

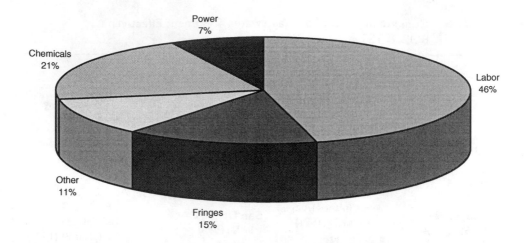

Figure 32.2. Average Expenditure Allocations for Wastewater Treatments Plant "Wet Operations" Among 31 Selected Utilities (Excluding Depreciation)
SOURCE: Adapted from Myron Olstein et al., *Benchmarking Wastewater Treatment Plant Operations: Collection, Treatment, and Biosolids Management* (Alexandria, VA: Water Environment Research Foundation, 1997), p. 9.9. Project 96-CTS-5. Reprinted by permission.

As with staffing for other municipal operations, the proper staffing of wastewater collection and treatment operations requires a reasonable sense not only of the anticipated workload but also of the human resources required to handle assorted tasks. Among the participating utilities in the Water Environment Research Foundation's benchmarking project, the average size of collection system repair crews was 3.4 members (Olstein et al., 1996, p. 9.10). Production ratios for a variety of wastewater services in several cities are provided in Table 32.21.

Note

1. Legislation affecting utilities often has pitted environmental and economic development proponents against one another. Local officials with direct managerial or general oversight responsibilities for public utilities are well advised to stay abreast of the latest developments and current standards.

Table 32.16 Frequency of Sewer Stoppages: Selected Cities

STOPPAGES PER 100 MILES OF SEWER LINE

Boise, ID
Number of sewer backups per 100 miles of sewer line: 5.6 (0.056/mi.) (1991)

Anaheim, CA
Target: No more than 33.3 stoppages per 100 miles of sanitary sewer line
Actual: 15 (1991)

Ocala, FL
Stoppages per 100 miles of sewer main: 16 (1996)

Duncanville, TX
Main line stoppages per 100 miles of sewer line: 21 (1998)

Greenville, SC
Stop-ups per 100 miles of sewer line: 26 (1990)

Wichita, KS
Sewer stoppages per 100 miles of sewer line: 29.7 (1998)

Corvallis, OR
Target: No more than 10 sewer backups per 100 miles (0.1/mi.)
Actual: 37.4 (0.374/mi.) (1992)

Winston-Salem, NC
Main stoppages per 100 miles of sewer main: 45 (1991), 48 (1992), 48 (1997)

Fort Collins, CO
Blockages per 100 miles of line: 53 (1991)

Durham, NC
Sewer blockage calls per 100 system miles: 95 (1998)

BACKUPS PER 1,000 SERVICE CONNECTIONS

Bellevue, WA
Public sewer system backups per 1,000 service connections: 0.61 (1998)

Fort Collins, CO
Backups per 1,000 customers: 0.7 (1996)

Table 32.17 Responsiveness to Sewer Emergencies and Complaints: Selected Cities

RESPONSIVENESS TO EMERGENCIES

Norman, OK
Targets: Average response time of 30 minutes or less during the day; 1 hour at night
Actual: 14.5-minute average during the day (1997); 14.5-minute average (1998)

San Clemente, CA
Percentage of responses to working hour emergencies within 15 minutes: 100% (1998)
Percentage of responses to non–working hour emergencies within 30 minutes: 100% (1998)

Denton, TX
Average response time: 15 minutes
Average downtime: 30 minutes (1991)

Sunnyvale, CA
Target: Respond to emergency malfunction calls and hazardous discharge notifications within 20
  minutes

Chesapeake, VA
Percentage of emergencies responded to within 30 minutes: 100% (1997)

Rockville, MD
Percentage of stoppages investigated within 30 minutes: 100% (1997)

Hurst, TX
Target: Respond to sewer main restriction complaints within 30 minutes
Actual average response time to reports of sewer main stoppage: Less than 30 minutes (1997)

Charlottesville, VA
Percentage of emergency calls responded to within 30 minutes: 74% (1998)

Cary, NC
Target: Respond within 45 minutes to pump station trouble alarms
Actual: 100% within 45 minutes (1999)

Lubbock, TX
Average response time: 48 minutes (1997)

Reno, NV
Target: Respond to sanitary sewer problems within 1 hour during the daytime; within 2 hours at
  night (1999)

Oklahoma City, OK
Target: Complete top-priority jobs in average of 1.75 hours or less
Actual: 1.37 hours (1999)

Calgary, Alberta
Percentage of responses to failures in the sanitary sewer system within 2 hours of notification:
  80% or greater (1992–1996)

Raleigh, NC
Percentage of sewer main obstructions cleared within 5 hours: 96% (1998)

New York, NY
Average response time to backups: 8.1 hours (1997), 6.2 hours (1998)
Percentage resolved within 24 hours: 97.7% (1997), 99.5% (1998)

Corpus Christi, TX
Target: Resolve at least 99% of customer service problems within 1 day (1999)

RESPONSIVENESS TO COMPLAINTS AND GENERAL MAINTENANCE

St. Petersburg, FL
  Target: Respond to 95% within 3 hours
  Actual: 95.0% within 3 hours (1995)

Orlando, FL
  Target: Respond to citizen complaints within 4 hours

Cary, NC
  Percentage of odor complaints responded to within 24 hours: 100% (1999)

Cambridge, MA
  Average number of days to close unscheduled requests for maintenance service (e.g., sewer cover
    repair, catch basin cleaning): 10.08 (1995), 8.0 (1996, estimated)

Kansas City, MO
  Percentage of Action Center complaints resolved within 2 weeks: 100% (1995)

Table 32.18   Routine Sewer Cleaning Cycles: Selected Cities

Peoria, AZ
  Targets: Clean entire system (lines under 12 inches) annually; clean all manholes and clean-outs
    within 18 months; clean larger mains as needed

Fort Collins, CO
  Percentage of wastewater lines hydrocleaned per year: 78% (1991), 70% (1993, projected)

St. Charles, IL
  Target: Clean all sanitary sewer lines and manholes at least once every 2 years

Upper Arlington, OH
  Target: Inspect and clean all sanitary sewer lines every 4 years

Table 32.19   Prompt Sewer Connections: Selected Cities

College Station, TX
  Percentage of new services installed within 5 working days of connection request date: 100% (1995)

Winston-Salem, NC
  Percentage of 4″ sewer connections installed within 10 days of permit issuance: 100% (1997)

Palo Alto, CA
  Target 1: Review service extension plans within 10 working days
  Actual: 10 working days (1997)

  Target 2: Install approved new customer services within 35 days of receipt of payment
  Actual: 40 days (1997)

Table 32.20   Industrial Water Pollution Inspection and Testing: Selected Cities

Tempe, AZ
  Percentage of industries inspected twice a year: 93% (1997); once a year: 7% (1997)
  Percentage of industries sampled 4 times a year: 82% (1997); twice a year: 18% (1997)
  Percentage of permitted industries in significant noncompliance: 5% (1997)

Portland, OR
  Percentage of industrial enforcement tests in full compliance: 94% (1998)

Table 32.21   Production Ratios for Wastewater Services: Selected Cities

POLLUTION CONTROL

Reno, NV
  3 worker hours per industrial user inspection; 2 worker hours per industrial user sample; 4 worker
  hours per treatment sample (1990)

SEWER LINE INSPECTION

San Clemente, CA
  Target: 1500 lineal feet video-inspected per crew day (2000)

Fayetteville, AR
  700 feet per 8 hours, using TV camera (3-person crew)

SEWER CLEANING

Duncanville, TX
  Average worker hours per mile of sewer main cleaned: 15.91

Winston-Salem, NC
  Worker hours per mile of sewer main cleaned: 18.74 (1991), 16.02 (1992), 18.38 (1997)

Santa Ana, CA
  Worker hours per mile of sewer main cleaned: 10.82 (1987)

San Clemente, CA
  Target: 2,500 lineal feet of collection line cleaned per worker day (2000)

Fayetteville, AR
  Sewer line cleaned per worker day: 2000 feet (4,000 feet for 2-person crew)

Pasadena, CA
  Miles of sewer cleaned per person day: 0.35 mile (assumes 7-person crew with equipment)

Richmond, VA
  Linear feet of sewer lines "jet cleaned" per crew day: 900 feet

Greenville, SC
  Miles of sewer line cleaned per worker per month: 1.5 (1987), 1.7 (1988), 2.3 (1989), 3.2 (1990),
    3.7 (1991)

Dunedin, FL
  Feet of sewer line cleaned per labor hour: 68.9 (1992)

MANHOLE REPAIR

Fayetteville, AR
5.3 worker hours per manhole repair

MAIN LINE REPAIR

Santa Ana, CA
11.82 worker hours per sewer main blockage/repair (1987)

Duncanville, TX
16.25 worker hours per repair

Fayetteville, AR
20 worker hours per repair

SERVICE LINE REPAIR

Duncanville, TX
14.77 worker hours per repair

SERVICE LINE STOPPAGE REMOVAL

Duncanville, TX
2.56 worker hours per stoppage

NEW SERVICE INSTALLATION

Duncanville, TX
15.74 worker hours per new service installation

33

Performance Milestones

A city may be judged on several different dimensions. Is it a good place to live? Is it a decent place to raise a family? How are the schools? Is the climate pleasant? What entertainment opportunities are available? Does the community support the arts? Is the local economy healthy? Is it growing? Is the community clean—physically and politically? Are municipal services good?

A city deemed to be outstanding on one dimension might be decidedly *un*remarkable—banished from the list of top cities—on another. Being number one is more than sheer luck, but it helps if the criteria used to establish the ranking coincide with the community's strengths and overlook points of weakness. On some lists, it also helps to have strong neighbors.

A city's reputation—favorable or unfavorable—may belong to an entire region. Reference to *city* often means *community* in a fairly broad sense. For instance, travelers who say they are from Dallas and report favorably on their hometown and its local government services actually might reside in suburban Mesquite or Plano and, without intending any deceit, may be reporting on living conditions and services in one of those cities. In such cases, the entire metropolitan area takes on a common identity symbolized by the central city. Some lists are no more exact than that.

Among local government officials, the definition of *city* is much more precise, and the relevance of *city rankings* is much less ambiguous. To them, city refers to the corporate municipal entity and its geographic territory. In this sense, it might be all right to judge a *community* on its climate, but a *city* should be judged only on political characteristics, economic and social conditions, service quality, and other factors that local government officials perhaps can influence.

Popular Ratings

Community ratings are popular. People like to know how their town stacks up against others. Some rankings are consistent with the narrower definition of city, but the authors of many lists have little concern with precise municipal boundaries and adopt a broader community view. The popular ranking found in the *Places Rated Almanac* (Savageau & D'Agostino, 2000; Savageau & Loftus, 1997) is like that. Although city names are used, the authors carefully explain that they are rating metro areas. The top "places" to live—Orange County, California, and Salt Lake City–Ogden, Utah, headed the lists in 1997 and 2000, respectively—are selected on the basis of climate, five types of facilities (health care, education, recreation, transportation, and the arts), and three economic and social indicators (crime, costs of living, and jobs). Rated as the best small metro area (population less than 250,000) in 2000 was Portland, Maine.

Similarly, recognition as an outstanding locale for business, retirement living, or affordable housing typically is an honor intended for the named city and its environs rather than strictly for that city alone. *Forbes* magazine, for example, has its list of "Best Places for Business Growth"; *Money* ranks "The Best Places to Live in America"; *Entrepreneur* names the "20 Best Cities for Small Business"; Reliastar Financial Corporation rates the 125 largest metropolitan areas for the financial security of residents; and Zero Population Growth ranks cities in terms of children's health, safety, and economic well-being. Although such lists are no doubt of interest to municipal officials —and especially to their chamber of commerce brethren—such ratings are of dubious value as scorecards for municipal government.

Several national programs do respect municipal boundaries, recognize the subtleties of local identity, and concentrate on civic matters, but even some of these focus on aspects of municipal government other than the effective delivery of traditional services. Citizens and local government officials desiring models of citizen participation, for example, may consult the honor roll of recipients of the National Civic League's (NCL) All-America Cities awards. If they wish to assess their own community's "civic infrastructure," they may do so by applying NCL's Civic Index, which emphasizes community vision, cooperation, communication, and leadership (NLC, 1999). The focus of the All-American Cities program extends well beyond municipal service delivery.

A tighter focus on municipal operations was apparent in the report cards prepared in 2000 for America's 35 largest city governments by a review team from Syracuse University and *Governing* magazine under the sponsorship of the Pew Charitable Trusts (Barrett and Greene, 2000). Reporters and researchers assessed the adeptness of the 35 cities in five crucial aspects of local government management—financial management, human resources, information technology, capital management, and managing for results—and assigned letter grades for each. Phoenix snared the only "straight A" report card among the rated cities. Others at the head of the class were Austin in financial managment and Minneapolis in information technology. The *Governing* assessments focused primarily on the management infrastructure of each local government, a crucial ingredient for operational success, rather than on the quality and efficiency of services themselves.

Other popular, third-party assessments of municipal performance extol innovation—especially innovation in a particular local government service. The Program Excellence Awards for Innovations in Local Government Management of the ICMA, the Solutions Awards sponsored by Public Technology, Inc., the Innovations in American Government Program of the Ford Foundation and the John F. Kennedy School of Government at Harvard University in partnership with the Council for Excellence in Government, and the Exemplary State and Local (EXSL) Awards Program sponsored by the National Center for Public Productivity all belong to this genre. Each provides important recognition to recipient governments and encouragement to others, but none of these esteemed programs purports to serve as an assessment device for judging municipal services in general.

Standards for Various Services

Standards and guidelines for several municipal functions, as described in previous chapters, have been offered by professional associations and others having a specific interest in those services. Although often developed by specialists with important insights regarding service delivery problems and possibilities, these guidelines are sometimes vulnerable to self-serving motives and, if adopted, should be applied with that caveat in mind.

Typically, single-service standards or guidelines emphasize process over product, or resource input over performance output. Even where standards for a particular specialty focus on performance objectives or prescribe reasonable performance expectations, they do so out of context, with little regard for the full mosaic of municipal services.

Multiservice Performance Guidelines

Even before Clarence Ridley and Herbert Simon (1943) raised the performance management profile of local government by encouraging comprehensive measurement of municipal activities, a series of halting steps signaled the fits and starts of various efforts to compare municipal performance meaningfully.[1] The rationale always had been that the performance of all units could benefit by the effective exchange of relevant performance data, but securing sufficient commitment and resources proved to be a recurring problem—until progress began to be made in the mid-1990s. Efforts by ICMA and by local governments and universities in individual states working cooperatively to compare performance information offered renewed hope for that strategy.[2]

In addition to the possible emergence of new databases of relevant performance statistics, a few promising initiatives focusing on good management and the delivery of basic municipal services have contributed to the growing interest in performance benchmarks. They have done so by turning attention toward the need for standards, heightening the profile of multicity comparisons, and renewing interest among municipal officials in the possibility of being recognized as a competent service provider or perhaps even as "one of the best." For example, local government practitioners have demonstrated a growing desire to know how their operations stack up against others and a willingness to be held accountable to reasonable standards. Evidence of that desire comes in several forms, not the least of which is the publication of workbooks and guidelines on standards and service levels for local governments in California, Washington, and Pennsylvania (LCC, 1994; *Level of Service Standards,* 1994; SPRPC, 1990).

Benchmarking

Benchmarking is a practice that has grown rapidly in popularity since the 1980s. Although variations are common, their progenitor—corporate-style benchmarking—involves the identification of best-in-class performers, the comparison of local performance outputs and results with those of top performers, the analysis of practices that account for any performance gaps, and the development and implementation of strategies to adjust the performance gap in one's favor.

This volume's collection of recommended standards, guidelines, performance records, and performance targets of individual cities offers a head start. These municipal benchmarks provide a pool of standards and performance data that allows initial assessments and sets the stage for cities wishing to undertake a more extensive, corporate-style benchmarking project.

Although leading municipal performers in several service activities have been identified, others remain untapped. The careful choice of benchmarking partners, chosen for their performance excellence, suitability on relevant characteristics, and willingness to cooperate, is vital to the success of a full-scale benchmarking project.[3]

A municipality can benefit from simply comparing its current performance with relevant benchmarks. In doing so, important strengths and weaknesses may be revealed and local officials may discover aspects of the operation that deserve detailed analysis. They may even decide to rearrange some priorities. To get full value from a benchmarking project, however, a municipality must take the next steps—supplementing the measures found here with others it collects itself, analyzing operations, and developing performance improvement strategies.

Performance Milestones

Resting on one's laurels is hazardous in city government. *Below average* performance is rarely acceptable; *good* performance can be made better; and those judged *the best* can hope to retain that status only through continuous improvement.

The purpose of identifying municipal benchmarks is to provide guideposts to city governments as they attempt to improve operations. Managers who desire easy answers and quick closure will find neither in benchmarking. Instead, they will find a means by which they can gauge the status of their performance and plot changes to enhance it. By comparing their own performance marks with those of other respected cities or with relevant standards, local officials can decide where improvements are needed and may identify models that could prove helpful as they design improvements.

City officials are encouraged to consult the full array of possible benchmarks for each service area. From that array, a subset of performance indicators has been assembled and labeled *performance milestones*. These are not benchmarks—the achievements of best-in-class performers. Benchmarks may be found in the individual chapters on municipal specialties. Instead, performance milestones denote standing just below best in class and serve as guideposts for cities on their way toward the upper echelon. In a sense, they are *near-benchmarks*. They represent good performance, and are worthy objectives for most municipalities (Table 33.1).

The performance targets identified in Table 33.1 are not destinations. They are milestones on the journey to optimum performance. Although a journey to an optimum place really has no end, some routes are more satisfying than others. Cities plotting their course by focusing on the performance milestones are traversing such a route. Those that have already passed the milestones must travel on, scouting out new guideposts on the basis of their own experience. As seasoned travelers know, there are few rest stops along this route.

Notes

1. For example, Downs and Larkey (1986) note that in perhaps the earliest of such initiatives the Census Bureau collected municipal performance statistics for a few years in the 1890s, until controversy over data interpretation and budget problems brought that effort to an end (p. 22).

(notes continue on page 469)

Table 33.1 Performance Milestones

Department/ Function	Performance Variable	Performance Milestone
Animal control	Percentage of emergency responses within 25 minutes, routine responses within 2 hours	90%
	Daily hours of shelter availability or other arrangement for receiving sick/injured animals	24 hours
City attorney	Percentage of oral opinions on routine matters within 24 hours; written opinions on routine matters within 10 days, written opinions on nonroutine matters within 30 days	90%
City clerk	Percentage of city council meeting minutes prepared within 10 days of the meeting	80%
	Percentage of requests for municipal records or information fulfilled within 1 day	95%
Courts	Percentage of traffic cases disposed within 60 days; within 90 days	70%/90%
	Collection rate for traffic fines	80%
Development administration	Average review time for subdivision plats and rezoning requests	30 days
	Percentage of commercial building permits processed within 10 days; residential, 3 days	95%
	Percentage of building inspections conducted within 1 day of request	95%
Emergency communications	Percentage of 911 calls answered within 10 seconds	90%
	Percentage of emergency calls dispatched within 2 minutes	75%
EMS	Average response time from initial call to arrival	6 minutes
Finance	Receipt of GFOA certificate for financial reporting	yes
	Percentage of monthly reports issued by 10th of following month	80%
	Amount by which return on investments exceeds average 3-month Treasury bill rate during the same period[a]	10%
Fire service	ISO public protection classification (fire insurance rate)	4
	Percentage of emergency responses within 6 minutes (including dispatch time)	95%
	Percentage of fires contained within room(s) involved on arrival	85%
Fleet maintenance	Equipment availability rate	94%
	Returns for rework	< 3%
Gas and electric services	Percentage of responses to reported gas leaks within 30 minutes	95%
	Retail customers per non–power generation employee	300+
	Line loss	gas, < 2%; electric, < 5%

(continued)

Table 33.1   Performance Milestones (continued)

Department/ Function	Performance Variable	Performance Milestone
Human resource administration	Percentage of positions filled within 90 days of vacancy	80%
	Percentage of position audits completed within 30 days of request	85%
	Percentage of vacancies with at least 3 qualified applicants	90%
Information systems	System availability during users' work hours	98%
	Percentage of responses to problems within 1 hour; resolution within 24 hours	90%
Library	Circulation per capita	7
	Percentage of not-immediately-available material provided within 30 days	85%
Management services— executive offices, budget, management audit	Percentage of recommendations from management audits/studies that are accepted/implemented	90%/70%
	Receipt of GFOA recognition for exemplary budget document	yes
Parking services	Revenues per program dollar spent	$3+
Parks and recreation	Developed/total park acres per 1,000 population	5 acres/7 acres
	Tennis courts per 10,000 population	2
Police	Crime rate as percentage of "projected norm" for population cluster and state or region[b]	90%
	Percentage of emergency responses within 6 minutes from call (includes dispatch time)	95%
	Patrol availability factor (percentage of time available for nondirected patrol)	40%
Property appraisal	Average assessment-to-sales ratio	90% to 100%
	Percentage of appraisals upheld at board of equalization appeal	> 50%
Public health	Local public health statistics as percentage of selected national targets[c]	< 100%
	Annual inspections per food service establishment	3
Public transit	Percentage on-time performance—peak/nonpeak periods	90%/95%
	Fares as a percentage of operating expenditures	35%[d]
Public works— engineering and miscellaneous services	Percentage of low bids for major projects within 15% of engineer's estimate	80%
Purchasing and warehousing	Average processing time from requisition to purchase order— sealed bids/other	60 days/10 days
	Out-of-stock ratio (inventoried items in warehouse)	6% or less

Risk management	Rate of occupational injuries (incidents per 100 workers)	< 8
	"Cost of risk"[e] as a percentage of the local government budget	2% or less
Social services	Percentage of adult job training clients placed in jobs and still employed after 90 days	> 50%
Solid waste collection	Routes completed on schedule	> 95%
	Missed collections per 10,000 scheduled stops	20 or less
	Stops per collection employee per week	1,100
Streets, sidewalks, and storm drainage	Percentage of potholes and road hazards repaired within 2 days of notification	90%
Traffic engineering and control	Percentage of centerlines, lane lines, and symbols on major streets repainted during the year	75%
	Average response time to damaged stop and yield signs; to malfunctioning traffic signals	3 hours; 1 hour
Utilities business office	Adjustments/rebillings due to error	< 0.1%
	Meters read per day per meter reader	350+
Water and sewer services	Compliance with AWWA standards of water quality[f]	100%
	"Unaccounted for" water	< 14%
	Percentage of BOD/TSS removed from wastewater	> 90%

a. *10% greater* does not mean *10 percentage points greater.* If the T-bill rate is 5%, this milestone would be met by a city having an investment return of 5.5%.

b. See Table 20.8.

c. See Tables 22.1 through 22.7.

d. 20%, if serving a population of less than 200,000.

e. A community's total "cost of risk" includes insurance premiums, risk management program cost, and other expenditures in support of risk management, as well as losses incurred in accidents, acts of nature, lawsuits, and other incidents.

f. See Table 32.2.

2. ICMA's Comparative Performance Measurement Consortium, the North Carolina Local Government Performance Measurement Project, and the South Carolina Municipal Benchmarking Project, for example, are designed to compile and share performance information among local governments.

3. Several excellent books have been published on the subject of benchmarking. Some of the best are Camp (1989), Spendolini (2000), Boxwell (1994), Fitz-enz (1993), Karlof and Ostblom (1993), Leibfried and McNair (1992), and Andersen and Pettersen (1996). For more specific application to the public sector, see Keehley, Medlin, MacBride, and Longmire (1997), Bruder and Gray (1994), and *Benchmarking Best Practices in the Public Sector* (1997).

References

Accident facts. (1998). Itasca, IL: National Safety Council.

Ahrens, M. (2000, January). *U.S. experience with smoke alarms and other fire alarms.* Quincy, MA: National Fire Protection Association.

Allan, I. J. (1990, November). *Unreserved fund balance and local government finance* (Research Bulletin series). Chicago: Government Finance Officers Association.

Almy, R. R., Gloudemans, R. J., & Thimgan, G. E. (1991). *Assessment practices: Self-evaluation guide.* Chicago: International Association of Assessing Officers.

American Association of State Highway and Transportation Officials. (1984). *An informational guide for roadway lighting.* Washington, DC: Author.

American Library Association. (1962). *Interim standards for small public libraries.* Chicago: Author.

American Library Association. (1967). *Minimum standards for public library systems.* Chicago: Author.

American national standard practice for roadway lighting (ANSI/IESNA RP-8-83). (1983). New York: Illuminating Engineering Society of North America.

American National Standards Institute. (1995). *American national standard for tree care operations: Tree, shrub and other woody plant maintenance—standard practice.* New York: Author.

American Public Health Association. (1981). *Public swimming pools: Recommended regulations for design and construction, operation and maintenance.* Washington, DC: Author.

American Public Health Association. (1991). *Healthy communities 2000: Model standards* (3rd ed.). Washington, DC: Author.

American Public Power Association. (2000, February). *Selected financial and operating ratios of public power systems: 1996.* Washington, DC: Author.

American Public Transit Association. (1998). *1996 transit fact book.* Washington, DC: Author.

American Water Works Association. (1992). *1992–1993 officers and committee directory.* Denver, CO: Author.

Americans With Disabilities Act of 1990, 42 U.S.C. § 12100 *et seq.*

Ammons, D. N. (1994, Spring). The role of professional associations in establishing and promoting performance standards for local government. *Public Productivity and Management Review, 17,* 281–298.

Ammons, D. N. (1995, January/February). Overcoming the inadequacies of performance measurement in local government: The case of libraries and leisure services. *Public Administration Review, 55*, 37–47.

Ammons, D. N. (1996). Local government standards via professional associations: How useful are they for gauging performance? In A. Halachmi & G. Bouckaert (Eds.), *Organizational performance and measurement in the public sector* (pp. 201–222). Westport, CT: Quorum.

Ammons, D. N. (2000a, Spring). Benchmarking as a performance management tool: Experiences among municipalities in North Carolina. *Journal of Public Budgeting, Accounting and Financial Management, 12*, 106–124.

Ammons, D. N. (2000b). *Performance benchmarks in local government risk management.* Fairfax, VA: Public Entity Risk Institute.

Ammons, D. N. (2000c, July). Taking the lead in risk management. *American City & County, 115*(10), 46–50.

Ammons, D. N., Coe, C., & Lombardo, M. (2000, January/February). Performance-comparison projects in local government: Participants' perspective. *Public Administration Review, 61*, 89-99.

Andersen, B., & Pettersen, P. (1996). *The benchmarking handbook: Step-by-step instructions.* London: Chapman & Hall.

Attanucci, J. P., Jaeger, L., & Becker, J. (1979, April). *Bus service evaluation procedures: A review—Short range transit planning* (Special Studies in Transportation Planning). Washington, DC: U.S. Department of Transportation.

Babachicos, P. F. (1998, December). *Report on NALGA's benchmarking and best practices survey for fiscal year 1998.* Lexington, KY: National Association of Local Government Auditors.

Baker, S. L., & Lancaster, F. W. (1991). *The measurement and evaluation of library services* (2nd ed.). Arlington, VA: Information Resources Press.

Bannon, J. J. (1976). *Leisure resources: Its comprehensive planning.* Englewood Cliffs, NJ: Prentice Hall.

Barr, R. C., & Caputo, A. P. (1997). Planning fire station locations. In A. E. Cote & J. L. Linville (Eds.), *Fire protection handbook* (18th ed., pp. 10.250–10.255). Quincy, MA: National Fire Protection Association.

Barrett, K., & Greene, R. (2000, February). Grading the cities: A management report card. *Governing, 13*, 22-91.

Barrett, K., & Greene, R. (1994, February 1). The state of the cities: Managing for results. *Financial World, 163*, 40–49.

Benchmarking best practices in the public sector. (1997). Research Triangle Park, NC: Southern Growth Policies Board. [A training module produced by the Southern Growth Policies Board and the Southern Consortium of University Public Service Organizations].

Benchmarking for quality in public service fleets. (1993). Iselin, NJ: National Association of Fleet Administrators. [A report prepared for the National Association of Fleet Administrators and the NAFA Foundation by David M. Griffith & Associates].

Bennett, J. (1993, August 2). Budget woes seize fleet buying cycles. *City & State, 10*, 9, 16.

Bens, C. K. (1986, November). Strategies for implementing performance measurement. *Management Information Service Report, 18,* 1–14. [Available from International City/County Management Association, Washington, DC].

Boxwell, R. J., Jr. (1994). *Benchmarking for competitive advantage.* New York: McGraw-Hill.

Brecher, C., & Mead, D. M. (1991). *Managing the department of parks and recreation in a period of fiscal stress.* New York: Citizens Budget Commission.

Broom, C., Jackson, M., Vogelsang Coombs, V., & Harris, J. (1998). *Performance measurement: Concepts and techniques.* Washington, DC: American Society for Public Administration.

Brown, Z., & Webster, C. (1913). *Buying list of books for small libraries.* Chicago: American Library Association.

Bruder, K. A., Jr., & Gray, E. M. (1994, September). Public-sector benchmarking: A practical approach. *Public Management, 76,* S9–S14.

Building Service Contractors Association International. (1992). *BSCAI production rate recommendations.* Fairfax, VA: Author.

Butler, R. F. (1985, March). Measuring productivity in local government purchasing. *Texas Town & City, 72,* 41–45.

Cable, S., Longoria, J., & Martin, E. (1996, March). *Performance management: An essential foundation for reengineering government.* Presented at the SAM International Management Conference on Reengineering: Businesses and Business Schools, March 14–16, 1996.

Camp, R. C. (1989). *Benchmarking: The search for industry best practices that lead to superior performance.* Milwaukee, WI: American Society for Quality Control Press.

Carpenter, V. L., Ruchala, L., & Waller, J. B., Jr. (1991). *Public health: Service efforts and accomplishments reporting—Its time has come.* Norwalk, CT: Governmental Accounting Standards Board.

Carter, L., & Vogt, A. J. (1989, Winter). Fund balance in local government budgeting and finance. *Popular Government, 54,* 33–41.

Center for Design Planning. (1979). *Streetscape equipment sourcebook 2.* Washington, DC: Urban Land Institute.

Center on Municipal Government Performance. (1998, September). *How smooth are New York City's streets?* New York: Fund for the City of New York.

Chilton Book Company. (annual). *Chilton's labor guide and parts manual.* Radnor, PA: Author.

Chute, A., & Kroe, E. (1999, February). *Public libraries in the United States: FY 1996* (NCES 1999-306). Washington, DC: U.S. Department of Education, National Center for Education Statistics.

Clawson, M. (1963). *Land and water for recreation.* Chicago: Rand McNally.

Code of federal regulations: Nondiscrimination on the basis of disability by public accommodation and in commercial facilities, 28 § 36 (1994).

Code of federal regulations, Protection of environment, 40 § 503.13 (1999).

Coe, C. (1999, March/April). Local government benchmarking: Lessons from two major multigovernment efforts. *Public Administration Review, 59,* 110–123.

Commission on Accreditation for Law Enforcement Agencies. (1993). *Standards for law enforcement agencies: The standards manual of the law enforcement agency accreditation program.* Fairfax, VA: Author. [Updated periodically]

Cope, G. H. (1987, Spring). Local government budgeting and productivity: Friends or foes? *Public Productivity Review, 41,* 45–57.

Cope, G. H. (1992). Walking the fiscal tightrope: Local government budgeting and fiscal stress. *International Journal of Public Administration, 15*(5), 1097–1120.

Couret, C. (1999, March). Let there be light. *American City & County, 114,* 51–61.

Craven, R. (1997). Fire department apparatus and equipment. In A. E. Cote & J. L. Linville (Eds.), *Fire protection handbook* (18th ed., pp. 10.201–10.214). Quincy, MA: National Fire Protection Association.

Crowell, A., & Sokol, S. (1993, May). Playing in the gray: Quality of life and the municipal bond rating game. *Public Management, 75,* 2–6.

Cummins, R. O., Ornato, J. P., Thies, W. H., & Pepe, P. E. (1991, May). Improving survival from sudden cardiac arrest: The "chain of survival" concept. *Circulation, 83,* 1832–1847. [American Heart Association Medical/Scientific Statement; approved by the American Heart Association SAC/Steering Committee, October 17, 1990]

Donahue, R. (Ed.). (1986). *Park maintenance standards.* Alexandria, VA: National Recreation and Park Association.

Downs, G. W., & Larkey, P. D. (1986). *The search for government efficiency: From hubris to helplessness.* New York: Random House.

Drebin, A., & Brannon, M. (1992). *Police department programs: Service efforts and accomplishments reporting—Its time has come.* Norwalk, CT: Governmental Accounting Standards Board.

Drucker, P. F. (1975). *Management: Tasks, responsibilities and practices.* New York: Harper & Row.

ElectriCities of North Carolina, Inc. (1997). *1997 Key Performance Indicators.* Raleigh, NC: Author.

Ferguson, E. J. (1980). Street maintenance. In G. J. Washnis (Ed.), *Productivity improvement handbook for state and local government* (pp. 677–728). New York: John Wiley.

Few, P. K., & Vogt, A. J. (1997, Winter). Measuring the performance of local governments. *Popular Government, 62,* 41–54.

Fiction catalog. New York: H. W. Wilson. [Published every few years, with supplements]

Fitch Investor Service. (1994). *Fitch ratings.* New York: Author.

Fitz-enz, J. (1993). *Benchmarking staff performance.* San Francisco: Jossey-Bass.

Folz, D. H. (1999, October). Recycling policy and performance: Trends in participation, diversion, and costs. *Public Works Management & Policy, 4,* 131–142.

Foster, W. S. (1978). *Handbook of municipal administration and engineering.* New York: McGraw-Hill.

Fowler, P. (1992, February). The establishment of health indicators in municipal health departments. *Texas Town & City, 79,* 20–24.

Fox, S. F. (1993, Spring). Professional norms and actual practices in local personnel administration: A status report. *Review of Public Personnel Administration, 13,* 5–28.

Glover, M. (1993, October 7). *Performance measurement: Local government efforts and successes.* Paper presented at the annual meeting of the Southeastern Conference on Public Administration, Cocoa Beach, FL.

Gold, S. M. (1973). *Urban recreation planning.* Philadelphia: Lea & Febiger.

Goldstein, N. (1977, April). Biocycle nationwide survey: The state of garbage. *Biocycle, 38,* 60-67.

Gordon, B. B., Drozda, W., & Stacey, G. S. (1969, December 31). *Cost-effectiveness in fire protection* (Working paper). Columbus, OH: Urban Studies Center, Battelle Memorial Institute.

Government Finance Officers Association. (1993). *Distinguished budget presentation awards program: Awards criteria.* Chicago: Author.

Governmental Accounting Standards Board. (1992, December 18). *Preliminary views of the Governmental Accounting Standards Board on concepts related to service efforts and accomplishments reporting.* Norwalk, CT: Author.

Governmental Accounting Standards Board & National Academy of Public Administration. (1997). *Report on survey of state and local government use and reporting of performance measures.* Washington, DC: Governmental Accounting Standards Board.

Granito, J. (1997). Evaluation and planning of public fire protection. In A. E. Cote & J. L. Linville (Eds.), *Fire protection handbook,* (18th ed., pp. 10.29–10.44). Quincy, MA: National Fire Protection Association.

Grizzle, G. A. (1987, Spring). Linking performance to funding decisions: What is the budgeter's role? *Public Productivity Review, 10,* 33–44.

Groves, S. M., & Godsey, W. M. (1980). *Evaluating financial condition.* Washington, DC: International City Management Association.

Groves, S. M., & Valente, M. G. (1986). *Evaluating financial condition: A handbook for local government* (2nd ed.). Washington, DC: International City/County Management Association.

Groves, S. M., & Valente, M. G. (1994). *Evaluating financial condition: A handbook for local government* (3rd ed.). Washington, DC: International City/County Management Association.

Guidelines for public libraries (IFLA Pub. No. 6). (1986). Munich, Germany: K. G. Saur for the International Federation of Library Associations and Institutions.

Haley, A. J. (1985). Municipal recreation and park standards in the United States: Central cities and suburbs, 1975–1980. *Leisure Sciences, 7*(2), 175–188.

Hall, J. R., Jr., & Cote, A. E. (1991). America's fire problem and fire protection. In A. E. Cote & J. L. Linville (Eds.), *Fire protection handbook* (17th ed., Sec. 1, Ch. 1). Quincy, MA: National Fire Protection Association.

Hall, J. R., Jr., & Cote, A. E. (1997). America's fire problem and fire protection. In A. E. Cote & J. L. Linville (Eds.), *Fire protection handbook* (18th ed., pp. 1.3–1.25). Quincy, MA: National Fire Protection Association.

Hall, J. R., Jr., Koss, M., Schainblatt, A. H., Karter, M. J., Jr., & McNerney, T. C. (1979). *Fire code inspections and fire prevention: What methods lead to success?* Boston: National Fire Protection Association.

Handbook for public playground safety (Pub. No. 325). (n.d.). Washington, DC: U.S. Consumer Product Safety Commission.

Handy, G. L. (1993, September). Local animal control management. *Management Information Service Report, 15,* 1–20. [Available from International City/County Management Association, Washington, DC]

Hatry, H. P. (1978, January/February). The status of productivity measurement in the public sector. *Public Administration Review, 38,* 28–33.

Hatry, H. P. (1980a, May/June). Local government uses for performance measurement. *Intergovernmental Personnel Notes,* 13–15. [Available from U.S. Office of Personnel Management, Washington, DC]

Hatry, H. P. (1980b, December). Performance measurement principles and techniques: An overview for local government. *Public Productivity Review, 4,* 312–339.

Hatry, H. P. (1999). *Performance measurement: Getting results.* Washington, DC: The Urban Institute Press.

Hatry, H. P., Blair, L. H., Fisk, D. M., Greiner, J. M., Hall, J. R., Jr., & Schaenman, P. S. (1992). *How effective are your community services? Procedures for measuring their quality* (2nd ed.). Washington, DC: Urban Institute and International City/County Management Association.

Hatry, H. P., Fountain, J. R., Jr., & Sullivan, J. M. (1990). Overview. In H. P. Hatry, J. R. Fountain, J. M. Sullivan, & L. Kremer (Eds.), *Service efforts and accomplishments reporting: Its time has come—An overview* (pp. 1–49). Norwalk, CT: Governmental Accounting Standards Board.

Hembree, H., Shelton, M., & Tyer, C. (1999, October). Benchmarking and local government reserve funds: Theory versus practice. *Public Management, 819,* 16–21.

Hitchcock, D. G., & Grossman, H. C. (1999, February 8). Benchmark general obligation ratios. Standard & Poor's *CreditWeek Municipal, 9,* 9–11.

Humane Society of the United States. (1982, April). *HSUS guidelines for animal shelter policies.* Washington, DC: Author.

Humane Society of the United States. (1986). *Responsible animal regulation.* Washington, DC: Author.

Humane Society of the United States. (1988, November). *HSUS shelter guidelines pertaining to potentially dangerous dogs.* Washington, DC: Author.

Humane Society of the United States. (1990, September). *HSUS guidelines for the operation of an animal shelter.* Washington, DC: Author.

Humane Society of the United States. (n.d.). *HSUS guidelines for responsible pet adoptions.* Washington, DC: Author.

In service to Iowa: Public library measures of quality. (1997). Des Moines, IA: State Library of Iowa.

Insurance Services Office, Inc. (1980). *Fire suppression rating schedule* (6-80 ed.). New York: Author.

Interim report on services for seven cities. (1998). Chapel Hill, NC: University of North Carolina, Institute of Government.

International Association of Fire Fighters. (1993). *Safe fire fighter staffing: Critical considerations.* Washington, DC: Author.

International City/County Management Association. (1999a). *Comparative performance measurement: FY 1997 data report.* Washington, DC: Author.

International City/County Management Association. (1999b). *Comparative performance measurement: FY 1998 data report.* Washington, DC: Author.

Jones, A. (1997, August). Winston-Salem's participation in the North Carolina performance measurement project. *Government Finance Review, 13,* 35–36.

Kansas City public works benchmarking report. (1998, May). Cincinnati, OH: Management Partners, Inc.

Kansas Library Association/Public Library Section. (1995). *Measurements of quality: Public library standards for Kansas.* Hutchinson, KS: Author.

Karlof, B., & Ostblom, S. (1993). *Benchmarking: A signpost to excellence in quality and productivity.* New York: John Wiley.

Karter, M. J., Jr. (1998). *U.S. fire department profile through 1997.* Quincy, MA: National Fire Protection Association.

Karter, M. J., Jr. (1999, November). *False alarm activity in the U.S., 1998*. Quincy, MA: National Fire Protection Association.

Karter, M. J., Jr., & LeBlanc, P. R. (1999, November/December). Firefighter injuries. *NFPA Journal, 93*, 47–56.

Keehley, P., Medlin, S., MacBride, S., & Longmire, L. (1997). *Benchmarking for best practices in the public sector*. San Francisco: Jossey-Bass.

Keillor, G. (1985). *Lake Wobegon days*. New York: Penguin.

Keillor, G. (1987). *Leaving home*. New York: Viking.

Kopczynski, M., & Lombardo, M. (1999, March/April). Comparative performance measurement: Insights and lessons learned from a consortium effort. *Public Administration Review, 59*, 124–134.

Kraus, R. G., & Curtis, J. E. (1982). *Creative management in recreation and parks*. St. Louis, MO: C. V. Mosby.

Krohe, J., Jr. (1990, December). Park standards are up in the air. *Planning, 56*, 10–13.

Kuzemka, K. (1997, November). Measuring your parking meter program. *The Parking Professional*, 16–23.

Lancaster, R. A. (Ed.). (1983). *Recreation, park and open space standards and guidelines*. Alexandria, VA: National Recreation and Park Association.

League of California Cities. (1994). *A "how to" guide for assessing effective service levels in California cities*. Sacramento, CA: Author.

LeGrotte, R. B. (1987). *Performance measurement in major U.S. cities*. Unpublished master's thesis, North Texas State University, Denton.

Leibfried, K., & McNair, C. (1992). *Benchmarking: A tool for continuous improvement*. New York: HarperCollins.

Level of service standards: Measures for maintaining the quality of community life. (1994, September). Kirkland, WA: Municipal Research and Services Center of Washington.

MacManus, S. A. (1984, Autumn). Coping with retrenchment: Why local governments need to restructure their budget document formats. *Public Budgeting and Finance, 4*, 58–66.

Massachusetts Water Resources Authority. (1997, July 23). *Management indicators for June 1997*. Boston: Author.

Massey, J., & Tyer, C. (1990, April-June). Local government fund balances: How much is enough? *The South Carolina Forum, 1*, 40–46.

McGowan, R. P., & Poister, T. H. (1985, February). Impact of productivity measurement systems on municipal performance. *Policy Studies Review, 4*, 532–540.

Mercer Management Consulting, Inc. (1992). *Benchmarking the quality of utility customer services: 1991 survey results*. Lexington, MA: Author.

Mitchell International. (annual). *Mechanical labor estimating guide*. San Diego, CA: Author.

Moody's Public Finance Department. (1993). *An issuer's guide to the rating process*. New York: Moody's Investor Service.

Moorman, J. A. (1997, January/February). Standards for public libraries: A study in quantitative measures of library performance as found in state public library documents. *Public Libraries, 36*, 32–39.

Moulton, C., Wright, P., & Rindy, K. (1991, April 1). The role of animal shelters in controlling pet overpopulation. *Journal of the American Veterinary Medical Association, 198*, 1172–1176.

National Arbor Day Foundation. (n.d.). *Tree City USA application.* Nebraska City, NE: Author.

National Civic League. (1999). *The civic index: Measuring your community's civic health* (2nd ed.). Denver, CO: Author.

National Commission on Accreditation for Local Park and Recreation Agencies. (1992, November). *Quick-start manual: Local park and recreation agency accreditation standards.*

National Commission on the State and Local Public Service. (1993). *Hard truths/tough choices: An agenda for state and local reform.* Albany, NY: Nelson A. Rockefeller Institute of Government.

National Council on Pet Population Study and Policy. (2000, February). *Summary of the projects of the National Council on Pet Population Study and Policy.* New London, MN: Author.

National Fire Protection Association. (1992). *NFPA 1500: Fire department occupational safety and health program.* Quincy, MA: Author.

National Fire Protection Association. (1999). *NFPA survey, update 9/99.* Quincy, MA: Author.

National Institute of Governmental Purchasing. (1994). *Results of the 1993 survey of procurement practices.* Reston, VA: Author.

O'Connell, G. B. (1989, June). Rate your city: Here's how! *Public Management, 71,* 7–10.

Olson, L. R., Shuck, D. C., Vinyon, M. J., & Lilja, J. B. (1980, March). The need for adequate fund balance levels. *Minnesota Cities, 65,* 9–12, 21.

Olstein, M., Blankenship, L., Burch, J., Freeman, N., Kingdom, W., Liner, B., Stevens, B., Koch, C., Eyre, K., Kochoba, T., & Vreeland, C. (1997). *Benchmarking wastewater treatment plant operations: Collection, treatment, and biosolids management.* Alexandria, VA: Water Environment Research Foundation.

Osborne, D., & Gaebler, T. (1992). *Reinventing government: How the entrepreneurial spirit is transforming the public sector.* Reading, MA: Addison-Wesley.

O'Toole, D. E., & Marshall, J. (1987, October). Budgeting practices in local government: The state of the art. *Government Finance Review, 3,* 11–16.

O'Toole, D. E., & Stipak, B. (1988, Fall). Budgeting and productivity revisited: The local government picture. *Public Productivity Review, 12,* 1–12.

Parry, R. W., Jr., Sharp, F. C., Vreeland, J., & Wallace, W. A. (1991). *Fire department programs: Service efforts and accomplishments reporting—Its time has come.* Norwalk, CT: Governmental Accounting Standards Board.

Paulsgrove, R. (1997). Fire department administration and operations. In A. E. Cote and J. L. Linville (Eds.), *Fire protection handbook* (18th ed., pp. 10.3–10.28). Quincy, MA: National Fire Protection Association.

Performance and cost data: Phase III city services. (1999). Chapel Hill, NC: University of North Carolina, Institute of Government.

Peters, T. J., & Waterman, R. H., Jr. (1982). *In search of excellence: Lessons from America's best-run companies.* New York: Harper & Row.

Peterson, G. E., Miller, M. J., Godwin, S. R., & Shapiro, C. (1984). *Guide to benchmarks of urban capital condition.* Washington, DC: Urban Institute.

Poister, T. H., & Streib, G. (1994, Winter). Municipal management tools from 1976 to 1993: An overview and update. *Public Productivity and Management Review, 18,* 115–125.

Poister, T. H., & Streib, G. (1999, July/August). Performance measurement in municipal government: Assessing the state of the practice. *Public Administration Review, 59,* 325–335.

Powell, N., & Bushing, M. (1992). *WLN collection assessment manual.* Lacey, WA: WLN.

Public Library Association. (1998). *Statistical report '98: Public library data service.* Chicago: American Library Association.

Public library catalog. New York: H. W. Wilson. [Published every few years, with supplements]

R. S. Means Company. (annual). *Means site work cost data.* Kingston, MA: Author.

Rider, K. L. (1979, May). The economics of the distribution of municipal fire protection services. *Review of Economics and Statistics, 61,* 249–258.

Ridley, C. E., & Simon, H. A. (1943). *Measuring municipal activities: A survey of suggested criteria for appraising administration.* Chicago: International City Managers' Association.

Rivenbark, W. C., & Carter, K. L. (2000, Spring). Benchmarking and cost accounting: The North Carolina approach. *Journal of Public Budgeting, Accounting & Financial Management, 12,* 125–137.

Rubin, M. A. (1991). *Sanitation collection and disposal: Service efforts and accomplishments reporting—Its time has come.* Norwalk, CT: Governmental Accounting Standards Board.

Savageau, D., & D'Agostino, R. (2000). *Places rated almanac.* Foster City, CA: IDG Books Worldwide.

Savageau, D., & Loftus, G. (1997). *Places rated almanac.* New York: Macmillan.

Schultz, G. R. (1997). Water distribution systems. In A. E. Cote & J. L. Linville (Eds.), *Fire protection handbook* (18th ed., pp. 6.31–6.49). Quincy, MA: National Fire Protection Association.

Shelton, B. R., & Bolen, B. (1999, February). Best economic development practices: Grand Prairie's streamlined permitting process. *Texas Town & City, 86,* 16–17.

Shivers, J. S., & Hjelte, G. (1971). *Planning recreational places.* Cranbury, NJ: Fairleigh Dickinson University Press.

Smith, G. N. (1992, February 18). Mayoral measures. *Financial World, 161,* 8.

South Carolina municipal benchmarking project: 1998. (1999). Columbia, SC: University of South Carolina, Institute of Public Affairs.

Southwestern Pennsylvania Regional Planning Commission. (1990, Spring). *Standards for effective local government: A workbook for performance assessment.* Pittsburgh, PA: Author.

Spendolini, M. J. (2000). *The benchmarking book* (2nd ed.). New York: AMACOM.

Standard & Poor's. (1983). *Credit overview: Municipal ratings.* New York: Author.

Standard & Poor's. (1994). *Municipal finance criteria.* New York: Author.

Standard & Poor's. (1995). *Municipal finance criteria: 1995.* New York: Author.

Standard on ratio studies. (1990). Chicago: International Association of Assessing Officers.

Standards for public library service in Ohio: 1998 revision. (1998). Columbus, OH: Ohio Library Council.

Stauffer, E. E. (1997). Fire department communication systems. In A. E. Cote & J. L. Linville (Eds.), *Fire protection handbook* (18th ed., pp. 10.191–10.200). Quincy, MA: National Fire Protection Association.

Stevens, B. J. (Ed.). (1984, June). *Delivering municipal services efficiently: A comparison of municipal and private service delivery*. New York: Ecodata, Inc. [Tech. rep. prepared for the U.S. Department of Housing and Urban Development, Washington, DC]

Terry, J. (1991, June 10). *Research project into performance budget measures of typical public service agencies used by select cities nationwide* [Memorandum]. Boston: City of Boston.

Thompson, L. H. (1991, May 8). *Service to the public: How effective and responsive is the government?* (Statement before the U.S. House of Representatives Committee on Ways and Means, GAO/T-HRD-91-26). Washington, DC: U.S. General Accounting Office.

U.S. Architectural and Transportation Barriers Compliance Board (Access Board). (1998). *Accessibility guidelines for buildings and facilities* (as amended through January 1998). Washington, DC: Author.

U.S. Department of Health and Human Services. (1999). *Health, United States, 1999*. Washington, DC: Author.

U.S. Department of Justice, Federal Bureau of Investigation. (1999). *Crime in the United States 1998: Uniform crime reports*. Washington, DC: Government Printing Office.

U.S. Department of Labor, Bureau of Labor Statistics. (1998, December 17). [News release].

U.S. Department of Labor, Bureau of Labor Statistics. (2000, February). Incidence rates of nonfatal occupational injuries and illnesses by industry and selected case types [Survey]. Available at www.bls.gov/oshstate.htm.

U.S. Department of Transportation. (1998, November). *Traffic safety facts 1997*. National Highway Traffic Safety Administration. (DOT HS 808 806). Washington, DC: Author.

U.S. Department of Transportation, Federal Highway Administration. (1987, December). *Highway performance monitoring system field manual*. Washington, DC: Government Printing Office. Available at www.fhwa.dot.gov/ohim/hpmsmanl/ hpms.htm (25 January 1999).

U.S. Department of Transportation, Federal Highway Administration. (1988). *Manual on uniform traffic control devices for streets and highways*. Washington, DC: Government Printing Office.

U.S. Department of Transportation, Federal Transit Administration. (1996a). *Transit profiles: Agencies in urbanized areas exceeding 200,000 population*. Washington, DC: Author.

U.S. Department of Transportation, Federal Transit Administration. (1996b). *Transit profiles: Agencies in urbanized areas with a population of less than 200,000*. Washington, DC: Author.

U.S. General Accounting Office. (1980, November). *Additional federal aid for urban water distribution systems should wait until needs are clearly established* [Report to Congress]. Washington, DC: Author.

U.S. General Accounting Office. (1993, February). *Performance budgeting: State experiences and implications for the federal government* (GAO/AFMD-93-41). Washington, DC: Author.

Usher, C. L., & Cornia, G. C. (1981, March/April). Goal setting and performance assessment in municipal budgeting. *Public Administration Review, 41*, 229–235.

Van House, N. A., Lynch, M. J., McClure, C. R., Zweizig, D. L., & Rodger, E. J. (1987). *Output measures for public libraries: A manual of standardized procedures* (2nd ed.). Chicago: American Library Association.

Walker, R. A. (1989). *Parks division quality standards manual*. Sunnyvale, CA: City of Sunnyvale.

Wallace, W. A. (1991). *Mass transit: Service efforts and accomplishments reporting—Its time has come*. Norwalk, CT: Governmental Accounting Standards Board.

Washburn, A. E., LeBlanc, P. R., & Fahy, R. F. (1999, July/August). Firefighter fatalities. *NFPA Journal, 93*, 54–70.

Weech, T. L. (1988, Summer). Small public libraries and public library standards. *Public Libraries, 27*, 72–74.

Withers, F. N. (1974). *Standards for library service: An international survey*. Paris: UNESCO Press.

Wright, P., Cassidy, B. A., & Finney, M. (1986, July). Local animal control management. *Management Information Service Report, 18*(7), 1–20. [Available from International City/County Management Association, Washington, DC]

Yates, D. (1977). *The ungovernable city: The politics of urban problems and policy making*. Cambridge: MIT Press.

Municipal Documents

City and County of Denver (CO). *Mayor's proposed 2000 budget.*

City of Albuquerque (NM). (1996, December). *Albuquerque progress report: December 1996.*

City of Albuquerque (NM). *Approved budget: Fiscal year 1997.*

City of Albuquerque (NM). (1998, October). *Goals in action: Goals, community conditions, service strategies, programs and objectives in the City of Albuquerque* (2nd ed.).

City of Alexandria (VA). *Fiscal year 1992–1993 approved budget.*

City of Alexandria (VA). *Fiscal year 1994–1995 approved budget.*

City of Alexandria (VA). *Fiscal year 1996–1997 approved budget.*

City of Alexandria (VA). *Fiscal year 2000 approved operating budget.*

City of Alexandria (VA). *Fiscal year 2000–fiscal year 2005 approved budget—Capital improvement program.*

City of Amarillo (TX). *Annual budget: 1992–1993.*

City of Ames (IA). *1998–99 program budget.*

City of Anaheim (CA). *Resource allocation plan: 1992/93.*

City of Ann Arbor (MI). *1996–1997 annual budget.*

City of Ann Arbor (MI). *Measures of success: First quarter FY 1996/97.*

City of Ann Arbor (MI). *1999/2000 budget.*

City of Auburn (AL). *FY93 annual budget.*

City of Auburn (AL). *FY97 annual budget.*

City of Bakersfield (CA). *Annual budget, 1989–90.*

City of Bellevue (WA). *1998 performance measures.*

City of Bellevue (WA). *1999–2000 budget* (Vol. 1).

City of Bellevue (WA). *1999–2000 budget* (Vol. 2).

City of Boca Raton (FL). *Approved budget, FY 1998–99* (Vol. 1).

City of Boise (ID). *Annual budget 1992–93.*

City of Boston (MA). *Mayor's management report: FY 1992* (Vol. 1).

City of Boston (MA). *Mayor's management report: FY 1992* (Vol. 2).

City of Boston (MA). *Mayor's management report: FY 1996* (Vol. 2).

City of Brentwood (TN). *Fiscal year 1996–97 annual budget.*

City of Burbank (CA) Fire Department. *Management reporting system: Year end report 1988–89.*

City of Burbank (CA). *1992–1994 biennial budget.*

City of Calgary (Alberta). *1998 operating budget: Volume V—Performance measures.*

City of Cambridge (MA). *Annual budget 1996–97.*

City of Carrollton (TX). *Comprehensive annual financial report: Fiscal year ended September 30, 1993.*

City of Chandler (AZ). *Annual budget 1992–1993.*

City of Chandler (AZ). *Annual budget: 1996–97.*

City of Chandler (AZ). *1999–2000 annual budget.*

City of Charlotte (NC). *FY88–FY89 objectives.*

City of Charlotte (NC) Budget and Evaluation Department. (1994, January). *Achievement of objectives: FY94 mid-year status report.*

City of Charlotte (NC). *FY95 operating plan.*

City of Charlotte (NC). *FY98 year-end corporate scorecard report.*

City of Charlotte (NC). *Corporate performance report: FY99 year-end.*

City of Charlotte (NC). *The corporate scorecard: Inventory, fiscal year 1999.*

City of Charlotte (NC). *FY99 business plan: Engineering and property management.*

City of Charlottesville (VA). *Adopted budget 1998–99.*

City of Chesapeake (VA). *Approved operating budget: 1992–93.*

City of Chesapeake (VA). *Fiscal year 1998–99 operating budget.*

City of Cincinnati (OH) Research, Evaluation and Budget. (1992, May 1). *Annual performance report: 1991 measures of success.*

City of Cincinnati (OH). *1993/1994 budget: Program information.*

City of Cincinnati (OH). *Biennial operating budget: 1995–1996.*

City of Cincinnati (OH). *Measures of success: 1995* (July 1996).

City of Cincinnati (OH). (n.d.). *City manager recommended benchmarks and current operation descriptions* (undated).

City of College Park (MD). (1992, May 26). *Adopted operating and capital budget for the fiscal year ending June 30, 1993 and five-year capital improvement plan.*

City of College Park (MD). *Adopted operating and capital budget for fiscal year 2000 and five-year capital improvement program.*

City of College Station (TX). *1992–1993 annual budget.*

City of College Station (TX). *FY 1996–1997 budget.*

City of College Station (TX). *FY 1999–2000 annual budget.*

City of Corpus Christi (TX). *Annual budget: Fiscal year 1992–1993.*

City of Corpus Christi (TX). *Annual quality plans: FY 1995–96.*

City of Corpus Christi (TX). *Performance plans: FY 98–99.*

City of Corpus Christi (TX). *Performance reports: December 1998.*

City of Corvallis (OR). *Quarterly operating report: Fourth quarter ending June 30, 1992.*

City of Corvallis (OR). *Fiscal year 1992–93 annual budget.*

City of Dallas (TX). (1989, August). *Dallas Police Department's recommendations to improve response time to citizens' calls for police service.*

City of Dallas (TX) Parks and Recreation Department. (1992, April). *City comparison study.*

City of Danville (KY). *Annual budget for fiscal year ending June 30, 1999.*

City of Dayton (OH) Office of Management and Budget. (1991). *1991 program strategies.*

City of Decatur (IL). *Annual performance budget, fiscal year 1992–1993* (Vol. 1).

City of Decatur (IL). *Annual performance budget: 1996–1997.*

City of Denton (TX). *Annual program of services: 1992–93.*

City of Denton (TX). *Annual program of services: 1998–99.*

City of Duncanville (TX). *Annual budget: Fiscal year October 1, 1991–September 30, 1992.*

City of Duncanville (TX). *Comprehensive annual financial report for the fiscal year ended September 30, 1996.*

City of Duncanville (TX). *Duncanville annual budget 1996–97.*

City of Duncanville (TX). *1998–99 annual budget.*

City of Duncanville (TX). *1999–2000 budget.*

City of Dunedin (FL). *Fiscal year 1992/1993 adopted annual operating and capital budgets.*

City of Durham (NC). *FY 1999–2000 budget.*

City of Edmonton (Alberta). *Business plan 1996–1998: Office of the city clerk.*

City of Eugene (OR). *Proposed budget: Fiscal year 1998.*

City of Eugene (OR). *Annual budget: Fiscal year 1999.*

City of Fairfield (CA). *Proposed 1992/93 budget and ten-year financial plan.*

City of Fayetteville (AR). (1989, January). *Public works maintenance management system: System manual.*

City of Fayetteville (AR). *Annual budget and work program: 1993.*

City of Fayetteville (AR). *Annual budget and work program: 1994.*

City of Fayetteville (AR). *1999 annual budget and work program.*

City of Flagstaff (AZ). *Annual budget and financial plan: 1998–1999.*

City of Fort Collins (CO). *1993 annual budget: Volume 1. Budget overview.*

City of Fort Collins (CO). *1993 annual budget: Volume 2. Program performance budget.*

City of Fort Collins (CO). *1998 and 1999 biennial budget, volume 2: Program performance budget.*

City of Fort Smith (AR). *Fiscal year 1999 budget.*

City of Fort Worth (TX). *1992–1993 annual budget and program objectives.*

City of Fort Worth (TX). *Annual budget and program objectives: 1998–99.*

City of Gainesville (GA). *Annual budget: Fiscal year ending June 30, 2000.*

City of Germantown (TN). *Fiscal year 1997 budget.*

City of Glendale (AZ). *1998–1999 annual budget.*

City of Grand Junction (CO). *Comprehensive annual financial report for the fiscal year ended December 31, 1991.*

City of Grand Junction (CO). *1992–1993 municipal budgets.*

City of Grand Prairie (TX). *Annual budget: Fiscal year 1996–1997 approved.*

City of Grand Prairie (TX). *Approved budget: Fiscal year 1997–1998.*

City of Grants Pass (OR). *Adopted operating/capital budget: Fiscal year 1996–1997.*

City of Greeley (CO). *1999 budget.*

City of Greensboro (NC). *Adopted budget 1998–1999 and projected budget 1999–2000.*

City of Greenville (SC). *1991 year-end performance report.*

City of Greenville (SC). *1992 adopted budget.*

City of Gresham (OR). *Adopted operating/capital budget document: Fiscal year 1992–93.*

City of High Point (NC). *Performance objectives and measures: Fourth quarter ending June 30, 1998.*

City of Houston (TX) Finance and Administration Department. *FY93 monthly status report for the period ended December 31, 1992.*

City of Houston (TX). *Adopted budget for fiscal year ending June 30, 1993* (Vol. 1).

City of Houston (TX). *Adopted budget for fiscal year ending June 30, 1993* (Vol. 2).

City of Hurst (TX). *Annual operating budget for fiscal year of 1991–92.*

City of Hurst (TX). *Annual operating budget for fiscal year 1998–1999.*

City of Iowa City (IA). *Comprehensive annual financial report for the fiscal year ended June 30, 1992.*

City of Iowa City (IA). *Financial plan for FY93–95.*

City of Irving (TX). *Annual operating budget: Municipal service programs 1992–1993.*

City of Irving (TX). *Annual operating budget: Municipal service programs 1996–1997.*

City of Irving (TX). *1999–2000 annual operating budget: Municipal services program.*

City of Jacksonville (FL). *Annual budget for the fiscal year ending September 30, 1996.*

City of Jacksonville (FL). *Annual financial plan for the fiscal year ending September 30, 1999.*

City of Kalamazoo (MI). *Comprehensive annual financial report: 1996.*

City of Kalamazoo (MI). *1999 proposed budget.*

City of Kansas City (MO). *Budget 1997.*

City of Kingsport (TN). *Annual budget 1997–98.*

City of Kingsport (TN). *FY 1998–99 budget.*

City of Knoxville (TN). *FY92–93 annual operating budget.*

City of Largo (FL). *Annual budget: Fiscal year 1993.*

City of Largo (FL). *Annual budget: Fiscal year 1996.*

City of League City (TX). *Annual budget for fiscal year 1992–93.*

City of Lee's Summit (MO). *Operating budget: Year beginning July 1, 1992.*

City of Little Rock (AR). (1990). *Management information systems manual.*

City of Little Rock (AR). *1993 annual operating budget.*

City of Little Rock (AR). *1997 annual operating budget.*

City of Long Beach (CA). *Adopted resource allocation plan: Fiscal year 1992–93/operating detail.*

City of Long Beach (CA). *Hitting the mark: City services and performance—Fiscal year 1995–1996.*

City of Long Beach (CA). *Annual budget 1996–97.*

City of Long Beach (CA). *Hitting the mark: City services and performance—Fiscal year 1996–1997.*

City of Longview (TX). *Annual budget 1992–1993.*

City of Longview (TX). *Annual budget 1993–1994.*

City of Loveland (CO). *1993 budget.*

City of Loveland (CO). *1999 adopted budget.*

City of Lubbock (TX). *Goals and objectives: 1992–93.*

City of Lubbock (TX). *Program of services: 1996–97.*

City of Lubbock (TX). *Annual budget 98–99.*

City of Macon (GA). *1997 annual budget.*

City of Milwaukee (WI). *1993 budget.*

City of Milwaukee (WI). *Special ADAP supplement to the proposed executive budget: 1993.*

City of Milwaukee (WI). *1994 budget.*

City of Milwaukee (WI). *1995 budget.*

City of Milwaukee (WI). *1999 plan and budget summary.*

City of New York (NY). (1992, September 17). *Mayor's management report.*

City of New York (NY). *Mayor's management report, fiscal 1997, volume 2: Agency and citywide indicators.*

City of New York (NY). *Mayor's management report, fiscal 1998, volume 1: Agency narratives.*

City of New York (NY). *Mayor's management report, fiscal 1998, volume 2: Agency and citywide indicators.*

City of Newburgh (NY). *The fiscal budget for the year 1993.*

City of Norfolk (NE). *Annual budget: Fiscal year 1996–1997.*

City of Norfolk (VA). *Building a community worth caring about: Approved budget July 1, 1998 through June 30, 1999.*

City of Norman (OK). *Budget: Fiscal year ending June 30, 1999.*

City of Norman (OK). *Budget: Fiscal year ending June 30, 2000.*

City of Oak Park (MI). *Annual budget, fiscal year 1998–1999.*

City of Oak Ridge (TN). *Goals and objectives program: Annual report fiscal year 1992.*

City of Oak Ridge (TN). *1993 budget.*

City of Oak Ridge (TN). *1994 budget.*

City of Oak Ridge (TN). *1999 budget.*

City of Oakland (CA). *1992–93 adopted policy budget* (Vol. 2).

City of Oakland (CA). *1996–97 adopted policy budget.*

City of Ocala (FL). *1992–1993 budget.*

City of Ocala (FL). *Fiscal year 1996–97 adopted budget.*

City of Oklahoma City (OK). *Annual budget: Fiscal year 1992–1993.*

City of Oklahoma City (OK). *Annual budget: Fiscal year 1999–2000.*

City of Orlando (FL). *Annual budget 1992/1993.*

City of Orlando (FL). *Annual budget 1995–1996.*

City of Orlando (FL). *Annual budget 1997–1998.*

City of Overland Park (KS). *1993 budget.*

City of Overland Park (KS). *1996 budget.*

City of Overland Park (KS). *1999 annual budget.*

City of Palo Alto (CA). *1998–99 adopted budget* (Vol. 1).

City of Palo Alto (CA). *1998–99 adopted budget* (Vol. 2).

City of Pasadena (CA) Management Audit Team. (1986, October). *Opportunities for improving efficiency and effectiveness: Department of public works.*

City of Peoria (AZ). *1991–92 approved budget and financial plan.*

City of Philadelphia (PA) Office of the City Controller. (1991, January). *Performance audit: Streets department/sanitation division.*

City of Philadelphia (PA). (1996). *Mayor's report on city services.*

City of Philadelphia (PA). *Mayor's report on city services (fiscal year 1998).*

City of Phoenix (AZ) Auditor Department. (1991, December 31). *Fire department performance indicators.*

City of Phoenix (AZ) Auditor Department. (1992, May 6). *Parks, recreation and library department performance indicators.*

City of Phoenix (AZ) Auditor Department. (1992, June 12). *Police department performance indicators.*

City of Phoenix (AZ). *The Phoenix budget 1992–93.*

City of Phoenix (AZ). *The Phoenix budget 1993–94.*

City of Plano (TX). *Program of services, 1998–99.*

City of Portland (OR) Office of the City Auditor. (1992, May). *An evaluation of city financial trends: 1980–91.*

City of Portland (OR). *FY 1992–93 adopted budget* (Vol. 1).

City of Portland (OR) Office of the City Auditor. (1993, January). *Service efforts and accomplishments: 1991–92* (Rep. No. 176).

City of Portland (OR) Office of the City Auditor. (1998, December). *City of Portland service efforts and accomplishments: 1997–98.*

City of Portland (OR). *Adopted budget fiscal year 1999–2000* (Vol. 1).

City of Prairie Village (KS). *1997 annual budget.*

City of Raleigh (NC). *Description of operating programs: FY 1992–93.*

City of Raleigh (NC). *Performance indicators: Fiscal year 1998–1999.*

City of Raleigh (NC). *Performance indicators: Fiscal year 1999–2000.*

City of Redondo Beach (CA). *Adopted budget fiscal year 1998–1999/adopted financial plan fiscal year 1999–2000.*

City of Reno (NV). *Program and financial plan: Fiscal year 1991–1992.*

City of Reno (NV). *1998/99 budget.*

City of Richmond (VA). *General fund budget 1993–1994: Proposed.*

City of Rochester (NY). *1992–93 budget.*

City of Rochester (NY). *1992–93 second quarter report: Program performance.*

City of Rochester (NY). *1996–1997 budget.*

City of Rock Hill (SC). *Budget and work program 1993.*

City of Rock Island (IL). (1993). *Budget for fiscal year 1993–1994.*

City of Rockville (MD). *FY99 adopted operating budget and capital improvements program.*

City of Sacramento (CA). (1992, September-November). *Budget workshop materials.*

City of Saginaw (TX). *Annual budget: Fiscal year 1992–1993.*

City of San Antonio (TX). *Adopted annual budget 1992–1993: Detail information.*

City of San Antonio (TX). *Proposed annual budget: FY 1996–97.*

City of San Antonio (TX). *Adopted annual budget, fiscal year 1998–1999.*

City of San Clemente (CA). *Annual budget and capital improvement program: Fiscal year 1999–2000.*

City of San Diego (CA). *Fiscal year 1999 semi-annual performance report.*

City of San Diego (CA). *Fiscal year 2000 annual budget: Volume 1—Citizens' budget.*

City of San Diego (CA). *Fiscal year 2000 annual budget: Volume 2—Department detail.*

City of San Diego (CA). *Fiscal year 2000 annual budget: Volume 3—Department detail.*

City of San Jose (CA). *Adopted operating budget 1989–90.*

City of San Jose (CA). *Proposed 1996–97 operating budget.*

City of San Luis Obispo (CA). *1991–93 financial plan and approved 1991–92 budget.*

City of San Luis Obispo (CA). (1992, July). *1991–93 financial plan supplement and approved 1992–93 budget.*

City of San Luis Obispo (CA). *1995–97 financial plan and approved 1995–96 budget.*

City of San Luis Obispo (CA). *1997–99 financial plan and adopted 1997–98 budget.*

City of Santa Ana (CA). *Annual budget 1988–89.*

City of Santa Ana (CA). *Annual budget 1996–97.*

City of Savannah (GA). Community, Housing and Economic Development Department. (1988, October). *Responsive public services programs.*

City of Savannah (GA). *1992 program of work.*

City of Savannah (GA). *1994 service program and budget.*

City of Savannah (GA). *Neighborhood quality benchmarks: 1996 report to the community.*

City of Savannah (GA). *1999 service program and budget.*

City of Savannah (GA) Park and Tree Department. *Condition standards, 1999.*

City of Scottsdale (AZ). *Performance report: 1999.*

City of Shreveport (LA). *1993 annual operating budget.*

City of Shreveport (LA). *Annual operating budget 1996.*

City of Sioux Falls (SD). *Comprehensive annual financial report: Fiscal year ended December 31, 1991.*

City of Smyrna (GA). *Annual financial plan: Fiscal year 1997.*

City of Southfield (MI). *Municipal budget: July 1992–June 1993.*

City of St. Charles (IL). *Budget plan 1992–1993.*

City of St. Petersburg (FL). (1992). *Approved 1992 program budget.*

City of St. Petersburg (FL). *Approved program budget: Fiscal year 1997.*

City of Sterling Heights (MI). *1992/93 annual budget.*

City of Sterling Heights (MI). *Annual budget 1996/97.*

City of Stockton (CA). *1996–97 annual budget.*

City of Sunnyvale (CA). (1993, June). *Performance indicators.*

City of Tacoma (WA). *Preliminary program biennial budget: 1993/1994.*

City of Tallahassee (FL). *Fiscal year 1993 approved detail budget.*

City of Tempe (AZ). *Annual budget 1998/99.*

City of Tucson (AZ). (1989, December). *City of Tucson parks and recreation master plan 2000: Planning our recreational future.*

City of Tucson (AZ). *Cost report summary: Fiscal year 1991–92.* [Prepared by S. Postil, Department of Budget and Research]

City of Tucson (AZ). *Fiscal year 1992–1993 budget.*

City of Tucson (AZ). *Adopted budget summary, fiscal year 1998–99* (Vol. 1).

City of Tucson (AZ). *Adopted budget operating detail, fiscal year 1998–99* (Vol. 2).

City of Upper Arlington (OH). *Annual budget: Fiscal year 1992.*

City of Vancouver (WA). (1991). *1991–1992 biennial budget.*

City of Victoria (TX). *Performance measurements pilot program,* September 2, 1997.

City of Washington, D.C. (1989a, March). *Improving ambulance operations in Washington, D.C.: A blueprint for change.* Washington, DC: Office of the City Administrator, Productivity Management Services.

City of Washington, D.C. (1989b, June). *Improving emergency medical services communications.* Washington, DC: Office of the City Administrator, Productivity Management Services.

City of Washington, D.C. (n.d.). *Improving vehicle maintenance productivity in the D.C. Department of Transportation* (Productivity Rep. No. 4). Washington, DC: Department of Transportation and Office of the City Administrator, Productivity Management Services.

City of Waukesha (WI). *Annual budget management plan: 1993.*

City of West Valley City (UT). *Annual budget: Fiscal year 1996–1997.*

City of White Plains (NY). *1996–1997 proposed budget.*

City of Wichita (KS). *1993/94 annual budget performance measures.*

City of Wichita (KS). *2000/2001 annual budget.*

City of Wilmington (DE). *Annual budget: Fiscal year 1993.*

City of Wilmington (DE) Office of Management and Budget. *MAPP: Management's administrative planning process, FY 1994.*

City of Winston-Salem (NC). *Annual budget program 1986–1987: Year-end report.*

City of Winston-Salem (NC). (1993, January). *Year-end reports for FY91–92.*

City of Winston-Salem (NC). *Annual budget program for FY92–93.*

City of Winston-Salem (NC). *Performance measurement objectives: Year-end reports FY 96–97.*

City of Winston-Salem (NC). *1997–1998 performance report and 1998–1999 business plan,* October 1998.

City of Winston-Salem (NC). *1998–1999 annual budget program and 1999–2003 capital plan.*

City of Winston-Salem (NC). (n.d.). *Fleet services project: Phase I report.* [Report prepared by Office of Organizational Effectiveness]

Lane Council of Governments. (1996, February). *City of Eugene progress report on 1995–96 city council goals.*

Metropolitan Government of Nashville and Davidson County (TN). *Fiscal year 1993 operating budget.*

Metropolitan Government of Nashville and Davidson County (TN). *Fiscal year 1998–99 operating budget: Departmental narrative* (Vol. 2).

Metropolitan Government of Nashville and Davidson County (TN). *Fiscal year 1999–2000 operating budget: Departmental narrative* (Vol. 2).

New Hanover County (NC), *1999–2000 budget.*

Sedgwick County (KS). *1991 budget.*

Town of Addison (TX). *Annual budget: Fiscal year 1992–93.*

Town of Avon (CT). *Annual operating budget 1998–1999.*

Town of Blacksburg (VA). *Adopted operating budget: Fiscal year 1998–1999.*

Town of Cary (NC). *Annual operating budget: Fiscal year 1999–2000.*

Town of Chapel Hill (NC). *First quarter report: 1996–97.*

Town of Chapel Hill (NC). *Annual report: 1998–99.*

Unified Government of Athens–Clarke County (GA). *FY 2000 annual operating and capital budget.*

Acronyms

AASHTO	American Association of State Highway Transportation Officials
ADA	Americans with Disabilities Act
ALA	American Library Association
ALS	advanced life support
ANI	Automatic Number Identification
ANSI	American National Standards Institute
APHA	American Public Health Association
APTA	American Public Transit Association
AWWA	American Water Works Association
BAN	bond anticipation notes
BLS	basic life support
BMI	body mass index
BOD	biological oxygen demand
CAFR	comprehensive annual financial reports
CALEA	Commission on Accreditation for Law Enforcement Agencies
CAMA	computer-assisted mass appraisal
CIP	capital improvement program
COD	coefficient of dispersion
CPR	cardiopulmonary resuscitation
CPS	Child Protective Services
CPSC	Consumer Product Safety Commission
cpu	cobalt platinum units
DTP	diphtheria-tetanus-pertussis
DUI	driving under the influence
DWI	driving while intoxicated
EAB	Emergency Ambulance Bureau
EEO	Equal Opportunity Employment
EMS	emergency medical service
EMT-D	emergency medical technician-defibrillation
EPA	Environmental Protection Agency
ESI	equivalent sphere illumination
EXSL	Exemplary State and Local
FBI	Federal Bureau of Investigation
FTA	Federal Transit Administration
FTE	full-time equivalent
FTMS	Financial Trend Monitoring System

GAAP	generally accepted accounting principles
GASB	Govenmental Accounting Standards Board
GED	general equivalency diploma
GFOA	Government Finance Officers Association
GO	general obligation
gpm	gallons per minute
Hib	Haemophilos influenzae type b
HR	human resources
HSUS	Humane Society of the United States
IAAO	International Association of Assessing Officers
IAFF	International Association of Fire Fighters
ICMA	International City/County Management Association
IFLA	International Federation of Library Associations and Institutions
ILL	interlibrary loan
ISO	Insurance Services Office
JTPA	Job Training Partnership Act
kPa	kilopascals
kWh	kilowatt-hours
LAN	local area network
LCC	League of California Cities
MBO	management-by-objectives
mph	miles per hour
MSA	metropolitan statistical area
NAFA	National Association of Fleet Administrators
NALGA	National Association of Local Government Auditors
NAFA	National Association of Fleet Administrators
NAPA	National Academy of Public Administration
NFIRS	National Fire Incident Reporting System
NFPA	National Fire Protection Association
NCALPRA	National Commission on Accreditation Local Park Recreation Agencies
NCL	National Civic League
NRPA	National Recreation and Park Association
NTU	nephelometric turbidity units
OSHA	Occupational Safety and Health Administration
PAA	Playground Association of America
PLA	Public Library Association
ppm; ppb	parts per million; parts per billion
psi	pounds per square inch
PSR	present servicability rating
PUD	planned unit development
ROW	publicly defined right-of-way
SPRPC	Southwestern Pennsylvania Regional Planning Commission
SEA	Service Efforts and Accomplishments
TDD	telecommunications device for the deaf
THM	trihalomethanes
TON	threshold odor number
TSS	total suspended solids
UCR	Uniform Crime Report

Index

About the Author

DAVID N. AMMONS (Ph.D., University of Oklahoma) is a Professor of Public Administration at the University of North Carolina's Institute of Government in Chapel Hill. He previously served in various administrative capacities in four municipalities—Fort Worth, Texas; Hurst, Texas; Phoenix, Arizona; and Oak Ridge, Tennessee—and has taught in the public administration programs of the University of North Carolina, the University of Georgia, and the University of North Texas.

His works include five previous books on local government—*Municipal Productivity, City Executives* (with Charldean Newell), *Recruiting Local Government Executives* (with James J. Glass), *Administrative Analysis for Local Government,* and *Accountability for Performance: Measurement and Monitoring in Local Government.* He consults with city and county governments on organizational and management concerns, including performance measurement, benchmarking, and productivity improvement.